NASA Technical Memorandum 4298

Fusion Energy for Space Missions in the 21st Century

Norman R. Schulze
NASA Office of Safety and Mission Quality
Washington, D.C.

National Aeronautics and
Space Administration

Office of Management

Scientific and Technical
Information Program

1991

ABSTRACT

Future space missions were hypothesized and analyzed and the energy source for their accomplishment investigated. The missions included manned Mars, scientific outposts to and robotic sample return missions from the outer planets and asteroids, as well as fly-by and rendezvous missions with the Oort Cloud and the nearest star, Alpha Centauri, and even beyond. Space system parametric requirements and operational features were established. The energy means for accomplishing missions where Δv requirements range from 90 km/sec to 30,000 km/sec (High Energy Space Mission) were investigated. The need to develop a power space of this magnitude is a key issue to address if the U.S. civil space program is to continue to advance as mandated by the National Space Policy. Potential energy options which could provide the propulsion and electrical power system and operational requirements were reviewed and evaluated. Fusion energy was considered to be the preferred option for most likely fulfilling the mission requirements and was analyzed in depth. Candidate fusion fuels were evaluated based upon the energy output and neutron flux. In addition to its mission enabling performance, fusion energy can offer significant safety, environmental, economic, and operational advantages. Reactors exhibiting a highly efficient use of magnetic fields for space use while at the same time offering efficient coupling to an exhaust propellant or to a direct energy convertor for efficient electrical production were examined. Near term approaches were identified. A strategy that will produce fusion powered vehicles as part of the space transportation infrastructure was developed. Space program resources must be directed toward this issue as a matter of the top policy priority.

SUMMARY: SPACE MISSIONS IN THE 21ST CENTURY

Critical to the implementation of NASA's space mission architecture in the 21$\underline{st}$ century and beyond is a high energy mission capability. Vehicles which have the energy performance permitting 100 km/sec to over 20,000 km/sec velocity changes, enable efficient manned and unmanned flights to all orbiting masses within the solar system – also, unmanned missions to the stars. The high energy capability is essential to:

- enable the missions
- provide safety
- permit economical space flight
- allow commercialization of space.

Constant 10^{-3}–10^{-4} g acceleration effectively accomplishes those missions. Jet power levels from 10 MW to 100 GW, produced by 1–10 kW/kg specific power propulsion systems, are necessary. Propulsion systems must deliver variable specific impulse, ~5×10^3 to 10^6 seconds, with firing durations of two months – ultimately 10's of years – with thrusts ranging from 1 N to ~1000 kN. Viable space-based vehicles demand ultrahigh reliability by today's standards. Reliability is essential to provide maintenance free, reusable operational flight systems as demanded by greater than ~10 AU range vehicles. Vehicles powered by "solid state" propulsion systems, where moving and erosive mechanisms are not mandated, offer an approach. Furthermore, the stellar distance communications, the permanent settlement of man on Mars where local planetary resources are necessary for safety and economics, and long duration manned spacecraft will require 10+ MW to 100+ MW, ultimately gigawatts, of electrical power using highly efficient direct converters.

Nuclear energy produces increases of >6 to 10 orders of magnitude in specific energy over chemical energy and is the only energy source capable of meeting those space power requirements. Fission thermal propulsion, while demonstrated, cannot. Fusion and matter-antimatter, while not demonstrated, theoretically can. Fusion inherently has significant advantages over both matter-antimatter and fission for safe, economical high performance space propulsion. Nonradioactive fuels, D-D or D-^{3}He, can be burned which eliminates global launch hazards and avoids impacting Earth's environment. D-^{3}He's charged particle energy converts directly to thrust for efficient propulsion and produces 1 to 5% neutron energy yields for minimal system mass – but is a more difficult fuel to burn at ~40 keV over D-T at ~10 keV, or D-D at ~30 keV.

Progress in magnetic confinement fusion (MCF) is being made on the tokamak, $Q \approx 0.8$, but not on a space confinement concept. Alternate confinement approaches are mandatory. Magnetic fields in MCF inherently offer the means for reliably meeting long firing durations. In space, some aspects of fusion are simplified. The expulsion of plasma particles for thrusting assists with resolving ash removal problems faced by terrestrial fusion reactors. Space's clean vacuum resolves terrestrial reactor vacuum issues. First-wall exposure is relieved by D-^{3}He's low neutron flux. *Confinement approaches exhibiting high reactor β are essential.* Options exist – For MCF, there are the Field Reversed Confinement (FRC) and dipole concepts. $\beta \approx 90\%$, illustrative of the required characteristics where less reliance is placed upon external magnetic fields. The FRC has inherent linear field properties desired for propulsion. Mechanisms to achieve plasma stability/heating and system thermal control require research. Inertial confinement (ICF) reportedly has shown positive fuel gains, but the concept's strict security classification seriously hampers an open evaluation. Drivers are a concern. Because no specific confinement approach can today be extrapolated with certainty for meeting the space requirements, we must accomplish the essential focused research. Testing is mandatory! But no space fusion research program exists world-wide. Alternate confinement programs for terrestrial applications were cancelled in 1991.

Applications for space are directly in front of us, but development of hardware for flight will not be quick. Thus, now is the proper time for NASA to initiate a space fusion research program to develop space applicable confinement designs. An expedient and cost effective program approach is to demonstrate first principles, then to proceed directly to full scale, net power reactors. An aptly funded program, $150-200M per full scale confinement approach, may make this capability available on the order of ~30 years. That investment will pay for itself based upon one manned flight to Mars alone.

In conclusion, we must take the necessary steps to reduce the requirements on mass placed into low Earth orbit as the strategic approach in preparing for the 21st century space program, whether fission or fusion. Fission fills a niche until fusion is developed. We need to move forward.

TABLE OF CONTENTS

Summary v

Foreword xiii

Preface... xv

Acknowledgements xix

Nomenclature xxi

1.0   Introduction 1-1

2.0   High Energy Mission Applications 2-1

      2.1   Mission Analysis 2-6

           2.1.1   Mission analysis calculations 2-6

           2.1.2   Mission categories considered 2-11

      2.2   Candidate high energy missions 2-12

           2.2.1   Manned Exploration 2-13

           2.2.2   Manned missions beyond Mars 2-38

           2.2.3   Unmanned outer planets and moons 2-41

           2.2.4   Asteroids and comets 2-60

           2.2.5   Inner planets 2-67

           2.2.6   Oort Cloud 2-69

           2.2.7   Near Earth Stellar Missions 2-83

           2.2.8   Solar 2-104

           2.2.9   Remote space based telescopes 2-105

           2.2.10 Lunar 2-107

           2.2.11 Stationary (lunar) power for spacecraft
                    propulsion and electrical power 2-108

           2.2.12 Mission life considerations 2-111

           2.2.13 Small space fusion reactors 2-113

           2.2.14 Aeronautical 2-113

           2.2.15 Fission reactor waste disposal 2-116

           2.2.16 Electrical power at space based facilities 2-117

           2.2.17 Other benefits 2-118

		2.2.18	Space commercialization and safety applications.	2-119
		2.2.19	Earth Orbital Applications	2-120
		2.2.20	Spin-offs	2-121
	2.3	Summary		2-121
3.0	High energy sources for space			3-1
	3.1	Candidate high energy sources		3-1
	3.2	Fusion		3-4
		3.2.1	Background	3-4
		3.2.2	Status	3-5
	3.3	Fission		3-7
		3.3.1	Nuclear Electric Propulsion (NEP)	3-7
		3.3.2	Nuclear Thermal Propulsion	3-8
		3.3.3	Gaseous core reactor	3-9
		3.3.4	Fission fragment rocket	3-10
	3.4	Matter-antimatter		3-14
		3.4.1	Background	3-14
		3.4.2	Status	3-17
	3.5	Strange matter		3-18
	3.6	Other		3-18
	3.7	Summary		3-19
4.0	General Discussion of Fusion Reactions			4-1
	4.1	Primary nuclear fusion reactions		4-1
	4.2	Other nuclear fusion reactions		4-7
	4.3	Summary		4-9
5.0	Theoretical Performance Capability of Fusion Energy Conversion for Space			5-1
	5.1	Summary		5-9
6.0	Flight System Considerations and Requirements			6-1
	6.1	Space reactors		6-2
		6.1.1	Specific power	6-2
		6.1.2	Thrust	6-2

	6.1.3	Specific impulse	6-2
	6.1.4	Fuel cycle	6-3
	6.1.5	Beta	6-3
	6.1.6	Ignition	6-4
	6.1.7	Throttle capability	6-4
	6.1.8	Plasma stability	6-4
	6.1.9	Power level	6-4
	6.1.10	Electrical power variability	6-5
	6.1.11	Dual mode operation	6-5
	6.1.12	Mass	6-5
	6.1.13	Efficiency	6-5
	6.1.14	Recirculation power	6-5
	6.1.15	Life	6-6
	6.1.16	Modes of operation	6-6
	6.1.17	Failure tolerance	6-6
	6.1.18	Space environment	6-6
	6.1.19	Summary	6-6
6.2	Interface of the fusion flight power reactor system with the flight vehicle		6-8
	6.2.1	Space restart capability	6-8
	6.2.2	Fuel storage capability	6-8
	6.2.3	Radioactivity	6-8
	6.2.4	Reuse	6-9
	6.2.5	Servicing	6-9
	6.2.6	Energy storage	6-9
	6.2.7	Size	6-10
	6.2.8	Life	6-10
	6.2.9	Mass	6-10
	6.2.10	Maintenance	6-10
	6.2.11	Pulsed versus steady-state operation	6-10
	6.2.12	Power conversion and transportability	6-11
	6.2.13	Emergency shutdown	6-11
	6.2.14	Heat balance and cooling	6-11

			6.2.15 Solid state propulsion	6-11

- 6.2.15 Solid state propulsion 6-11
- 6.2.16 Self diagnostics and corrections 6-12
- 6.2.17 Operations 6-12
- 6.2.18 Safety 6-12
- 6.2.19 Redundancy 6-12
- 6.2.20 Space Station compatibility 6-13
- 6.2.21 Environment 6-13
- 6.2.22 Reliability 6-13
- 6.2.23 Testing and qualification 6-13
- 6.2.24 Space based vehicle design 6-13
- 6.2.25 Disposal 6-14
- 6.2.26 Economics 6-14
- 6.2.27 Status 6-14

- 6.3 Summary 6-15

- 7.0 Fuel and Design Options for Space Fusion Reactors 7-1
  - 7.1 Fuel selection 7-2
    - 7.1.1 Space fuel options 7-2
    - 7.1.2 Summary of the fuel discussion 7-10
  - 7.2 Reactor concepts 7-10
    - 7.2.1 Magnetic Confinement Fusion (MCF) 7-10
    - 7.2.2 Inertially Confined Fusion (ICF) 7-20
    - 7.2.3 Other confinement concepts 7-24
  - 7.3 Evaluation of space applications of fusion energy 7-35
    - 7.3.1 Propulsion: fusion engine 7-35
    - 7.3.2 Electrical Power 7-39
  - 7.4 Summary 7-41

- 8.0 Status and Performance of Potential Space Fusion Reactors 8-1
  - 8.1 Terrestrial program status – general background 8-1
  - 8.2 Comparative Research Maturity, Fusion Reactor Designs 8-3
  - 8.3 Space fusion reactors (SFR) 8-9
  - 8.4 Energy conversion for propulsion 8-10

8.5	Electrical power generation	8-22
8.6	Summary	8-23

9.0 Acceptability: Safety, Environmental, Reliability, and Space Maintenance Considerations ... 9-1

    9.1 Safety ... 9-2
        9.1.1 Public safety and environmental considerations ... 9-3
        9.1.2 System Safety ... 9-7
        9.1.3 Operations safety ... 9-10

    9.2 System Reliability ... 9-13
    9.3 Space maintenance and life cycle approaches ... 9-15
    9.4 Summary ... 9-16

10.0 Space Program Operational Economics and Program Implementation ... 10-1

    10.1 Purpose and approach of the economic-program implementation evaluation ... 10-1
    10.2 Reactor and Flight Vehicle Operations ... 10-7
    10.3 Comparative Agency Economic Issues ... 10-9
    10.4 Manned Mars Missions ... 10-13
    10.5 Science Missions ... 10-22
    10.6 Looking Forward ... 10-25
    10.7 Summary of Operational Advantages and Suggested Developmental Cost Inferences ... 10-26

11.0 Broad Issues ... 11-1

    11.1 Philosophy on advanced space research ... 11-1
    11.2 Roles of NASA and DOE ... 11-5
    11.3 Environmental impact to extensive lunar mining ... 11-7
    11.4 Involved parties ... 11-7
    11.5 Priority of net power ... 11-7
    11.6 Reasonable payload mass ... 11-8
    11.7 Timeliness of the stellar mission ... 11-8

12.0 Conclusions ... 12-1

13.0 Space Program Options ... 13-1

    13.1 Option 1: analytical and systems study tasks ... 13-2

13.2 Option 2: Space Fusion Experiments Support Program 13–6

13.3 Option 3: Dedicated NASA Space Fusion Research Program 13–9

13.4 Option 4: Expedited Development of a Prototype Space Fusion Flight System 13–11

13.5 Summary 13–12

14.0 Recommended Space Fusion Strategy and Program Plan 14–1

14.1 Basis for the strategy 14–2

14.2 Assumptions 14–3

14.3 Fusion Program to Address Key Assumptions 14–5

14.4 Program Definition: Funding Level 14–44

14.5 Program Management 14–45

14.6 Program Summary 14–46

15.0 Recommendations for a High Energy Space Mission Program Using Fusion Energy 15–1

16.0 References 16–1

Appendix A A–1

Appendix B B–1

Appendix C C–1

FOREWORD

The goal of this work is ultimately to assist in the achievement of space science and exploration missions and thus to assist in the implementation of the National Space Policy.

This analysis commenced with a self initiated inquiry concerning the importance of and the need for more advanced forms of space mission energy conversion, thereby permitting the accomplishment of more ambitious space science objectives. In a sense it began with discussions with the Associate Administrator of the Office of Aeronautics and Space Technology, Dr. Colliday, in 1987 who took an interest in this subject.

As one set of questions led to another during the initial phase of the inquiry, a drive developed to conduct an independent investigative analysis on where the space program is headed with regard to energy conversion. It ultimately has culminated in this report, the scope of which increased substantially in response to interest shown and questions raised as the study progressed.

The work was thus initiated by the author as an unfunded endeavor with the exception of travel cost reimbursements by Code Q and by funding from the Headquarters Solar System Exploration Division for SAIC, Schaumberg, IL to perform mission performance analyses. Actually, nearly all of the work was accomplished during off duty time, including some of the travel.

The approach has been to coordinate closely with and to solicit the opinion of others in their fields of expertise to obtain the best possible response to any given topic as well as to perform a comprehensive literature search. The report is not intended to be interpreted as necessarily representing the position of any organization or of any individual with whom the study content was discussed, nor that of NASA.

It is intended to be a presentation of all key issues which need to be addressed as part of a comprehensive space strategy. These include mission objectives, the energy options available, the preferred approach-and why, the system requirements, status of the fusion program and probability of achieving results, confinement preferences, safety, reliability, maintainability, economics, operations, program options, and recommended strategy. Various levels of depth are selected with greater emphasis placed upon the more important topics for the purpose of this report. Emphasis is placed upon confinement options and more depth upon the preferred approach. Part of the depth also reflects the technology status in some situations.

PREFACE

Only on rare occasions does man have the opportunity to truly alter the way of technically accomplishing missions in a manner that significantly affects history. Current technologies become well embedded, and the initiation of wholly new technical concepts is a rare event, making most changes more of a technological refinement process rather than a radical departure from the norm. Now we face one of those rare opportunities — and challenges — relative to the space propulsion and power and thereby in drastically changing space flight:

> Imagine the intensity and anxiety as an Alpha Centauri Stellar Class Spacecraft first reaches its destination and successfully concludes the final stage engine burn, a mission which the stellar spacecraft had been executing faithfully for nearly three centuries — one which was accomplished using the spacecraft's native intelligence and a very highly efficient energy conversion system. The historic news does not arrive at Earth until 4.3 years later, but its arrival had been well anticipated since the Alpha Spacecraft had been faithfully reporting new science data ever since departing from Earth three centuries earlier. Then, several months later, consider with awe as it pauses during its search of the heavens surrounding its new stellar home to concentrate studiously upon one particular location, causing the 3 meter telescope to focus on a thermal anomaly noted by a delicate IR scanner. The first images reveal an intriguing planet, with the presence of an atmosphere. Could it be an oxygen and water atmosphere? The next data stream back to mother Earth provides the long sought after information...

Is the above description fact or fiction? To answer that question, think how the most respected scientists of 300 years ago would have described today's science and technology!

If high energy systems, such as fusion, can be satisfactorily manipulated to man's purpose and benefit in space, then exciting new space endeavors, such as stellar exploration, Oort Cloud science exploration, and major sample return missions from all planets, the asteroids, and many comets, can be mastered and space travel performed in a manner that man has only dared to dream. Energy can be made available for man's permanent presence on Mars as well as for his exploration of other bodies within the solar system. Given a sufficient amount of controlled energy, terraforming can even become a reality.

With the ever optimistic belief that man can achieve the once considered "not achievable," a dedicated analytical investigative study was undertaken primarily in 1988-89 to examine where the space program is headed with respect to meeting the high energy requirements that could be anticipated for future missions. During the course of the study I specifically made a concerted effort to determine whether or not fusion energy holds any promise for accomplishing missions of this nature; and if it does, what are the ramifications? Should NASA conduct research on fusion technology to develop its potential for space flight?

The conclusion reached is that the conversion of dream into the reality is achievable. Those missions are achievable, however, only provided that a concerted effort is made to adopt fusion energy for application to space. Advanced high energy missions involving the Alpha Centauri stellar mission category, and beyond, will require very long missions and some new thinking toward mission and spacecraft design. The task to develop fusion energy is an extraordinarily demanding endeavor, requiring NASA's total commitment and attention now, if we are to achieve that goal within a reasonable period of time. Otherwise significant programmatic, safety, environmental, and economic advantages, will be forfeited.

The present study – an effort conceived in 1987 – first analyzed advanced alternative space energy conversion techniques, and the practicality thereof, for purposes of:

- improving manned space exploration,
- gathering unique space science data,
- the conduct of space science research,
- safety enhancements,
- enhancing mission success.

To expand on this report's topical point – high energy space missions – one must first necessarily consider the types of missions projected and then consider the energy options for accomplishing those missions. The first program criteria in order to proceed will always be, "Can it be done within a reasonable time frame?" followed by, "What will it cost?" The next consideration should be, "What is the total safety impact?" The environmental impact should follow then, assuming affirmative responses.

Mission objectives are of utmost importance in these deliberations. Given the desire to conduct a manned Mars program, what is the best means to accomplish either one mission, multiple missions, or an extended presence there? The answer is a function of the question. For the first objective, i.e., the single Martian mission, the capability made available by the current chemical energy option would certainly economically trade more favorably over another more advanced approach. But for a continual presence, there are optional, safer propulsion means that will permit a continual Martian operational basis.

Assembled herein, then, is the space high energy mission story from the perspective of an all encompassing range of NASA programs including space and aeronautical activities, manned, unmanned, science, applied research, research programs, and program assurance. With regard to the latter, this study, in part, advances a new concept – advanced mission program assurance. It amalgamates research planning with flight program mission applications from a program assurance perspective. In that sense it is unique. Program assurance does not concern itself with advanced mission thinking, planning, and research, except to typically to provide oversight safety in the conduct of experiments.

The purpose for any program assurance activity is to obtain a high degree of confidence that a particular program will succeed. There are many reasons for program failure. Many revert to whether or not the proper planning and theoretical basis were sound at the beginning. Then, where sound, the degree of commitment and capability of the implementing staff enter into program assurance considerations. The amount of "up-front" funding is a measure of commitment and a key determining factor whether flight programs will meet objectives within budget and on schedule. The concept forwarded here is that to minimize program risk and to maximize program success, consideration of advanced research as a part of NASA's total program assurance activity is of utmost importance. This is a most appropriate phase for program assurance since it is one that offers the potential for problem elimination – the strongest assurance tool available. Hence, to consider program assurance for future missions much emphasis is placed in this report on theory, research, application, planning, strategy, and a program systems approach as the means to meet with success of NASA's advanced missions. The strategy offered is based upon that perspective.

Actually, this study reflects the interest and desire of several individuals who assisted, to see the United States' space program realize its fullest potential for the benefit of space science and space exploration by the means of the greatest and most expeditious utilization of national and natural resources possible. It was essentially an unfunded effort (with the exception of some of the mission performance computations performed by SAIC).

Authorization to proceed was granted late in the spring of 1988. The approach taken was to visit extensively with those individuals who have dedicated their life to fusion science and technology and to pure space science. That personalized approach was emphasized, as opposed to solely relying upon reviews of copious documentation without interpretation. One significant task performed in this study involved the integration of scientific background into space systems concepts and experience to be applied toward future operational mission considerations.

The first week was in residence at the University of Illinois with Dr. George Miley, Fusion Studies Laboratory. That was followed by two weeks at the University of Wisconsin with Dr. Gerald Kulcinski and Dr. John Santarius, Fusion Technology Institute. Seminars were held or attended at the universities. The site visits concluded with two additional weeks at the Lawrence Livermore National Laboratory with Dr. Grant Logan, Magnetic Confinement Fusion Laboratory. There discussions were held with individuals in the Inertial Confinement Program at Livermore as well. In addition, the Eighth Fusion Topical Meeting at Salt Lake City was attended to obtain a current perspective of the United States' terrestrial fusion program. Other activities included serving as one of the lecturers for Dr. George Miley's sponsored "Space Fusion Minicourse," reviews of documents, and discussions with fusion staff and space scientists. In addition, a trip was made to the Los Alamos National Laboratory where the status of their FRC fusion programs were reviewed with Drs. R. Siemon, M. Tuszewski, and others. Later, a short visit was made with Dr. Furth at the Princeton Physics Laboratory. Discussions were held with Drs. Roth, Reinmann, and Englert who had been previously involved with the Lewis Research Center's fusion research program during the 1960's and 1970's, the subject of Appendix A which is published in *Fusion Technology*, January 1991 issue (Sch91). During these visits, numerous individuals were consulted; the concept of *space fusion* forwarded; and the study's objectives were presented. Due to manpower/funding constraints, use was mainly made, where possible, of existing documentation rather than on the conduct of new analyses.

ACKNOWLEDGEMENTS

I am particularly very appreciative of the following individuals whose assistance and support for this study made it a viable accomplishment:

1. Dr. George Miley: Director, Fusion Studies Laboratory, University of Illinois

2. Dr. John Santarius: Senior Scientist, Fusion Technology Institute, University of Wisconsin

3. Dr. Gerald Kulcinski: Director of Fusion Technology Institute, Grainger Professor of Nuclear Engineering, University of Wisconsin

4. Dr. Grant Logan: Deputy Associate Director for MFE Development and Plans, Lawrence Livermore National Laboratory.

Special thanks and recognition are due to Dr. Santarius, who served on the Technical Planning Activity (TPA), referred to in this report, for his valuable assistance in the section dealing with plasma physics issues and status and his review of many drafts. Dr. Arthur Code, astronomer, at the University of Wisconsin was very instrumental in assisting on the subject of science missions, especially astronomy. Many in the fusion and science community have taken the interest and donated time to review the report and offer suggested changes to improve the contents and ensure that, where possible, the points made are technically sound. The list is long, but I would like to acknowledge their contribution and that their efforts are deeply appreciated.

The study would not have been as meaningful nor as expeditiously conducted without the valuable experience of those individuals consulted at the National Laboratories and the fusion research universities, plus many others not listed. Their insight, farsightedness, and experience were crucial in understanding the processes involved with fusion, the critical issues to address, how the program advanced to its current stage, and most importantly, in projecting where one can reasonably expect future advances. The manner by which this study was conducted is considered to be a sufficiently successful process that the following recommendation is made. For any future endeavors involving space fusion, the investigating personnel should further cultivate the relationship by directly interfacing with the fusion community which this activity has initiated.

Two key individuals at Headquarters, whose authorization and support permitted the undertaking of this study and who share interests in bold new endeavors were Dr. William Ballhaus, Associate Administrator for the Office of Aeronautics and Space Technology and Mr. George Rodney, Associate Administrator for Safety and Mission Quality. The encouragement and interest expressed by Dr. Geoffrey Briggs, Director of the Solar System Exploration Division, is deeply appreciated. Mr. Alan Friedlander and Mr. Jim McAdams of SAIC, Schaumburg, Illinois were very helpful in very quickly supporting this study on mission performance matters, including many subsequent consultations on propulsion performance. Mr. Friedlander presented a summary of the results of the mission analysis at the AAS/GSFC International Symposium on Orbital Mechanics and Mission Design in April 1989 at the Goddard Space Flight Center. This entire group of supporters all share a common interest in space and the advancement of man's knowledge.

The work would not have been possible without the kind understanding and assistance of my wife, Joan, who in my absence was left with the thankless job of performing many of my tasks at home while I was on travel or at work during odd hours.

NOMENCLATURE

LIST OF SYMBOLS

^{3}He	= helium-3, isotope of helium
^{11}B	= boron-11, isotope of boron
$\langle$Isp$\rangle$	= average specific impulse, seconds
$\langle\sigma v\rangle$	= reactivity parameter, cm^3/s
a	= radius of the spherical heavy-metal tamper shell (MICF), m
a	= the radius of the laser-compressed fuel mass (ICF), m
a_o	= mean acceleration, m/sec^2
A_i	= average ion mass number, a.m.u.
AU	= astronomical unit = 1.5×10^{11} m
Au	= gold
B	= magnetic field, tesla
C	= ratio of total plasma pressure to ion pressure
c	= velocity of light = 3×10^8 m/s
c*	= characteristic exhaust velocity, m/s
C_F	= thrust coefficient
cm	= centimeter
D	= deuterium, isotope of hydrogen
D-T	= deuterium-tritium
D_f	= photovoltaic foil (collector) receiver diameter, m
D_r	= laser receiver diameter, m
E	= energy
E*	= charged particle energy release per fusion
E_a	= alpha particle energy, joules
E_{fus}	= total (neutron + ion) energy released per fusion reaction, joules
E_{ign}	= plasma energy at ignition point, joules
E_M	= additional magnetic energy, joules
E_p	= proton energy
F	= thrust, newtons
f_b	= fuel burnup fraction
f_C	= $E_{charged,\ eff}/(E_{fusion})$, effective charged fraction of fusion energy yield
f_m	= lunar mass utilization factor
f_{mp}	= lunar mass utilization factor for propellant
F_{rep}	= IFEL pulse repetition rate
G_{fom}	= total energy gain figure-of-merit
G_{ideal}	= ideal fusion gain
GW	= gigawatts (10^9 watts)

Nomenclature

h	= Planck's constant, 6.6×10^{-34} joule seconds
hν	= photon energy quanta, joules
H	= henry
Isp	= specific impulse, seconds
J	= energy, joules
J	= mission difficulty parameter, m^2/sec^2
K	= torus elongation
k	= ratio of specific heats, c_p/c_v
keV	= kiloelectron volts
kg	= kilograms
m	= mass
m	= meters
MA	= megamps
MeV	= million electron volts
M_{laser}	= laser mass, MT
M_o	= initial vehicle mass, MT (= propellants + inert vehicle + payload)
M_{of}	= vehicle mass at completion of thrusting, metric tons
M_p	= propellant mass, MT (includes fuels and diluent)
$M_{reactor}$	= reactor mass, MT
MT	= metric tons
MW	= megawatts
MW/m^3	= fusion plasma power density
N	= thrust, newtons
n	= neutrons
n	= ion density, number of ions per cubic centimeter
n_e	= number of electrons
n_i	= fuel ion density, ions/m^3
nm	= 10^{-9} meter
n_t	= Lawson parameter, $cm^{-3}s$ (fusion plasma = plasma losses)
p	= plasma pressure, pascals
p	= proton
p	= the fusion plasma power density
P_j	= jet power, kW
P_l	= average laser power
P_{rad}	= radiation loss
R	= range between transmitter and foil (collector) receiver
R	= universal gas constant, 8×10^3 J/kmole K
R/a	= torus aspect ratio
RF	= radio frequency
s	= ratio of a FRC's radius to an average gyroradius in the device

s	= seconds
T	= temperature, K
T	= Tesla
T	= tritium, isotope of hydrogen
t	= round trip flight time, years
T_e	= electron temperature, keV
T_i	= plasma's ion temperature, K or keV
$T_{ign(ideal)}$	= ignition temperature with Bremmstrahlung loss only
T_{ign}	= ion temperature at ignition
T_p	= duration of thrusting
v_i	= mean ion velocity, m/s
V_m	= magnetic field volume, m^3
V_p	= plasma volumes, m^3

GREEK

α_{IFEL}	= IFEL specific power, kW/kg
α_p	= specific power, kW/kg
α_{p1}	= specific Power, where α_p = 1 kW/kg
α_{p10}	= specific Power, where α_p = 10 kW/kg
α_{p100}	= specific Power, where α_p = 100 kW/kg
α_r	= reactor specific power, kW/kg
β	= ratio of plasma pressure to magnetic field pressure [$p/(B^2/2\mu_o)$], %
χ_E	= thermal diffusivity, m^2/s
Δv	= velocity incremental change, km/sec
γ	= payload mass fraction, % (payload mass/initial vehicle mass)
η_a	= electrical efficiency of the auxiliary plasma heating system
η_c	= coupling efficiency of energy to raise plasma to ignition
λ	= laser wave length
m	= H/m
m_o	= 1.6×10^{-6} henry(H)/meter
v	= vibration frequency, vibrations/sec
ρ	= gyroradius, cm, (characteristic radius of a charged particle's orbit gyrating around field lines in a magnetic field)
ρ_w	= shell mass density, kg/m^3
σ	= nuclear cross section, cm^2
τ	= fusion reaction time, seconds
τ_E	= cross-field thermal conductivity loss time for plasma energy
τ_E	= energy confinement time, seconds
τ_ι	= fuel ion confinement time during the burn at peak compression, seconds

ACRONYMS

AAS	= American Astronomical Society
CTOR	= Compact toroid reactor
DARPA	= Defense Advanced Research Projects Agency
DOE	= Department of Energy
FEL	= Free electron laser
FRC	= Field Reversed Configuration, magnetic confinement experiment
GEO	= Geosynchronous orbit
GSFC	= Goddard Space Flight Center
HEPS	= High Energy Power System
HESM	= High Energy Space Mission
ICF	= Inertial Confinement
IFEl	= Induction FEL
ISAM	= In-situ Analysis Mission
JET	= Joint European Torus, magnetic confinement experiment
LEO	= Low Earth Orbit
LLNL	= Lawrence Livermore National Laboratory
LSX	= Large Scale "s" Experiment (FRC), magnetic confinement experiment
MCF	= Magnetic Confinement Fusion
MEM	= Mars Excursion Module
MFE	= Magnetic Fusion Energy
MICF	= Magnetic Inertial Confinement Fusion
RFP	= Reversed Field Pinch, magnetic confinement experiment
RTE	= Return to Earth Abort
RTG	= Radioisotope Thermoelectric Generator
NERVA	= Nuclear Engine for Rocket Vehicle Application (fission thermal rocket)
OTV	= Orbital Transfer Vehicle
PLT	= Princeton Large Torus, magnetic confinement experiment
SAIC	= Science Applications International Corporation
SDI	= Strategic Defense Initiative
SFP	= Space Fusion Program
SFR	= Space Fusion Reactor
SOAR	= Space Orbiting Advanced Fusion Power Reactor
TFTR	= Tokamak Fusion Test Reactor, magnetic confinement experiment
TPA	= Technical Planning Activity
VISTA	= Vehicle for Interplanetary Space Travel Applications

1.0 INTRODUCTION

A new class of space missions – those requiring high energy levels – and the means for their accomplishment is addressed. Other new concepts are introduced. One major underlying consequence of this study is to forward the concept of *Advanced Mission Program Assurance*. Program assurance is treated in NASA as the activity which concerns itself with (1) the successful and (2) the safe implementation of current programs. It is, therefore, concerned with the conduct of programs in a manner that minimizes risk from the viewpoint of both objectives – mission success and safety. It has not been concerned with advanced missions. One key measure of program risk, however, is the degree of technology understanding at the commencement of any program. Program risk, in that light, is a function of the research and development status of the technology used. Whether the proper research steps are being taken to assure that the space program advances with the proper perspective for mission success and safety considerations is a reasonable question to address as part of a risk assessment task since it determines the foundation of a program's technical state. Safety is a direct function of the nature of the energy sources, including both the inherent and the designed-in degree of control. The higher energy levels have been typically considered the more hazardous. New thought is given herein to that adage. This report then advances a new concept, advanced mission program assurance, and the means for reduction of risk through technical advances, not only prior to program commencement, but as an input to research programs.

Looking then with many visions to the future, what is the focus that NASA should be making regarding missions requiring high energy sources? How can program and safety risk levels be reduced? Does one energy source offer any inherent advantages over another? What is the state of NASA's space related energy conversion research, and how well will the agency be prepared to meet the visions and the challenges of the forthcoming programs? These are some of the fundamental questions that motivated this analysis.

The key element in the accomplishment of our future missions is energy. It shown that fusion energy, if available, will serve an essential role in the implementation of NASA's space mission. Fusion will advance the state of risk reduction as an added benefit to the mission enabling capabilities which it provides.

The study flow that addresses the above and other topics is presented in Fig. 1.1.

1.0 Introduction

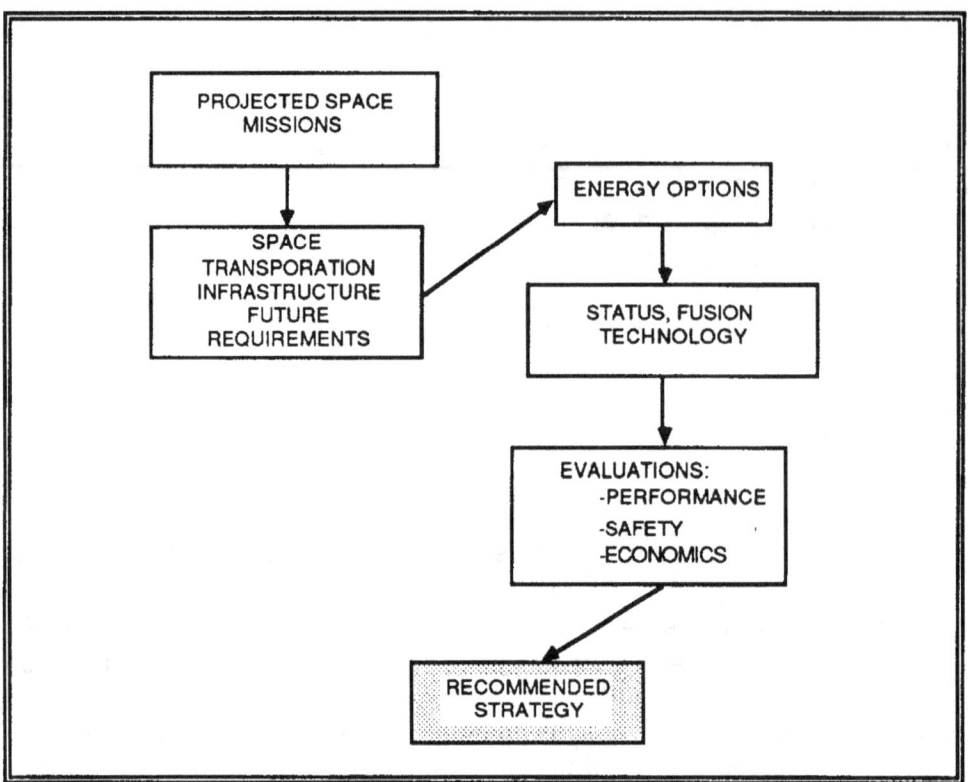

Fig. 1.1. Study flow and content.

During the course of this study two different opinions were voiced. One stated that NASA can accomplish all objectives using current propulsion infrastructure technology. The development of an engineering program like fusion will detract from today's science missions. The other point made was that there is no doubt that we would use fusion if it were available. The question is, "How viable is fusion energy?" This study, therefore, treats both points – diversion of science funding and system viability – with equal interest and concern.

The current United States program for fusion energy, managed by the Department of Energy (DOE), has appropriately been dedicated to address terrestrial power applications to help respond to the critical energy shortages that can be anticipated in the not very distant future. Now is an appropriate time, however, to focus attention on a different application, namely space, where the high specific energy yield from fusion fuels makes it a potentially highly desirable source of space power, perhaps the most desirable of all known energy sources. But, the mission requirements for NASA's civilian space fusion applications, particularly propulsion, differ in some important aspects which will not necessarily be addressed in the terrestrial electrical power program. Hence, research results are not anticipated to be accomplished for space applications as a consequence of differences in program goals between the two agencies.

This document, thus, reports on the results of an analysis, the objectives of which are:

- to determine the nature of bold new missions in space science and exploration by initiating a new High Energy Space Missions (HESM) class – one which has not been addressed by NASA,
- to determine the energy means for accomplishing that mission class with emphasis on the preferred source of energy,
- to examine the flight system requirements, operations, and other pertinent aspects, such as, feasibility, safety, and the economics involved in the energy conversion means of accomplishing the missions,
- to enhance manned space flight exploration and safety,
- to compare the advanced high energy sources, primarily fusion and fission, and
- to present a high energy conversion program strategy for NASA to implement to accomplish high energy space missions and in particular, to synthesize an advanced energy program for space.

To accomplish the above objectives, the analysis integrated space missions with the means by which various sources of energy including fusion and fission would be employed to accomplish those missions. Specifically, sections in the report are included to:

- evaluate space missions where high energy can be an enabling or enhancing technology,
- examine fusion reactors and system requirements for possible space applications,
- show the technological means for the enhancement of safety and the reduction of operational costs of space flight missions, and
- to show why it is important to provide an energy option to NASA's space program.

The results of the analysis show many advantages of "space fusion energy."

Throughout this report a clear distinction has been made to differentiate the two applications of fusion energy as indicated by the phrases "space fusion" and "terrestrial fusion." Those phrases are used to distinguish the individual, distinctive features which permit a reactor to meet the requirements for their

1.0 Introduction

respective applications. A "space fusion reactor" is, therefore, one which meets the criteria presented in Section 6.0, "Flight System Considerations and Requirements" to accomplish space missions of the type defined in Section 2.0, "High Energy Mission Applications."

Considerable emphasis is placed on missions in this report. One dividend of this study is to bring forth new ideas and to make projections where the space program might be headed and specifically to examine and expand options for both unmanned and manned programs. The advantages offered apply equally to both.

Science missions in particular have received emphasis. The rationale is quite clear, namely, the return from space research is much greater using unmanned spacecraft with science instruments. There is a perceived need for this level of mission enabling energy, particularly for missions beyond the solar system — the next step in space, which has not been addressed.

While humans offer the opportunity to perform real-time in-situ decision making, the institution of man as a science instrument is performed at a high cost level taking into consideration the return from the investment. On the other hand, it is man's innate character to explore and to expand his presence beyond his current environment. The economic and safety trade-offs of high performance propulsion will favor the manned missions due to the intensity of operational requirements and due the fact that humans are on board. The Manned Mars Missions will require high power levels to deliver more massive payloads. We will eventually look beyond Mars for exploration. High performance propulsion beyond our currently researched technology is essential for those missions.

The Agency objective is to accomplish a good balance between science and manned missions such that one complements the other, also an objective of the approach presented herein.

Traditionally space science has operated in a "energy-responsive" mode as opposed to setting requirements for advancing technology. During a review of this report near its completion, the judgement was offered that mankind traditionally refrains from the pursuit of endeavors of a science nature, but instead advanced energy activities are directed toward strategic defense applications. This report suggests that we transcend that time honored philosophy to advance toward more humane, sublime principles. Knowledge is, in the end, the instrument of advancing civilization and mankind's well being. That, in the end, is the result of science.

There are many reasons, the history of which is discussed in Section 2. A change to a more active role is suggested, however, and several points are worthy of note here.

The importance to NASA of greater science program emphasis is made in the Summary and Principle Recommendations of the Advisory Committee on the Future of the U. S. Space Program (anom90, p. 9).

Principal Recommendations Concerning Space Goals

It is recommended that the United States' future civil space program consist of a balanced set of five principle elements:

- A science program, which enjoys highest priority within the civil space program, and is maintained at or above the current fraction of the NASA budget

- A mission to Planet Earth (MTPE) focusing on environmental measurements

- A mission from Planet Earth (MFPE), with the long-term goal of human exploration of Mars, preceded by a modified Space Station which emphasizes life-sciences, an exploration base on the moon, and robotic precursors to Mars focusing on environmental measurements

- A significantly expanded technology development activity, closely coupled to space mission objectives, with particular attention devoted to engines

- A robust space transportation system.

While the Committee's recommendations above did not address the energy means to accomplish their recommendations, the conclusions drawn in this study and recommendations are fully supportive of those recommendations.

The importance of abundant energy for the conduct of science missions is well illustrated by the National Research Council (NRC) Space Science Board's report on *Planetary and Lunar Science Exploration* (Don88). Although the subject of high energy classes of missions was not one of the board's considerations, discussion was provided on the technical developments needed by the future science programs:

> Many of the recommended investigations will be enabled or enhanced by technical developments beyond those of the mid-1980's.

Seven recommended technical developmental areas were listed, three of which directly relate to energy, or at least to the need for energy augmentation pertaining to:

- low thrust propulsion to permit serious study of comets, asteroids, and the solar system beyond the inner planets;

1.0 Introduction

- power sources;
- on-orbit staging, assembly, and fueling to offer new capabilities for the more ambitious missions, especially returning samples from Mars.

Those statements give a clear statement of the importance of higher energy needs for the conduct of more ambitious, energy intensive space science missions.

Even the four other NRC recommended technical developmental areas have some ancillary appropriateness for high energy discussions. These included: hard or semihard lander technology to reduce spacecraft costs; further developments in robotics and artificial intelligence (AI); radiation and high temperature electronic improvements; and support for data analysis and data interpretation from transmissions to the ground from spacecraft.

The additional magnitude of vehicle performance increase, as provided by fusion energy, for example, would directly benefit science by enabling missions of greater difficulty. Furthermore, more intensive science objectives and missions of greater duration are achievable, including transporting multiple, heavier landers of the type referred to in this report – while using less massive orbital launch vehicles and a reduced number of launches. In missions where spacecraft traverse very long distances which involve lengthy flight times, like greater than a century, fusion creates a new meaning and challenge for advanced artificial intelligence capabilities. As that high energy mission class is implemented, there will be a need to design the capability for spacecraft systems to revise science objectives during real time flight operational situations as well as to conduct preplanned in-situ data analysis. For example in the stellar mission, the magnitude of the distances involved dictates that we employ AI rather than have a spacecraft wait 9 years at Alpha Centauri for data to travel to Earth and back for the spacecraft to receive Earth's return instruction. Self diagnosis, self analysis, and even repair, in that context assume an entirely new meaning and an exciting challenge. The augmented science in the form of more electrical power for additional experiments and provisions for higher data rates that high energy missions can provide would also indicate the need for even greater improved data support for missions than was envisioned. In addition, there would be the challenge to conduct meaningful on-board real-time science data analysis and automated decision making. In a sense, the spacecraft's "brain" will in some respects have to become trained to think like that of a scientist!

What is the motivating factor to consider initiating a NASA Space Fusion Research Program at this time? The analysis shows significant advantages could be accrued if fusion energy conversion were available.

An overview of the newly issued "U. S. National Space Policy" is presented in the "Executive Summary." Fusion energy fits well within the context of the policy goals and that capability should ultimately become a part of the NASA space transportation infrastructure. To examine that point, consider the policy statements further (p 4):

> The goals of United States space transportation policy are: (I) to achieve and maintain safe and reliable access to, transportation in, and return from, space; (2) to exploit the unique attributes of manned and unmanned launch and recovery systems; (3) to encourage to the maximum extent feasible, the development and use of United States private sector space transportation capabilities; and (4) to reduce the costs of space transportation and related services. (p. 4)

Consistent with the first goal, this paper examines and places great emphasis upon the safety advantages of fusion as well as the impacts and hazards which can be defined at this time. Also, there is great emphasis placed upon the 4th goal – means to reduce costs of doing business in space. Goals 2 and 3 are also relevant to this study as discussed later in the report.

One of the key points made in this document is the need to have an option for an energy source to the only other potential contender for high energy missions – fission. By accomplishing fusion energy development, which is considered to potentially provide a greater energy capability, reliance is not placed upon a single energy source; there is an option. Other aspects of fusion are likewise examined: safety to the public and to the flight crew, reduction in the costs of doing business in space, reliability, and ultimate possible commercial uses.

An activity similar to this one in scope but performed for Air Force missions by the Air Force Studies Board's Committee on Advanced Fusion Power was conducted in 1987 (Mil87). In the Executive Summary,

> The committee concluded that the prospect for achieving aneutronic fusion [free of neutrons] is doubtful, but the use of advanced fuels like D-^{3}He appears more feasible and offers many advantages for space applications. ... Based on the fact that no other insurmountable technical problems are envisioned and using projections of competitive performance parameters (such as power/unit mass) the committee concluded that fusion is potentially attractive for select Air Force applications.

The committee then recommended that a comparison be made with fusion and non-fusion technologies and that, if the Air Force intends to pursue fusion, a follow-on study be formulated to define a research and development program plan for using fuels with reduced neutron yields.

1.0 Introduction

Subsequent to that study's completion, the Air Force concluded that no Air Force mission in the fusion energy level existed and decided that the recommended follow-on work would not be pursued. Later, however, a one year, $150K contract with McDonnell Douglas Astronautics was awarded through the Air Force Rocket Propulsion Laboratory, the purpose of which was to advise the Air Force on the prospects of using fusion for space propulsion for Air Force missions. That report has been recently completed (Hal89). Many of the conclusions made in the the Air Force study and this study are similar although there was no collaboration between the two.

Taking today's fusion research status into account, we note that a number of new developments have taken place, particularly with regard to the viability of the low neutron producing fusion fuel reactions. Those fuels are particularly suited to space program missions and, in fact, they have greater applicability and advantage to the space application than to the commercial power plant fuels.

Furthermore, the widely ranging mission applications, the timeliness, the progress in the DOE program, and the terrestrial program's developmental status stress the importance of dealing with this matter now. Both the DOE Magnetic Confinement Fusion (MCF) program and the Inertial Confinement Fusion (ICF) program have made some very significant advancements. The development activity for space fusion energy will involve a considerable period of time. If fusion energy were available now, substantial program operational savings, amounting to tens, even 100's, of billions of dollars, would be realized by reduced operational costs.

Next, the value of high energy to the manned interplanetary missions with respect to the safety of the flight travelers is apparent when one considers that the radiation exposure hazards, the psychological difficulties, and the physiological problems are all reduced by shorter trip times. Cosmic radiation, integrated over a period of time as experienced during a chemical propulsion powered Manned Mars Mission, is perceived to be a serious safety concern at the present time.

Another advantage is that the time available for the conduct of science is lengthened by the added mission capabilities, made possible by the additional payload mass. The vehicle's payload mass fractions are increased, making more massive payloads possible and thereby effecting economies of missions. Safety also requires greater mass for shielding of the flight crew and to provide options for safe flight operations.

In addition, renewed interest in advanced planning in NASA was stimulated at the start of this endeavor by the President Reagan's request and from the charge by Congress, to "formulate a bold agenda to carry America's civilian space enterprise into the 21st century." (Pai86) What could be more bold than to undertake new energy developments leading to the settlement of the solar system and to missions to the stars? A related review of new space endeavors culminated in the report, *NASA, Leadership and America's Future in Space*, the results of which were presented to NASA during March 1987 (Rid87). An advanced missions planning office, The Office of Exploration, was established to continue the planning. On July 20, 1989 President Bush requested Vice President Quayle to chart a new course for the nation's space program, one which looks to the moon, Mars, and beyond. The NASA response was prepared by a task force during a 90-day study of a human exploration program (Anom89). A draft of the *Executive Summary* of this high energy mission report was provided to that effort and used as an input. A revised United States National Space Policy was written, November 2, 1990, affirming the United States commitment to space science and exploration (Anom89).

The availability of any large, specific source of energy, such as fusion, will open new horizons for NASA. We can conduct more bold missions than ever previously dreamed. Just consider some of the various thinking starting to come to the forefront, like the Office of Exploration's sponsored Space Enterprise pilot study which examined the use of lunar resources for large scale commercial applications, including the mining of helium-3 fuel for terrestrial applications and the construction of large solar power satellites using lunar materials. Consider the importance to a Martian settlement of the availability of megawatts to gigawatts of electrical power and high performance propulsion for logistics support.

Fusion energy, or any high specific power system, can contribute to cost savings by carrying out multiple mission objectives on any given flight which otherwise could never be economically feasible and/or technically possible. Fusion energy can potentially become a very significant factor for NASA's space mission architecture in the future mission planning equations now underway. Under the current mission performance restrictions due to low energy systems, a large number of small payloads are required using a multitude of launches by chemical propulsion systems. A high performance fusion powered vehicle has the advantage of reducing that number and from that perspective, substantially lowering mission costs and improving safety by reducing the launch operational requirements.

Further, consider that the technologically relatively easy, low energy science program mission objectives can be expected to be accomplished in the not too distant future. Where will we look for energy beyond those missions? Is Earth remote sensing to become the final phase of "space flight," or are we going to get involved with exploration further out than our current energy potential allows? After the near Earth missions have been flown, the demand will be for the more difficult, high energy missions. Attention will turn toward higher specific energy sources which are capable of providing higher specific power. But that development will be time consuming, requiring preparation now.

From the results of this analysis, fusion is the most viable high energy candidate source to consider; and in view of the above rationale, now is the appropriate time to consider it. The key issue to deal with is not what it can accomplish as

1.0 Introduction

much as the question of how viable is fusion as an energy source for space missions. That is the issue on which space resources must be focused.

Under the current strategy, no effort exists for space fusion research. Actually, a fusion program would not commence a new discipline to the agency. Earlier in NASA's history, 1958-1978, a modest fusion research program was established and research performed at the Lewis Research Center, in the Advanced Concepts Branch, Electromagnetic Propulsion Division, to pursue fusion energy conversion and applications for space. That activity was terminated in 1978 in the wake of the many cost reductions that NASA underwent during that decade. The program, contributions, and bibliography are discussed in *Fusion Technology*, January 1991 issue which is provided as Appendix A (Sch91) for ready reference.

In the light of the advances in the terrestrial fusion program, the renewed interest in advanced space missions, and the belief that fusion energy does offer a potentially enabling and enhancing space technology which should be researched by NASA for application to the space exploration and science programs, a concerted study effort was made by the author to carefully analyze the application of fusion energy to space. This was accomplished using the expertise of fusion scientists at the National Research Laboratories and universities as well as from the available literature. Specifically, site visits were made to the University of Illinois, University of Wisconsin, and the Lawrence Livermore National Laboratory during the latter part of August and in September 1988, to the Los Alamos National Laboratory in October 1988, and to the Princeton Plasma Physics Laboratory in 1989. This report reflects the results of an independent study commenced then plus continuing reviews of the literature and discussions with interested individuals in the space science and fusion disciplines.

As an enabling technology, fusion accomplishes strategic objectives for NASA missions:

- NASA's goal of manned exploration of the solar system including the establishment of a Mars colony which can be implemented in a more economical manner.

- Attainment of new science missions not otherwise practical.

- More intensive science research return as a result of more massive payload capabilities.

- Accomplishment of new manned space exploration missions, those beyond Mars, that otherwise could not be considered due to power and trip time limitations.

- Improved safety of flight. It is clear from fusion powered mission flight times that safety is enhanced by reduced flight times, a topic of discussion later. Although it is too early to provide proof, it appears that fusion technology and its machinery holds promise for significantly enhancing the inherent safety and reliability of manned spaceflight equipment.

- Major economic improvements in the cost of doing business in space.

- Free enterprise potential for space. If there ever is to be any hope that free enterprise will play a role in the wide development and utilization of space, there must be an abundance of energy at an affordable cost. Fusion has to be considered the prime contender, at least for the foreseeable future. For that to occur, substantial increases in specific power must be achieved to increase the payload mass fraction and to make available less manpower intensive vehicle system designs. Higher efficiencies are mandated.

- Reduce requirements for large numbers of heavy lift launch vehicles.

With the study objective in mind to evaluate whether NASA should undertake a space fusion program, two key fusion related topics must be addressed from a management perspective, namely,

(1) the missions for space science, exploration, and other applications where fusion can provide benefits and

(2) its technical viability as a credible energy source for space applications.

Those two subjects consequently comprised the focus of this report. The report is completed with a statement on what we should do about it.

In the context of this report, "space" is taken in a broad sense, that is, it encompasses space operations commencing at the Earth's surface, i.e., the transportation to and the return from space, as well as the trip times to and the visitation time at extraterrestrial bodies. The space missions presented in Section 2.0, "High Energy Mission Applications," comprise a new mission category not previously considered. Some mission types for which NASA has already planned science objectives have been extrapolated to high energy applications, plus other new missions not considered. In part, some missions were inspired by the 1988 National Academy of Science report (Don88) outlining a map for an exciting science program, particularly with regard to astronomy, physics, the planets, sun, and lunar science.

The rationale for mission considerations is quite clear. Fusion must either enable man to accomplish mission objectives that would not otherwise be achievable and allow the accomplishment in a more cost effective manner, or alternatively, realize higher yields such as safety and economic enhancements.

1.0 Introduction

Ultimately it must "pay" for itself, and that is the vein in which space fusion, or any other endeavor, should be pursued.

The technology advances which have been made toward the production of net fusion energy provide us with a better understanding of the effort involved with space fusion. While fusion energy conversion systems have not advanced to the degree that nuclear fission has, there is merit in evaluating reasonable candidate fusion concepts for performance and costs estimates for obtaining rough comparisons. That comparison process is particularly appropriate since the more currently advanced researched fission thermal propulsion and power technology has not been performance and cost substantiated for flight operations either. Fusion has a more extensive technological developmental background in comparison with gaseous core reactors and is well in advance of matter-antimatter energy systems.

With the conviction that both of the two key issues – cost effectiveness and reactor developmental physics – will be proven, consideration was further given to the system aspects of fusion powered vehicles. System considerations ultimately must be taken into account in the development of a flight program. Without the capability to effectively implement the system requirements, the concept of fusion energy becomes only of academic interest for space, or, alternatively, the magnitude of the flight program efforts becomes grossly understated, a common program error. The system considerations ultimately are the driving criteria for critical parameters like fuel selection, and it is most important that system considerations be pursued at the earliest stage to achieve an optimal program.

Finally, taking into account the aforementioned topics, a recommended strategy is offered as part of NASA's overall strategic planning. With recognition that balanced budgets and balanced research program priorities are a part of the research management decision process in NASA's aeronautical and space research programs, program options, including the recommended strategy, for space fusion are provided.

The aforementioned comments were provided to acquaint the reader with the study objectives, content, and the approaches taken. Some comments are now offered with regard to the reader of this document. It is intended to provide both a program mission analysis and a technical analysis to focus on a thoughtful articulation of the key managerial and technical issues for consideration, principally by NASA management as part of its decision process. But the report is meant to be more than a technical management summary. It is intended to serve as a stand alone technical report for one who is unacquainted with the field of fusion and the application of fusion energy to space missions. A very brief fusion tutorial is provided in sufficient depth to permit one unacquainted with fusion technology to understand the report's contents. Although technical data and descriptions are included to substantiate the report's conclusions, it is not intended to be a treatise for the expert in the technical fusion field. Consequently, there may be and probably are particular, important parameters to fusion experts that were not discussed here. As with many broad, encompassing works of this nature, many specific supportive details cannot be elaborated upon. References are given to substantiate the presented data. Since the report is lengthy, some sections have been written as stand-alone, so there is repetition of some of the study's major points and themes between sections.

The study focuses heavily upon the importance of fusion energy to space. That focus should not be misinterpreted as a narrow, no-option approach. Indeed, the opposite is intended. That is, the goal is to uncover and forward an articulated discussion of an energy option which has been neglected. Comparative evaluations of energy sources have been presented at the request of individuals with whom the report was given early reviews. This overall topic is brought forward as one which must be given senior attention:

> The importance of performing developmental research to provide larger, safe energy sources for the continuation of an advancing space program—one that can deliver all that is being requested of the agency in the 21st century— is such that it cannot be neglected any longer.

2.0 HIGH ENERGY MISSION APPLICATIONS

The current planning for space science missions has focused upon that science which can be gleaned from energy capabilities using the existing space transportation propulsion and power capabilities. The preferred approach is obviously to pursue science based upon the establishment of mission requirements which originate solely from the need to fill voids in science understandings. The current chemical energy systems characteristically yield inherently low specific energy which can only lead to energy conversion systems having a low performance design, that is, systems which can deliver ~7 km/sec. Of interest are systems which can deliver Δv's from 90 to 20,000 km/sec. Using a higher magnitude of specific energy from nuclear processes where the inherent energy releases are much greater per unit mass, new space science and exploration programs became viable while others can be accomplished more quickly and with greater science yields and mission reliability than otherwise possible. The means for achievement comprises the thesis and pursuit of this study.

The current low energy level approach to science missions, i.e., the consideration of missions falling within chemical propulsion capabilities, is not unreasonable and could certainly be anticipated, particularly in view of the Shuttle tragedy which resulted in the deferral of space science missions. Immediately following the accident, the science community had been primarily occupied with and oriented towards launching the currently built spacecraft rather than considering advanced new space missions of the energy level analyzed here. The impact of the deferrals on space science and the concern over mission launches are particularly understandable when one takes into account that the last major scientific payload launched prior to Magellan in 1989 was Pioneer Venus in 1978. It has been sufficiently difficult to seize the available science which is obtainable within the bounds of current technology without requesting any new ventures requiring the use of advanced energy systems, particularly of the magnitude that is not available. Even with the current specific energy sources and their energy conversion technology limitations, NASA has been able to accomplish tremendous improvements in science yield through ingenious mission planning using gravity assists and through improvements in science instrumentation by the principle investigators of space experiments. Further, it would not be prudent to place the ability to conduct one's science program on a non existent capability. But there are significant limitations that can be broadened and science horizons increased using high energy flight systems just as the current chemical propulsion technology has broadened our knowledge over the Earth based science instruments and systems. Tremendous gains have been made and are also yet anticipated from space based science instrumentation and systems which examine γ-rays, x-rays, ultra violet and infrared radiation, atmospheric distortion-free visible light, in-situ science outposts, and locally observing

2.0 High Energy Mission Applications

particles and fields data, imaging sensing systems, plus others. Much science remains, but access will be increasingly difficult and time consuming as missions extend further into space. The conclusion drawn is that science, while it should in general be the most forward looking of any discipline, has recently in the case of space science been placed in a conservative position of concentrating on science objectives which consider only current energy conversion technology.

Let us examine approaches and policies over the past decade to examine the where program planning has led us into today's space science program activities. From that experience let us then proceed with guidance for the future. Lowering the costs for the conduct of space science has been a key factor in limiting space science to low energy missions. The programmatic means to reduce costs was addressed by the SSEC (Solar System Exploration Committee) (Anom83).

> The Committee believes that the planetary programs have grown in costs because of three dominant factors. ... The Committee's Core program recommendations concerning this and their implications, are summarized as follows:
>
> 1. Maximize hardware and software inheritance ...
>
> 2. Control scientific scope of missions
>
>   • Restrain and focus scope of missions;
>
>   a. Payloads limited to highest priority objectives.
>
>   • Judicious separation and combination of mission objectives;
>
> 3. Minimize changes after original mission definition
>
>   • Forego missions where technology developments are of an enabling nature;
>
>   a. No requirements for launch capability beyond that already available;
>
>   b. No missions requiring solar electric propulsion system. (p 82)

The Committee considered that in order

> To maintain the tightest control possible over costs, the missions of the Core program **should impose no requirements for enabling technologies** (for example, new upper stages, low-

2.0 High Energy Mission Applications

> thrust propulsion systems, mobile lander systems, intact sample return capability). (p 71)

On the other hand, the committee also recognized that mass margin is a cost driver.

> Mass reduction programs required by inadequate initial mass margins have been very expensive, and, therefore, the *Mariner Mark II* approach makes a conservative allowance for such growth. (p 79).

Recognition was also given to the economic advantage of combining mission objectives:

> Furthermore, combining as many objectives as are technically feasible into a single mission to lower the cost of achieving all the objectives can provide a cost-effective mission that is not affordable.(p 70)

Clearly there are counter economic and performance forces at work. The mass margin and performance margin and the multiple mission objectives are best served by greater space propulsion power. The alternative is to severely limit mission objectives, and that restraint is also suggested in recommendation 2 above. Hence, in order to be affordable, the space science program by policy in essence has been restrained from considering high energy advanced missions of the type considered herein, creating a dichotomy in terms of the overall program economics.

The payload cost information presented on page 69 shows that the average cost of science payload programs is on the order of $400M, excluding the more expensive Viking, Voyager, and Galileo programs. These costs exclude the launch costs which can nearly double the total costs to place a science payload into LEO. Refer to Fig. 2.1 which is Fig. 6 in the reference.

2.0 High Energy Mission Applications

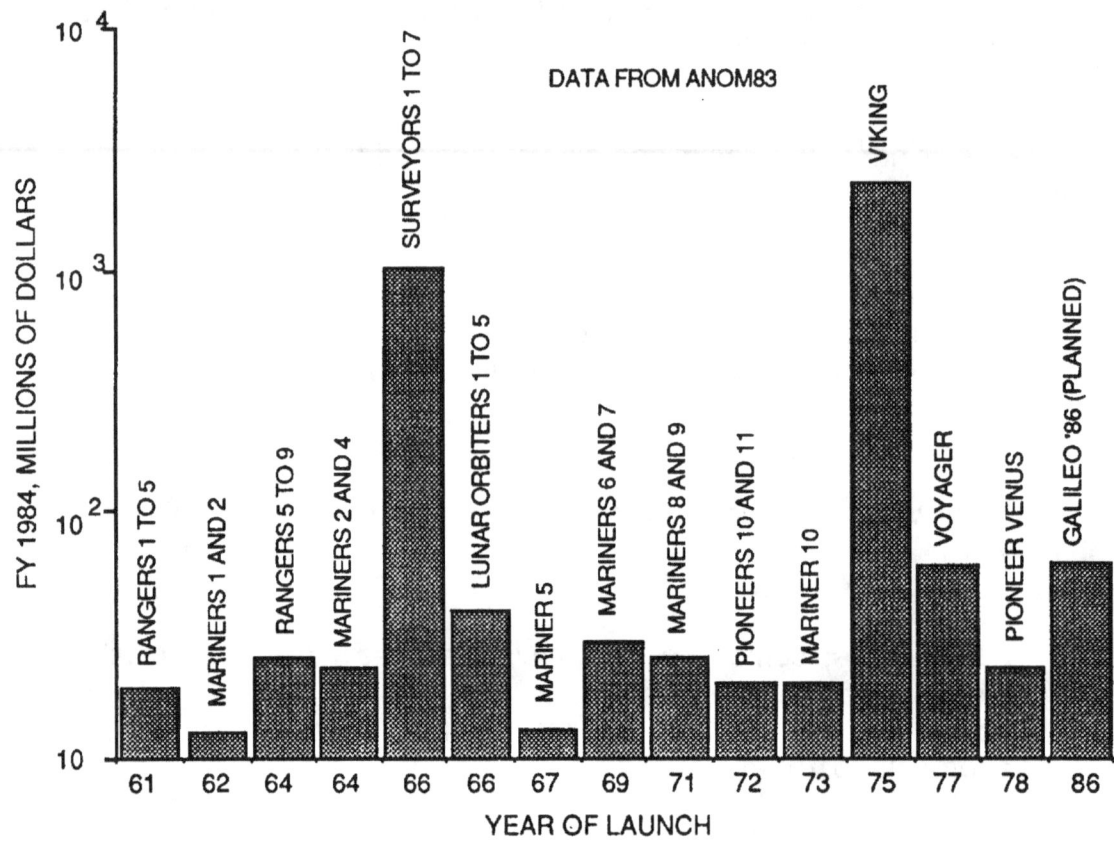

Fig. 2.1. Science payload program costs (Anom83).

Where repeated missions are conducted to obtain the additional science at the same target, the launch costs are obviously additive. The advantage of greater payload delivery capability is apparent, but funding limitations have resulted in the implementation of the aforementioned performance limitation measures in order to conduct a space science program. The Committee was concerned about the limitations imposed as indicated by the following:

> Inevitably, some of the highest priority science goals recommended by the Space Science Board cannot be accommodated within the Core program. Specifically, the return of samples from Mars and the exploration of the Martian surface by mobile laboratories are goals whose scope inevitably must lead to costs that are not affordable given current NASA priorities. (p 70)

This approach has been technologically restraining with respect to new developments concerning higher energy solutions. Mission requirements have not been forwarded for high energy missions, and consequently a high energy capability has not been researched. But without new high energy technology, the program will ultimately stagnate. As a result, costs for the missions today are higher than necessary if a high performance space energy conversion capability were available.

The point is that an entire systems approach must be taken to arrive at an optimal solution. We cannot examine missions, the transportation means to accomplish those missions, the research, the safety, the reliability, the economics, the timeliness, and the management of those programs all as separate entities. The tendency is to focus too narrowly on one objective, i.e., compartmentalization occurs. This activity attempted to take an overall systems approach to examine all aspects of space missions – the above elements plus others as well.

This study activity initiates consideration of suggested high energy missions, those which advanced high energy propulsion could best serve – a category not given consideration in view of the past constraints. The energy source options to carry out those missions are elaborated upon in Section 3.0 "High Energy Sources for Space." Those missions are intended to advance the space program by making available the capability to conduct exploration and science programs at greater distances using greater payload masses and by providing quicker trip times for attaining a greater and more rapid return of science data and conduct of space exploration.

What could be a stronger motivating factor for encouraging enterprising young scientists to enter the space science field than to provide a quick return on data and to provide greater mission operational flexibility, including more advanced missions and the capability to alter mission objectives in real time to a greater extent than presently available? Under the current technology, a space scientist could come very close to spending the person's entire professional career on one mission just as is occurring on the Galileo mission which will exceed ~20 years from concept to data retrieval. The length of time for return of science data will only become more lengthy as the frontiers of space science expand further and further from Earth. Hence, one major motivation of and objective for this endeavor was to evaluate the shorter flight times and greater return of science that could be achieved by the use of high energy. The science program should be capable of being conducted under more flexible auspices.

The first step in considering high energy sources is to evaluate the requirements placed upon the flight systems as established by space mission requirements. Therefore, in this section, high energy mission classes are hypothesized and examined. These missions are not intended to be all inclusive of the high energy science mission objectives – an ever expanding frontier – but are examined to consider and expand the wide variety of potential applications to illustrate the significance of the potential of high energy missions and to infuse inspiration for further in-depth pursuit.

2.0 High Energy Mission Applications

The determination and establishment of mission objectives and requirements are fundamental steps in establishing requirements for increased propulsion and electrical energy capabilities over our current visions. If no mission requirement exists for which the high energy yield has an application, then clearly the matter should be pursued no further.

2.1 MISSION ANALYSIS

To comparatively quantify the mission advantages of high energy missions and to provide an indication of the new mission capability offered, mission analyses were conducted and are reported in this section. That work also established some key system requirements. Selected mission performance analyses were calculated by SAIC, Schaumburg, Illinois, under contract to the Headquarters Solar System Exploration Division to perform advanced mission analyses. Specifically, those responsible for the calculations were Messrs. A. Friedlander and J. McAdams. A meeting, held at SAIC Schaumburg in September 1988, resulted in agreement to examine the high energy missions presented in this report, namely, manned Mars, robotic sample return missions from each of the outer planets, multiple asteroid visits with sample returns, the Oort Cloud, and the nearest star – Alpha Centauri – missions. Additional missions were subsequently included which are presented in this report. There it was decided that for purposes of this study the fast calculation technique, discussed below, would be sufficiently accurate for the comparative analysis and for approximations of key mission parameters. Fusion performance information parameters were as provided by the author as reported herein. Other input parameters and performance characteristics were provided by SAIC. The results were presented by Mr. Friedlander at the American Astronomical Society, AAS/GSFC International Symposium on Orbital Mechanics and Mission Design in April 1989 (Fri89).

2.1.1 MISSION ANALYSIS CALCULATIONS

The mission analysis program algorithms were based upon the techniques employed by Dr. W. E. Moeckel (Moe72). Because the objective was to provide an indication of the performance capabilities of high energies of the fusion class rather than to conduct precision mission planning trajectories, the results from the mission performance calculation approach were considered to be adequate. As a result, the computational time was considerably shortened. The data presented provide, thus, good parametric comparisons but would require modification where the influence of gravity becomes a significant factor. For example, with respect to planetary phasing, the calculations are considered representative of actual trajectories provided that the phasing reflects the distances as used in the calculations. But if the inertial data were obtained reflecting 180° out of phase distances of Earth-Mars where the effect of gravity

2.0 High Energy Mission Applications

would affect the "real" trajectory, significant errors would result as a consequence of the gravity-free assumption. Using today's techniques of gravity assists, but applied to a fusion powered spacecraft, the flight times presented could, for many missions, be considerably shortened over those presented in this study.

Since these are not "real" trajectories, but are distance determined values under constant acceleration, the accuracy of the calculations was of great interest and was therefore pursued. SAIC had made a comparison of detailed trajectory calculations with the Moeckel rapid estimation technique for low thrust and high thrust systems. The results are presented in Fig. 2.2 which shows this estimation technique to be slightly pessimistic for the case evaluated, i.e., the "actuals" will be less than presented in this report (Anom86).

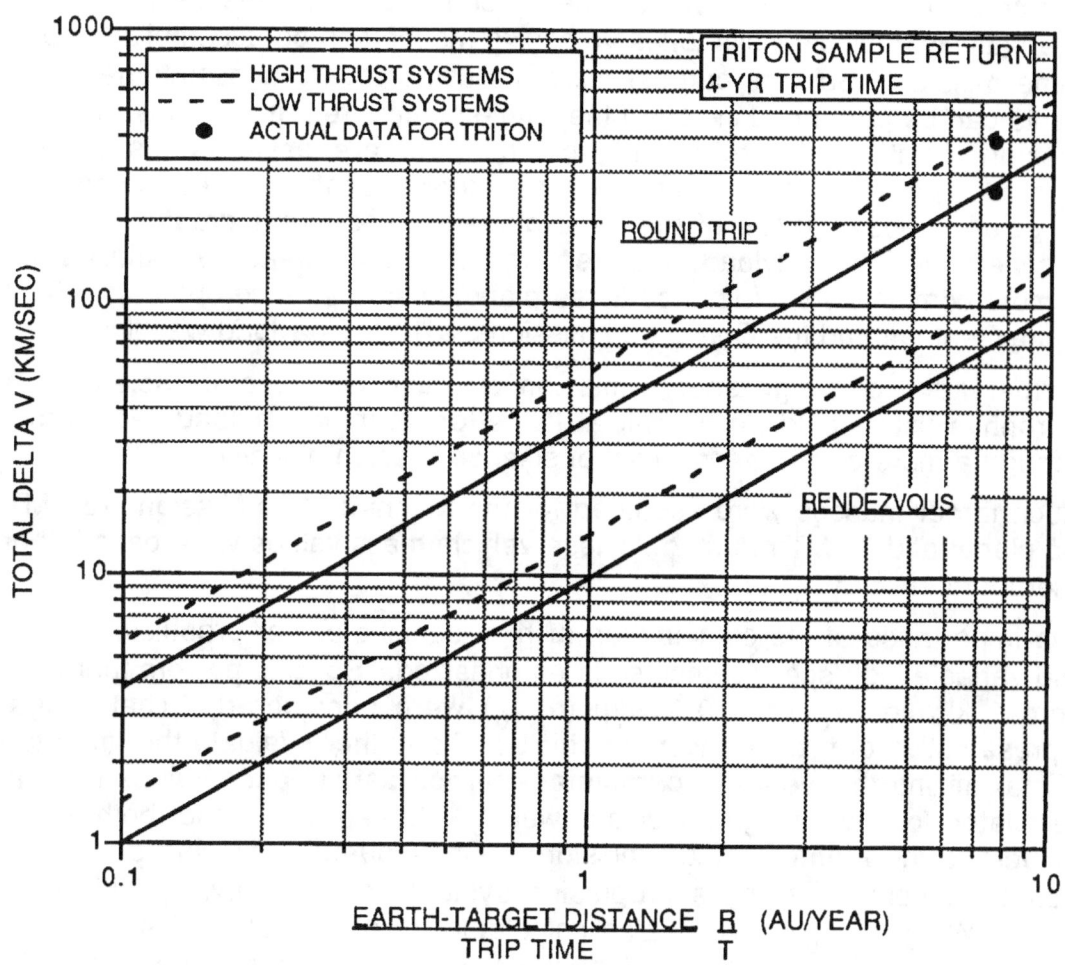

Fig. 2.2. Comparison of trajectory calculation techniques.

The figure compares a high thrust system with a low thrust one for the particular set of mission parameters considered. The question of high thrust versus low

2-7

2.0 High Energy Mission Applications

thrust advantages must be considered on a case-by-case basis. Where the effects due to gravitational losses are low and where the advantage of integration of velocity increments can be seized as in the situation where firing durations are lengthy, then low thrust systems are clearly preferred.

For the system inputs to the trajectory program, a consistent set of spacecraft data was used for the outer planet science missions. That is, the outbound payload mass of 20 metric tons (MT) for the unmanned scientific payloads was estimated to represent a reasonable mass advancement along with a 10 MT return payload, bringing to Earth a significant soil sample, atmospheric sample, and data. That mass is greater than an order of magnitude over current payloads for today's outbound payloads which of course are energy constrained permitting no sample return capability. The capability as discussed herein permits single sample returns from the most massive moons, and, therefore, these are the greatest energy demanding sample return missions. Alternatively, multiple visits to less massive moons can be accomplished on the same mission. Science outpost missions using rovers and stationary power plants can be performed in lieu of the single sample return missions. Extended mission durations for retrieving science data are essential as programs progressively advance toward the outer planets since the desired period of time for science observations of the local planetary seasons increase. Electrical surface power and payload mass requirements will increase as science tends to more completely characterize those planets using science outposts. For the Alpha Centauri and the Oort Cloud missions, a 10 MT payload was flown.

Further advanced high energy mission planning by planetary scientists and astrophysicists would be beneficial in better defining payload requirements through a more comprehensive list of science mission objectives.

Much larger masses were assumed for the Manned Mars Mission: 133 MT to the planet and 61 MT returned. These vehicle mass values were based upon a prior Martian study.

The importance of the combination of high specific power propulsion systems and variable high specific impulse is emphasized. Specific powers that ranged from 1 kW/kg (α_{p1}) to 10 kW/kg (α_{p10}) were considered. That range is representative of fusion powered vehicles. To further evaluate the importance of maximizing that very key parameter, mission performance values were also calculated for a very high specific power, 100 kW/kg, (α_{p100}) for both the Oort Cloud and the Alpha Centauri missions. Consequently, mission performance values for vehicles flying a propulsion system having a low specific power, 0.067 kW/kg ($\alpha_{p0.067}$), were also calculated for all planetary missions. That value represents an advanced nuclear electric propulsion system (NEP) ($\alpha_{p0.067}$) target. Refer to Fig. 2.3 (Rie88).

2.0 High Energy Mission Applications

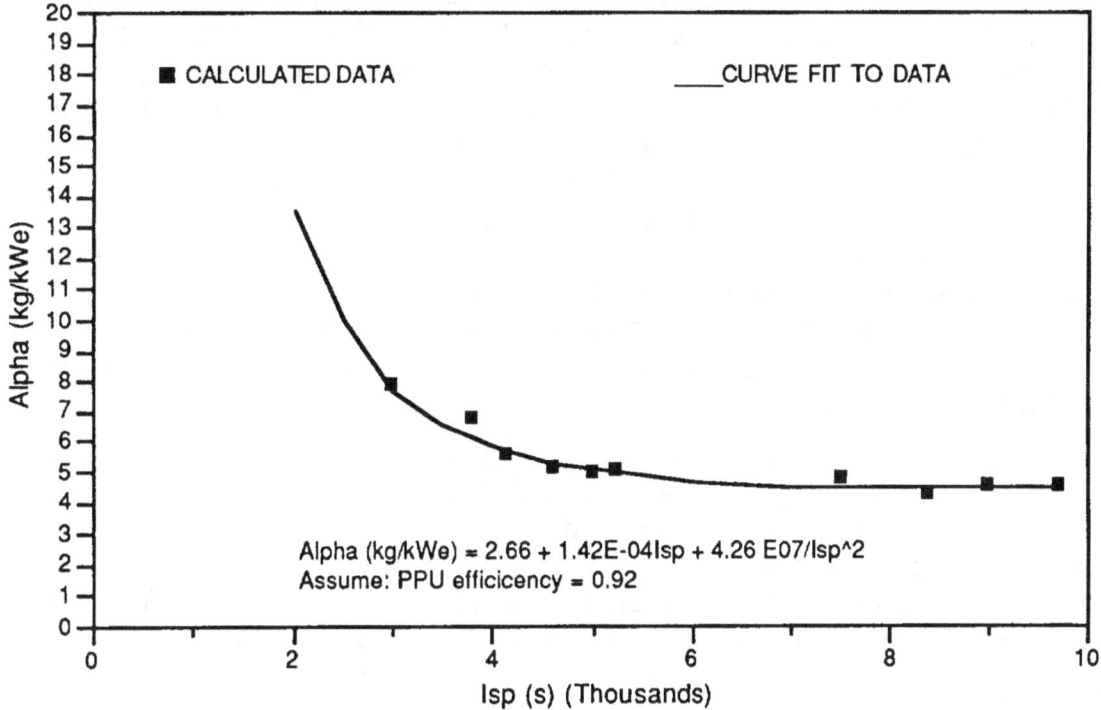

Fig. 2.3. NEP performance projections (Rie88).

Because these mission performance calculations are specific power normalized calculations, any energy conversion system having those specific powers properties, would of course yield similar results, provided that the same propulsion and power performance characteristics can be attained using alternate energy sources.

The missions beyond the solar system were not considered practical for NEP, and therefore, $\alpha_{p0.067}$ powered trajectories were not calculated for those missions. The physical constants were obtained from a JPL program (Anom), the Oort cloud data from the "Comets," Wilkening (Wil82). Specific data used for the analysis are further discussed in each of the sections below in the "Mission Performance Analysis" discussions.

The program optimized the vehicle mass by firing the propulsion system for a duration of two thirds of the heliocentric flight time and 100% of the planetocentric time. That approach minimizes the sum of the propellant and propulsion system mass. It assumes that the specific impulse can be varied without limits. There are, of course, limits; and those limits are discussed in the text rather than shown on the mission curves. The set of data presented in the report include the initial vehicle mass, propellant mass, delta velocity, power, and specific impulse. The specific impulse values shown are averaged values over the mission firing duration, the method chosen to present the data in this

2.0 High Energy Mission Applications

report as a figure of merit. Vehicle mass is minimized for those flights where specific impulse and thrust are optimally varied. It is important to note that the mission performance calculations included planetary gravitational forces for the spacecraft while attaining escape velocities from the respective planets during the planetocentric phase, but those losses were excluded as negligible during the heliocentric phase of the mission and of course during the interstellar missions. Trip times for missions within the solar system, therefore, included the time to spiral out during acceleration and to spiral in during braking maneuvers as well as the traverse time. The time interval for mission activity at the target bodies for the round trip missions was not included since stay time determinations are independent of the energy source used for placing mass at a target (destination). Flight times shown in the report's figures are, therefore, the total flight time, not the total mission time which would include time for activity at the target.

The type of thrust assumed was low, Type II, and a constant jet power in a gravity free field was assumed as mentioned earlier (Moe72). Specific impulse is a variable, the average value of which is reported in the text unless otherwise noted. The Δv term for energy imparted is used in this text, it being the more familiar parameter for the measure of energy supplied to the vehicle. Another mission performance calculation parameter is J, the difficulty parameter, where

$$J_n = \int_0^{T_{pn}} a^2 dt \equiv a_0^2 T_{pn}, \quad m^2/sec^3. \tag{9}$$

The mission difficulty parameter as a function of specific mass and payload ratio is given by:

$$\alpha J \equiv N\alpha J_n = 2N\left[1 - (m_{pay}/m_{0,1})^{\frac{1}{2N}}\right]^2. \tag{13}$$

The distances traversed as a function of time for a fly-by mission is:

$$R = (2N/\alpha)^{1/2}\left[1 - \left(\frac{m_{pay}}{m_{0,1}}\right)^{1/2N}\right](\tfrac{2}{3}T)^{3/2} = \Delta v(\tfrac{2}{3}T). \tag{16}$$

The total thrusting time as a function of J is given by:

$$J \equiv NJ_n = Na_0^2 T_{pn} = a_0^2 T_p. \tag{14}$$

2.0 High Energy Mission Applications

The Δv as a function of the mission difficulty parameter is shown below:

$$\Delta v = a_0 T_p = \left(J T_p\right)^{1/2} = \left(2 N T_p / \alpha\right)^{1/2}\left[1-\left(m_{pay} / m_{0,1}\right)^{1/2N}\right]. \tag{15}$$

The above equations are referenced to the Moeckel paper "Comparison of Advanced Propulsion Concepts for Deep Space Exploration" (Moe72). Additional information concerning the method of calculations used by SAIC for this study can be obtained from Fri89, "Performance of Advanced Missions Using Fusion Propulsion," presented at the AAS/GSFC International Symposium on Orbital Mechanics and Mission Design.

2.1.2 MISSION CATEGORIES CONSIDERED

The mission class considered herein required large energy levels: manned Mars; manned missions beyond Mars; multiple asteroid rendezvous with a sample return; sample return from the outer planets' moons of Jupiter, Saturn, Uranus, Neptune, and Pluto; multiple asteroid sample return; fly-by and rendezvous with the Oort Cloud; and fly-by and rendezvous with the nearest star, Alpha Centauri. A fly-by mission to Barnard's star was also briefly examined. Multiple stage performance was considered for only the latter three missions. Other high energy missions are suggested in the text, but energy requirements were not determined.[1]

Interest in sample return missions was expressed by the Space Science Board's report. (Don88, *Planetary and Lunar Exploration,* p. 16) "Many crucial types of chemical and isotopic analysis can only be made on samples returned to Earth. Such studies bear not only on the present state of crustal material, but also on its origin, age, and history. ... For comets the main consideration is to preserve the original physical state of the material."

The optimal technique for preserving the original state of matter is to perform in-situ analysis. The large payload delivery offered by high energy powered vehicles could be a very significant contributor toward providing mankind with a comprehensive in-situ science data analysis capability. In-situ analysis

---

[1]The Advisory Committee on the Future of the U. S. Space Program has suggested that the role of science be increased and that NASA advance the technology needed for space missions in their report dated December 10, 1990. The purpose of this document on advanced high energy missions is not to provide emphasis on manned versus unmanned programs but instead to call attention to the need to develop a high performance propulsion system which can provide Δv's on the order of 90 to 10,000 – 20,000 km/s regardless of any particular program activity focus. The content herein is in complete accord with the Advisory Committee's position.

2.0 High Energy Mission Applications

techniques are beyond the scope of this endeavor; but with a high energy capability, a new class of missions (ISAM - In-situ Analysis Mission) can be considered since we can now consider the delivery of large masses required for robotic labs. These are referred to as scientific "Outposts" in the context of this report.

2.2 CANDIDATE HIGH ENERGY MISSIONS

Why high energy missions?

A high energy propulsion capability is mission enabling for the missions discussed below. High energy performance optimization and long duration propulsion infrastructure performance capabilities will be uniquely enabling, particularly for the stellar and Oort Cloud missions which have not been addressed. The high energy missions described below are therefore considered to be new missions and are vital applications of high energy flight systems. This report's mission performance analyses for more advanced missions were conducted to quantify the benefits for some selected difficult missions.

Consider "NASA's Goals."

(1) advance scientific knowledge of the planet Earth, the solar system, and the universe beyond;

(2) expand human presence beyond the Earth into the solar system;

(3) strengthen aeronautics research and develop technology toward promoting U. S. leadership in civil and military aviation.

Successful pursuit of these major goals requires commitment to the following supporting goals:

(2) develop facilities and pursue science and technology needed for the Nation's space program. (Anom89, p. II-9)

The goal of this report is, thus, to assist in the accomplishment of the three major NASA goals by suggesting that fusion energy be included in new technology research which should be pursued as a supporting commitment.

One recent document which was useful in illustrating the power levels for current advanced missions is the JPL report D-3547 (Man87). Missions evaluated typically in that study exhibited power levels ranging between 80-300

2.0 High Energy Mission Applications

kW$_e$. The largest power consuming application was 7 MW$_e$, a cargo-carrying Interplanetary Transport Vehicle (ITV).

An excellent source for science mission objectives, representing the most advanced scientific mission objective thinking, was the NRC Space Science Board Report. (Don88 *Space Science in the Twenty-First Century*, Imperatives for the Decades 1995 to 2015).

We can consider three space mission categories that can benefit from a high energy capability:

HIGH ENERGY SPACE MISSION CATEGORIES

- I. Manned Solar System Exploration
- II. Space Science-
  - Solar system
  - Interstellar
- III. Applications

I. Manned Solar System Exploration Missions

2.2.1 MANNED EXPLORATION: ENERGY OPTIONS AND ADVANTAGES OF HIGH ENERGY

Mars–and ultimately beyond

Mission Description

Unmanned science missions in support of manned exploration missions to Mars are currently being planned and examined in depth for early in the 2000 year time frame. The ultimate goal now is manned Mars exploration and settlement, an event which, if it actually materializes or not, perhaps will depend upon factors about which we are probably unaware today.

2.0 High Energy Mission Applications

The extent of NASA's exploration missions will be determined by the availability of high specific power and performance propulsion and power systems. The availability of high specific power systems will establish whether a permanence of man at Mars will become practical or whether it is possible to perform only a very small number of visits for short durations of weeks. A permanence of man there will be very energy demanding from a space logistics requirements perspective. High specific power systems permit the design of vehicles to high payload factors, thereby enabling an economical logistics capability. In that key necessity, high specific energy will undoubtedly play a major role in determining whether the manned permanence endeavor is technically viable. Logistics must be mandated as a critical infrastructure element in the future space mission architecture.

Active planning for Mars exploration commenced in the 1960's. To investigate the in-flight transportation energy requirements, a study was conducted to establish the mass of the lander vehicle. The 1967 Manned Mars Excursion Module (MEM) study by North American Rockwell showed that chemical propulsion could accomplish a reasonable round trip Manned Mars Mission (Can68). Refer to Fig. 2.4 to observe the effects on performance of the two propulsion systems – fission powered Nuclear Engine for Rocket Vehicle Application (NERVA), and chemical propulsion, in combination with aerobraking.

Mission Options	Earth Orbit Escape	Mars Orbit Capture	Mars Orbit Escape
	Δv KFPS (KM/SEC)		
	16 (4.9)	15 (4.6)	16-10.5 (4.9-34.2)
CASE 1	NUCLEAR (ISP = 800 SEC)	CHEMICAL (ISP = 800 SEC)	CHEMICAL (ISP = 800 SEC)
CASE 2	NUCLEAR	NUCLEAR	NUCLEAR
CASE 3	CHEMICAL	AEROBRAKE	CHEMICAL
CASE 4	NUCLEAR	AEROBRAKE	CHEMICAL

2.0 High Energy Mission Applications

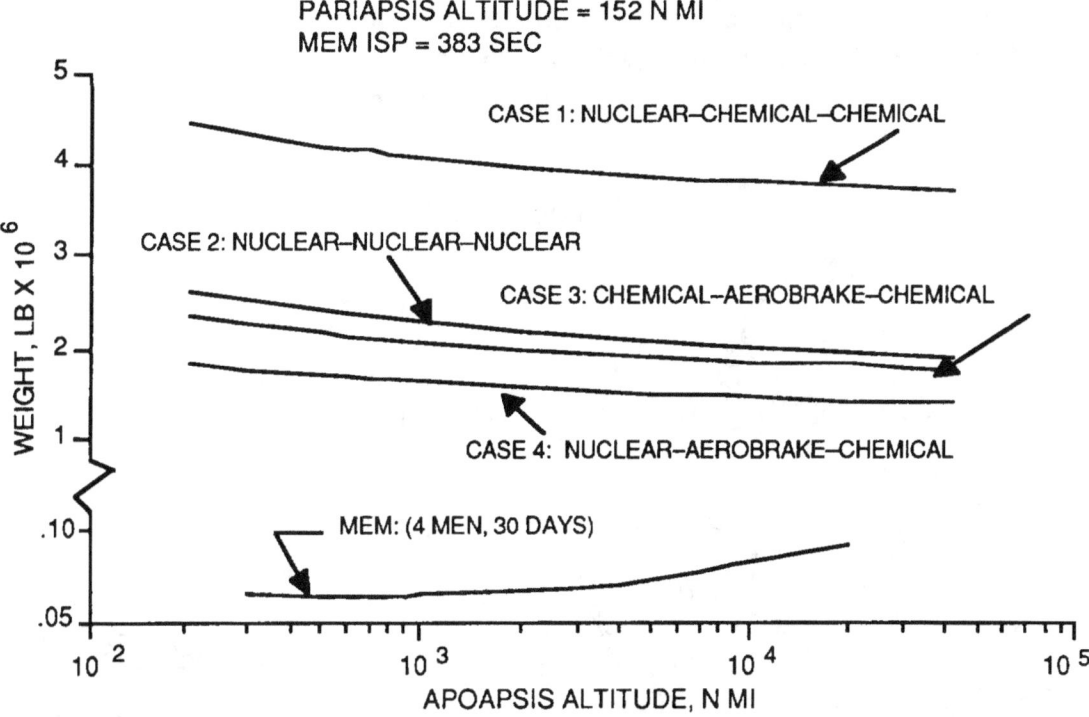

2.4. MEM study results showing the mission performance for combinations of energy sources – nuclear, chemical, and aerobraking. (Can68, Fig. 3-11).

Two missions were considered in that study, a 4-man/30-day Martian stay and a 2-man/4 day stay. Aerodynamic braking was selected in lieu of retro propulsion at Mars. The initial vehicle mass was ~ 910 MT. A 50 MT payload was placed into a Martian orbit. The Δv requirement to escape from Earth orbit to Mars was 4.9 km/sec. For this mission, use of the fission engine for Earth orbit escape reduced the initial vehicle mass by only 16 MT in comparison with the chemical propulsion system's performance (400 seconds specific impulse). Chemical propulsion (400 seconds specific impulse) provided the return trip's Δv also. The return engines burned FLOX/methane. Aerocapture at Earth was used in lieu of a retro burn. The value of 800 seconds specific impulse performance was determined by the contractor to be a reasonable performance for the technology then. Since the NERVA program was concluded subsequent to the Rockwell Mars mission analysis, that value is still considered to be a reasonable performance goal although current studies are extrapolating the nuclear engine's performance beyond as, for example, up to 900 or even 1000 seconds. While that, too, may be attainable, the question is the degree of risk incurred as a result of pushing the heat exchanger's material limits to a greater stressed thermal load. Clearly for quicker, more massive payloads the higher performance of fission will be advantageous. We can also state the advantage of greater operational windows using fission. The answer to chemical versus fission is a function of the mission defined. If the goal is simply to perform a

2.0 High Energy Mission Applications

single manned Mars mission, then chemical systems can accomplish it and without the costs associated with qualifying fission.

A description of the lander vehicle is provided in Fig. 2.5.

Recommended Design

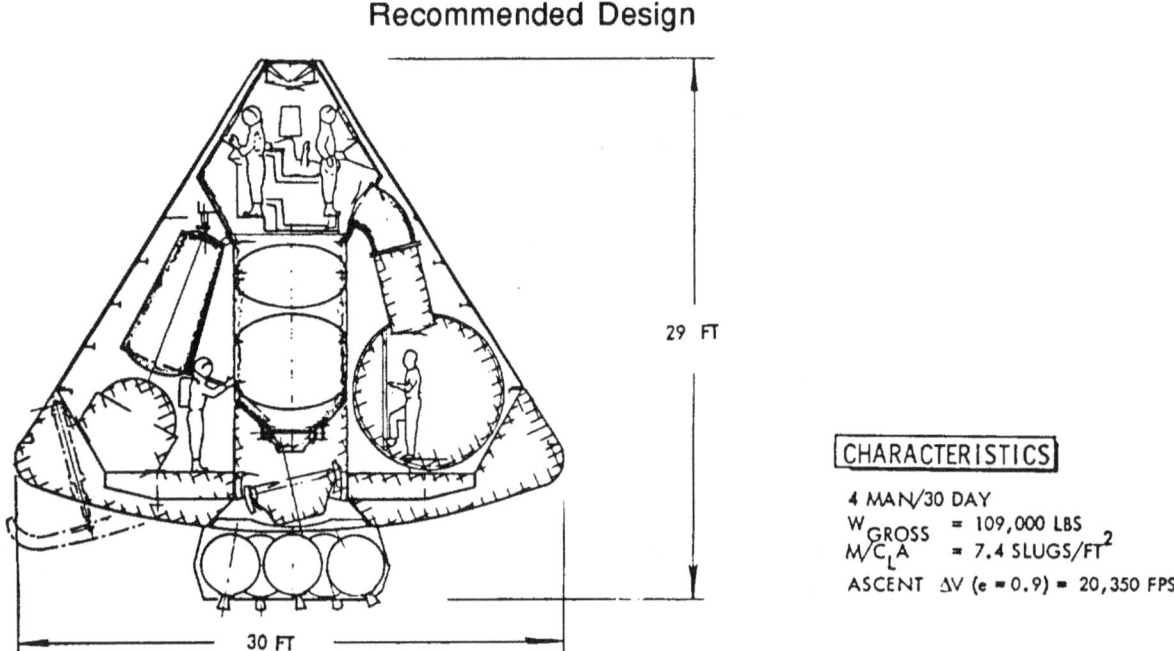

CHARACTERISTICS

4 MAN/30 DAY
W_{GROSS} = 109,000 LBS
$M/C_L A$ = 7.4 SLUGS/FT^2
ASCENT ΔV (e = 0.9) = 20,350 FPS

Fig. 2.5. MEM configuration. (Rockwell chart SD 67-755-4).

The MEM is that portion of the payload to be carried by the trans-Martian spacecraft to accomplish the descent and ascent mission phases similarly as accomplished in Apollo by the Lunar Module. The mission profile is described in the following manner by Dr. Canetti.

> The MEM would be passive and unmanned during the Earth orbital operations and interplanetary transit phases except for scheduled checkout and maintenance operations. After the Mars capture orbit has been achieved, the subsystems would be checked out and activated, and the MEM manned and separated from the spacecraft. Low thrust de-orbit motors would be fired at a predetermined time and position to effect entry and landing in a prescribed landing area. Entry generally would occur with the lift vector up; roll control would be employed for minor navigational adjustments. To decelerate the MEM, retropropulsive thrust would be initiated at the equilibrium velocity of about 300 fps (0.9 km/sec) and applied so that the vertical component balances the difference between weight and lift. Portions of the heat shield would be

jettisoned to reduce weight. Touchdown would occur after a short hover period over the final landing site. Until touchdown, the crew would occupy the control cabin atop the vehicle.

A laboratory and living quarters, connected to the control cabin by a tunnel and airlock, would be provided for surface operations. At lift-off, much of the control equipment and structure would be left on the surface; propulsion tankage would be staged during ascent. Normally, the MEM would ascend to an intermediate phasing orbit and, after appropriate phasing with the spacecraft orbit, effect rendezvous and docking. After the crew and scientific payload are transferred to the spacecraft, the MEM would be abandoned in Mars orbit.

Abort capability exists before entry, before landing, and on the Mars surface; there is no abort capability during entry. The most critical requirement is imposed just before touchdown when the ascent stage must be separated, the ascent engine ignited, and a turn-around maneuver performed to correctly orient the thrust vector.

The study concentrated on the Martian operations to establish whether the MEM vehicle mass would be sufficiently low such that current chemical propulsion would meet energy requirements or whether higher propulsion system performance would be required. The conclusion was that fission propulsion is not required to accomplish a Manned Mars Mission. The existing chemical propulsion/aerobraking technology was better from a performance standpoint for the mission considered and the projected performance levels assumed.

The other proposed nuclear propulsion mission under consideration at that time was for Earth-moon transfer. As a consequence of the operational complexities and safety aspects associated with nuclear propulsion of the NERVA design class for trips to the Moon, NASA removed nuclear propulsion from future space application considerations, and research on it was terminated. At that point the feasibility of the thermal fission technology propulsion was well demonstrated.

Fusion power and propulsion have not yet been inserted as a requirement for application to the planning equations for Mars and other advanced missions. Some thoughts are provided below in that regard.

> Advantages of high energy fusion systems over other energy sources:

  1. Reduced trip times
  2. Safety

2.0 High Energy Mission Applications

3. More massive payloads
4. Environmental benefits
5. Economy of space travel
6. New missions enabled
7. Conduct of more science more quickly
8. Sustenance of man for a permanent presence of man beyond Earth
9. Space manufacturing
10. High payload mass fraction vehicles
11. Greater operational flexibility in mission planning.

1. <u>Reduced trip times</u>

   One recent study shows that a fusion powered round trip mission to Mars, including a two week stay, can be accomplished within a total time of 90 days (Anom88: "Fusion Propulsion Study"). Short trip times have obvious enabling advantages from a safety viewpoint as well as from a cost and performance perspective. A 90-day mission provides substantial flight operational savings over a 1-2 year flight time mission. Trip times as a function of vehicle masses for manned and unmanned payloads are presented later in this section.

2. <u>Safety</u>

   Safety is discussed further in Section 9.0. Some major points follow. The Mars one-way trip time of up to nine months is a very significant time penalty paid for the use of lower performance systems. Under the current planning which use low performance propulsion systems, there are significant health and safety hazards for a manned Mars program which may be acceptable for a one flight mission but which may be unacceptable for settlement.

   These then constitute the major environmental hazards to control:

   Space environmental hazards

   1- galactic radiation

   2- solar events

   3- psychological

   4- physiological.

   High specific power/impulse reduces the first two hazards by virtue of shorter trip times. Two sources of natural radiation hazards exist, galactic cosmic ray exposure and solar flares. For cosmic rays the dosage severity is determined by the exposure duration. From exposure to cosmic rays plus secondary radiation, the flight crew can be exposed to their dose limit within a short time, i.e., at the dose of 0.1 rem per day a space traveler will receive in 50 days the 5 rem limit established for the clean-up crew during the Three Mile Island accident. Protection from solar flares can be accomplished but with a mass penalty. A severe flare will otherwise be fatal.

What can be done to protect the crew? A small shielded safe haven is perhaps the optimal solution. It is certainly a strong design option for emergency use in the event of an unexpected solar event. Limiting launches to periods of anticipated cycles of low solar activity cycles will surely be exercised. This subject is explored further in Appendix B which presents another approach to the use of fusion energy, referred to as "LASERPATH."

The effects of being enclosed in a small space for nine months causes great concern for the psychological aspects, or mental health, of the flight crew. There are questions concerning the psychological fidelity of ground testing and its true capability to simulate the realism of not being able to rapidly return to Earth in the event of a contingency. Test subjects are obviously aware that they can immediately "return to Earth" in a very short time, if necessary. A sufficiently high power system can alleviate the concern by providing propulsive braking and power for return to Earth abort (RTE). RTE abort is a very important concept for safety that will be mandated if the vehicle performance capability were attainable.

Long duration trips impose a severe penalty upon the crew due to the physiological aspects of extended periods of weightlessness, and it is a subject under considerable research, especially by the Soviet Union where the conduct of a series of orbital weightlessness experiments, now in excess of one year's exposure to man, have been methodically under way for many years.

3. <u>Massive payload transportation</u>

High energy allows the consideration of manned payloads on the order of 130 MT outbound to Mars and 60 MT returned to be considered on a reasonable flight time scale. Science payloads to the planets can be more massive, on the order of 20 MT outbound and 10 MT returned mass while a 10 MT payload mission to the nearest star can be examined. The results are presented later in this section.

2.0 High Energy Mission Applications

4. <u>Environmental benefits</u>

Three separate, distinct categories of environmental concerns are reduced by high specific power performance – the naturally present radiation environment, the provision for a suitable environment for man's habitability, and the environmental impact due to man's generated waste products.

    Environmental elements:

        1- space environment

        2- habitability for man on a planet

        3- waste products.

The space environment elements have been discussed from the safety perspective.

High specific power aids the second environmental element, habitability, by enabling an adequate life supporting environment on an extraterrestrial body. For the second phase of manned Mars exploration, extended duration stay times are anticipated (Pai86). Self sufficiency will be a primary program objective. The key enabling technology making this practical is an abundance of energy. This phase of the mission will require a sizable habitat rather than the confines of a space suit or small landing craft cabin as used in Apollo or designed for the MEM. Essential features for an advanced habitat would include: a self contained, recyclable, breathable atmosphere; plant growth; water; and temperature conditioning, the implementation of which will depend upon a large energy source. In addition, there will be power demands to accomplish work related functions. Power estimates for Mars could not be found, but a lunar base mission power level of 2 MW was determined for supporting an independent 24-person lunar base including the conduct of human tended science, materials processing, and mining (Fri88). Materials processing appears to dominate the power requirements in the studies reviewed.

The third environmental factor concerns man's generated environmental waste. Consider the mission operational scenarios described in the discussion of the MEM at the beginning of this section. The penalty for the use of chemical energy is in the generation of significant debris, both in a Martian orbit and on the Martian surface. Each landing and launch at Mars will result in jettisoned debris. A wasted vehicle stage will remain in orbit around Mars for each return flight to Earth, creating a space debris problem. There is a concern about space debris in Earth orbit now, without resolution. Obviously the preferred approach is to avoid the problem in the first place. The high specific energy mission capability to transport greater mass will permit mission designers the use of space flight equipment that will avoid the generation of orbital and surface debris. The goal would be to provide a sufficient mass-to-Mars capability that will permit the use of a reusable Martian lander and ascent vehicle. To continually transport Martian landers will obviously be a very expensive method.

Because of the great distances and long flight times, and therefore program costs, the advantage of in-situ materials processing will ultimately be developed and become refined. For that to happen, the power level for future missions will of necessity increase substantially over today's predictions. Our understanding of the state of in-situ technology utilization and science requirements for Mars is in its infancy at the present time, and it is possible to hardly more than hazard a crude guess regarding the total power consumption. The current per capita consumption in the US is estimated at a steady state power level of 250 GW used by nearly 250 million citizens which is approximately 1,000 watts per person. To extrapolate that value linearly as a figure of merit to a colony of 1,000 persons for example, would indicate 1 MW power requirement. That requirement will be higher in space since the environmental parameters are more demanding at Mars than on Earth. Extrapolated to the Friedlander lunar study, it would indicate ~80 MW level. As another mission function, materials processing will be a part of future surface operations, and that can be expected to be highly energy intensive.

To provide the energy, this much we know. To use its naturally occurring energy sources at Mars, there will be available only thermal gradients, solar, and wind unless exploration research reveals a source from mining that offers an alternate energy source. The wind could be used as supplemental energy, but it clearly is not a dependable source unless the means to store very significant quantities of electrical power can be developed. Significant gains in solar cell efficiency are required before they can act as more than buffers to provide only small quantities of electrical power. A study would have to be conducted to examine the viability of thermal power conversions. If fusion is unavailable, that leaves chemical and/or fission systems to be transported to Mars, a very expensive proposition. An unknown/undefined alternative, pending the outcome of additional science on Mars geology, is to develop in-situ processing of chemicals for energy.

Fission power plants will, by nature of their neutron emissions, require a significant cooling capacity, and on Mars that will probably require the use of radiators for coolant purposes because of the absence of water and the thin atmosphere. Underground water may offer an alternative, if proven. There are, consequently, significant mass factors and cost parameters to be taken into consideration in trade studies between fusion and fission concerning heat rejection. A key advantage for fusion is realized if direct electrical energy conversion is used, especially for D-^{3}He, where waste heat is minimized. For Martian ground operations, the fusion reactor has the advantage of being capable of safe, simple well-defined refueling technology performed by the transfer of cryogenic liquids unlike the fission

2.0 High Energy Mission Applications

reactors which will require very special handling or abandonment at the conclusion of useful life. That will raise attendant environmental issues.

Ultimately after settlement, similar environmental issues will have to be addressed for the Martian environment that we must now face here on Earth – "How are the radioactive and other environmental wastes to be eliminated?" The advanced fusion fuels will not totally eliminate that problem, but they should decrease the issue by a substantial factor. Perhaps by that time we will become more knowledgeable of the reactor's physics so that the reaction could become more purely aneutronic.

Environmental issues are discussed in Section 9.0.

5. Economy of space travel

This topic is discussed in Section 10.0. The high performance capability reduces the mass requirement in low Earth orbit which effects enormous cost savings. The use of lunar volatiles from the mining of ^{3}He is anticipated to reduce the costs of doing business in space through the use of local planetary resources.

6. Enable new missions

That is the topic of this section, the details of which follow.

7. Conduct more science more quickly

From the science mission perspective, the abundant power provided by fusion allows for transmitting more data at a higher bit rate and for transporting heavier payloads which permits the delivery of more science instruments for both surface and orbital missions. The higher power transportation system delivers the science payload more rapidly as discussed in Section 2.2.3.

8. A sustenance of man for a permanent presence of man beyond Earth

The OAET (Office of Aeronautics and Exploration Technology) NASA University Space Engineering Research CULPR (Center for the Utilization of Local Planetary Resources) is especially important when one considers that it is in the most fundamental essence, exploring the means by which a lifeless planet can be made hospitable, not just from the life support aspects, but by providing the material necessities. Another application is the use of lunar volatiles produced as a by-product of the mining of helium-3 as mentioned earlier. The volatiles include the important life support gases as well as propulsion, including oxygen and hydrogen. That capability is being studied by the University of Wisconsin in the Wisconsin Center for Space Automation and Robotics. From an energy requirements perspective, research on the means to provide in-situ life support is very important. Planetary energy exploration is another topic which is in need of research and development in order to make available the power for extraterrestrial habitability. Mars is clearly an excellent location, beyond the moon, on which to accomplish extraterrestrial utilization research and to prove principles. It is absolutely essential that we learn the skills that develop the technology which uses extraterrestrial materials and resources. They are needed, not for transport back to Earth, but for local exploration and settlement purposes. By today's space transportation cost and performance standards, transporting extraterrestrial processed materials will not be affordable, except in rare situations, such as, the mining of ^{3}He on the moon or on other bodies in the solar system.

Conversion of an inhospitable planet's environment to one that is life sustaining and supporting is more than a science fiction fascinating subject; it is a crucial, fundamental technology if man is to press his presence beyond the bounds of Earth on a permanent basis – a third phase of the Martian mission. That mission can be anticipated to be of great benefit to us in understanding our planet better, a topic particularly important to "Mission To Planet Earth." Martian exploration technology can be expected to aid in the understanding and management of Earth's environmental problems. Then too, as we ultimately look beyond this solar system for settlement, it is conjectured that if a planet in another solar system is found hospitable to man, there is a reasonable probability that the planet will already have been occupied by a similar natural process of evolution with some type of native inhabitants. If this is indeed a correct conjecture, and then we obviously have either the option of passing it by or sharing it, depending upon the phase of its evolution. Planets capable of supporting life are considered to be very rare by some scholars of the subject, but that is speculative. Barnard's star, at only 6.0 light years away, has held great interest as the promise of perhaps another planetary system. Some thought has been given to the subject of the production of habitable planets by Oberg, a process referred to as "terraforming" (Obe82). These and other missions discussed in the reference Hart and Zuckerman paper are of an advanced nature.

The presence of life beyond Earth is a deep rooted, fundamental question in analytical minds. Two NASA programs have devoted resources in pursuit of the question of extraterrestrial life, the Viking and SETI (Search for Extraterrestrial Intelligence). Much more can be done. The only known life exists on Earth; and the intelligent life here is, by astronomical standards, an event which happened only yesterday. In terms of astronomical dimensional scales, life is merely a surface phenomenon, that is, it exists in only a miniscule portion of the universe – a transition region, a boundary layer between the very dense massive regions, Earth, and the very tenuous but predominant region of the universe, space. Even within that sub-atomic dimensional scale, it is confined to a very narrow energy level of a few degrees. Yet there are billions of stars per galaxy and billions of galaxies within the universe. Are we really alone as intelligent beings in the universe? Are the Earth species the only type of life form? Solar system science objectives discussed in Section 2.2.3 include the gathering of basic

2.0 High Energy Mission Applications

information concerning life forming situations such as is proposed with Titan and the comets. But the Titan mission for life form evaluations is very limited – it must be. While these objectives cannot answer the fundamental question of the probability of planetary formation, they are important to better understand life. The stellar mission category pursues the topic further; it is intended to address planetary formation in star systems and to address formations with older single stars as well as the most probable source for extraterrestrial life. The NASA goals in Section 2.2 pursue scientific knowledge of the universe, an expansion of man into the solar system, the conduct of aeronautical research. Forwarded here, then, is an additional thought reflecting the question of life.

> A major new NASA space goal, in addition to understanding the origin of the solar system and universe and their ultimate destiny, is suggested, namely, the determination of extraterrestrial life, particularly including the presence of planets having habitability characteristics and the detection of life beyond Earth.

That goal is compatible with "US National Space Policy" which states that:

> NASA ... will conduct a balanced program to support scientific research, exploration, and experimentation to expand understanding of: ... (6) the factors governing the origin and spread of life in the universe. (pp. 5,6).

In that context, as a first step, a visit to Alpha Centauri is analyzed in Section 2.2.5. As a corollary to placing emphasis upon space life forms and for better understanding life's origins on Earth, missions to the comets, the asteroids, and Titan are particularly significant. These are discussed in Sections 2.2.2 and 2.2.3.9.

9. <u>Space manufacturing</u>

   The CULPR is conducting research dedicated to developing the technology for space derived materials including propellants, structural metals, and shielding for permanent settlements. The benefits of this capability is also discussed in Appendix B. The objective is to develop an in-situ manufacturing capability to reduce dependence on at least those bulky, massive low technology materials which would otherwise require transportation from Earth. This self-sufficiency technology is also required in order to develop a high degree of outer space autonomy, thereby dramatically decreasing the cost of space operations and space logistics.

2.0 High Energy Mission Applications

The University of Arizona was granted approval to proceed with that program in 1988. That research program comprises significant technology for accomplishing the permanent expansion of man beyond Earth into the solar system. The utilization of local planetary resources is a very important consideration for remaining in space where logistic costs will be substantially higher than those on Earth.

10 <u>Provide high payload mass fractions</u>

Payload mass variations are decoupled to a large extent by the high mass fractions as permitted by high performance flight systems. That can achieve economies in terms of a decreased probability of weight reduction programs as well as to provide a higher unit loading of the launch vehicle. The values gained are more than an order of magnitude improvement. This topic is discussed as part of the missions later in this section.

For the Earth's moon, Dr. Logan, Lawrence Livermore Laboratory, has looked at the use of a laser driven ablation thruster as well as free electron lasers (see Appendix B). Now if we extrapolate that concept for use on Mars, it can transmit energy by a fusion powered laser to an interplanetary vehicle for plasma thruster propulsion at a sufficient "space power" level to send substantial payloads to either of the Martian moons. This concept is discussed further in Section 2.2.9. It offers a very high payload mass fraction since the energy source remains stationary.

11. <u>Greater operational flexibility</u>

Additional mission flexibility and science objectives are gained by providing large energy sources in local planetary orbits to conduct science and exploration missions on extraterrestrial soil. A large number of **multiple** Martian science objectives can be accomplished through a central power source with energy beamed directly to the power consuming device. These devices include long duration rovers, science surface outposts on moons, weather stations, orbiters, etc. Since some are not large consumers of power, a means of storage could be provided for intermittent energy transmissions. With higher power capabilities, the transmitted data rates can be substantially increased for a higher rate of science return to Earth from Mars. Imaging, for example, is a high power consuming function.

Early contingent crew returns to Earth (aborts) is a new element in operational flight planning which is propulsive energy intensive. This, too, has obvious direct relevance to safety.

Where the use of chemical propulsion is mandated, stay times at Mars will be determined more by planetary phasing reflecting the propulsion system's capabilities rather than by pure mission objectives. Mission analyses will be required to quantitatively define the performance penalties for initial mass and payloads. There are low mass launch opportunities to Mars that the chemical propulsion systems can meet. The consequence of missing a launch window is to delay a launch for several years. One

2.0 High Energy Mission Applications

launch window missed has the consequence of extending the program for a minimum of 2 to 3 years, probably the cost equivalent to develop fusion energy.

Mission performance analysis

Staff at the University of Wisconsin have been analyzing the space application of fusion energy. In a recent study, Dr. Santarius used the low thrust calculations of Dr. Stuhlinger to compare fusion and chemical propulsion system payload deliveries for the same flight time to each of the destinations: the moon, Mars, and Jupiter, Fig. 2.6,

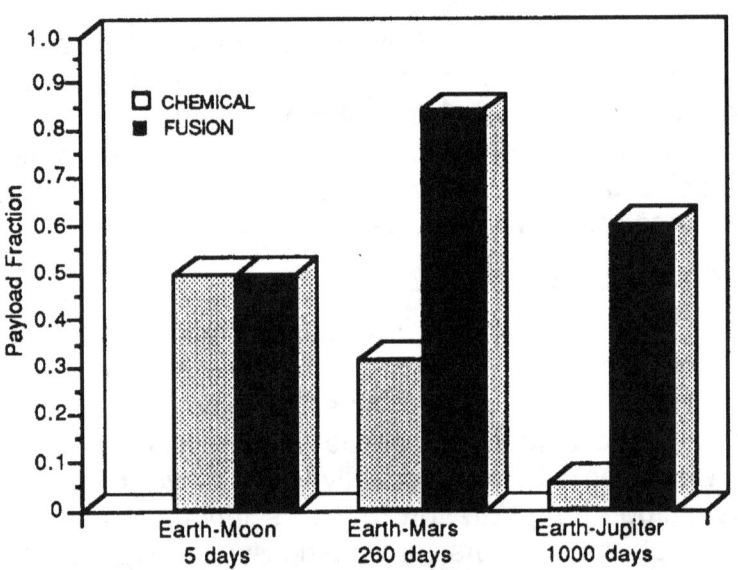

Fig. 2.6. Fusion and chemical propulsion system payload deliveries for the same flight time.

and for the reduction in mission time for the same payload, Fig. 2.7 (Stu64).

2.0 High Energy Mission Applications

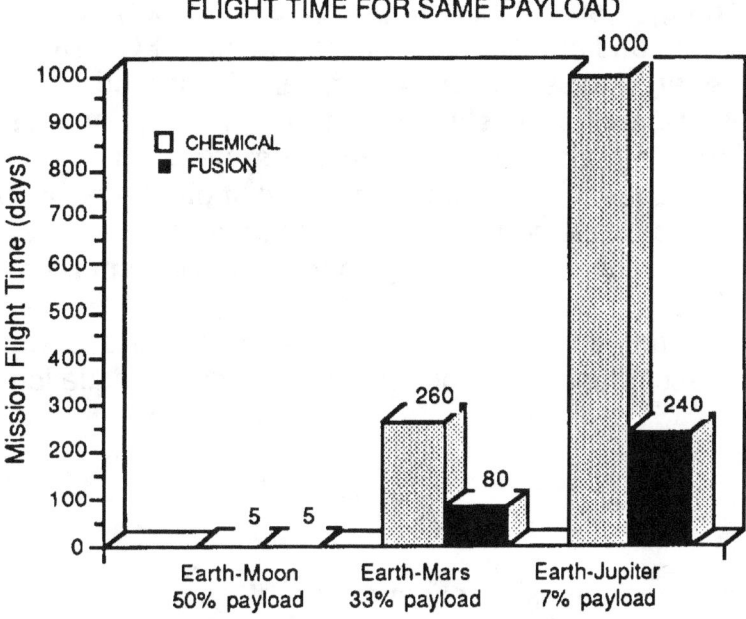

Fig. 2.7. Fusion and chemical propulsion system payload delivery time for the same payload (San88).

The payload mass fraction trade as a function of flight time for a trip to Mars is shown in Fig. 2.8.

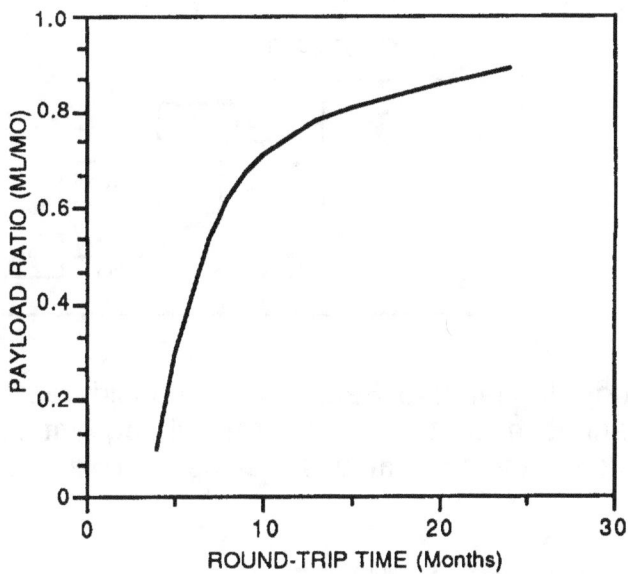

Fig. 2.8. Payload mass fraction as a function of flight time for a Mars trip.

The above calculations assumed a specific power of 1 kW/kg.

2.0 High Energy Mission Applications

These brief energy system analyses point the way toward substantial savings in mission flight time and/or the mass transported into LEO. The benefits of fusion for the high energy missions are graphically illustrated. For the moon, the performances are approximately the same for chemical and fusion; for Mars, a big improvement is realized from the fusion system; for Jupiter, a tremendous improvement is acquired. Figs. 2.6 and 2.7 point out the basis for the statement that fusion can be expected to economically transport large payloads over long distances. Fusion achieved a large gain in the payload mass fraction, i.e., from 0.1 to 0.6 without a large penalty in flight time. Large mass fractions are essential in lowering costs in any transportation mode and exponentially so in space flight in accordance with the rocket performance equation:

$$m_f = m_o e^{-\Delta v / g I_{sp}}$$

To provide a comprehensive examination of possible energy requirements, these are the missions that can be considered for manned exploration of the solar system:

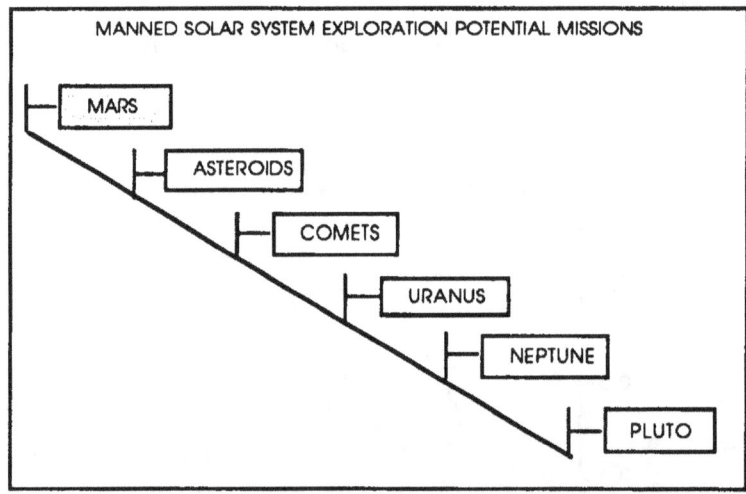

Early in this study Jupiter and Saturn were considered to be too radiation intensive for manned exploration although subsequent discussions indicate possible exploration of the outer moons may be feasible.

MARS

Manned Mars Mission performance data were calculated for a design point as part of this study. Mission performance data from the SAIC calculations (Fri88 & McA88) for a manned rendezvous and sample return mission are discussed below. The results of the calculations are inclusive of planetocentric and

2.0 High Energy Mission Applications

heliocentric times but exclusive of stay times, a propulsion independent variable for the purpose of this study. For these manned Mars mission performance calculations, an outbound vehicle mass of 133 MT was used, and a return mass to Earth orbit of 61 MT. These **round trip** missions were considered to depart from a 1000 km altitude Earth circular orbit and to park in a 500 km altitude Martian circular orbit. No aerodynamic braking was used in these missions, only vehicle propulsion, a safer mission operational mode. In this Martian mission performance analysis, only single stage vehicles are flown. This concept, therefore, eliminates the space debris issues discussed earlier. The payload masses were based upon the results of a Marshall Space Flight Center study (Anom87).

The results of these low thrust calculations ($\sim 10^{-3}$ to 10^{-4} g) using the aforementioned design data are presented in Figs. 2.9 through 2.12. Included are the results for M_o, M_p, Δv, and Isp. Curves which show trends and tables which provide single data points are provided. The same format is used for discussions of all missions.

The text for each of the mission performance figures includes single data points for simple reference. Basically the trade is between trip time and initial mass in low Earth orbit (LEO). The power level establishes the reactor size requirement while specific impulse shows the performance level necessary to meet the flight time. The propellant mass is important in establishing the operational costs for launch as discussed later in Section 10.0. The curves are provided to give the reader a broad perspective of the performance levels achieved and the performance trends.

Benefits

Here we discuss the advantages for manned space missions that can be realized from high performance energy propulsion systems, as exhibited by the expectations from the use of fusion energy.

1. Reduced flight time and initial vehicle mass

Fig. 2.9, the initial vehicle mass versus flight time, shows the significance to the mission flight time of high performance specific power and impulse, as considered to be within the domain of fusion reactor's capability.

2.0 High Energy Mission Applications

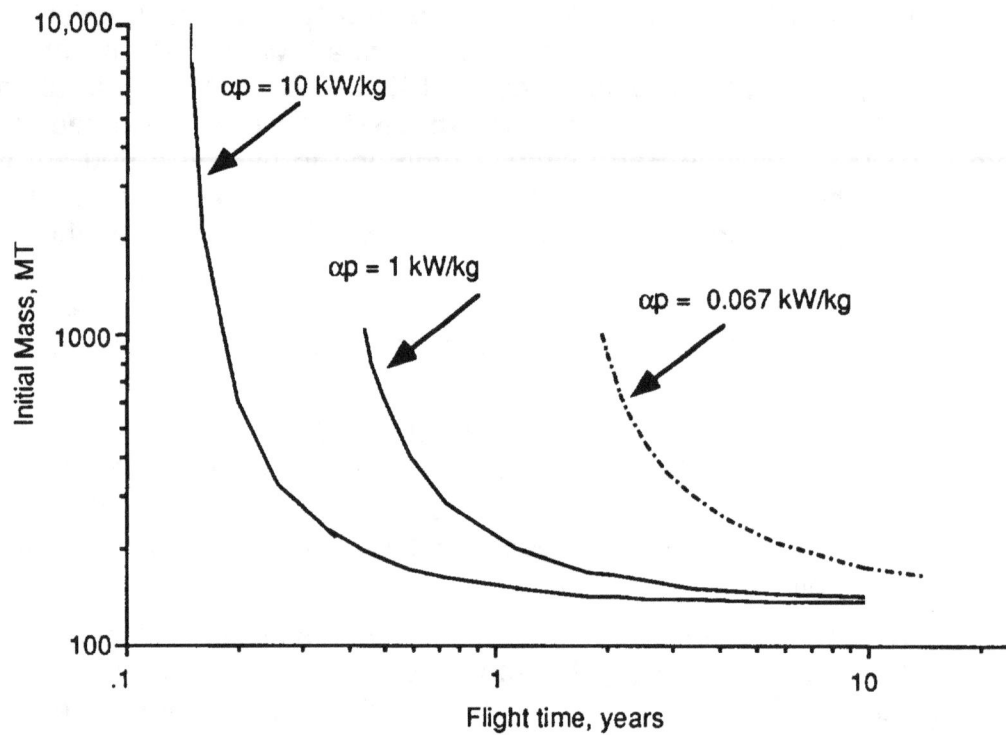

Fig. 2.9. Initial vehicle mass variations with flight duration for a Manned Mars Mission.

In fact, the same point is made for the mission performance parameters for all missions considered herein. That parameter alone distinguishes a most significant difference between an Earth based commercial electrical power generation plant and a space bound propulsion system.

Using a propulsion system designed with a specific power of 10 kW/kg (α_{p10}), NASA could perform a Manned Mars Mission round trip flight time of 110 days with a total initial vehicle mass of 274 MT in LEO, inclusive of the payload, fuel, propellant, and propulsion system/vehicle mass.

A more realistic performance goal for early developmental space fusion reactor designs is those propulsion system designs having a specific power of 1 kW/kg (α_{p1}). For that design, round trip times as short as 160 days were calculated. The large initial vehicle mass, 1,041 MT, and large quantity of propellant consumed, 681 MT, Fig. 2.10, for the $\alpha_{p1}/0.44$ round trip mission, and its attendant costs, would indicate that a more practical trade would be to extend the total flight duration to 6 months, 3 months each way, to achieve a propellant reduction of 210 MT.[2]

---

[2] For brevity, the nomenclature α_{pxx}/YY is used in this report to designate first the propulsion system's specific power [xx] and secondly the round trip time flight time [YY] nominally in years unless otherwise noted.

2.0 High Energy Mission Applications

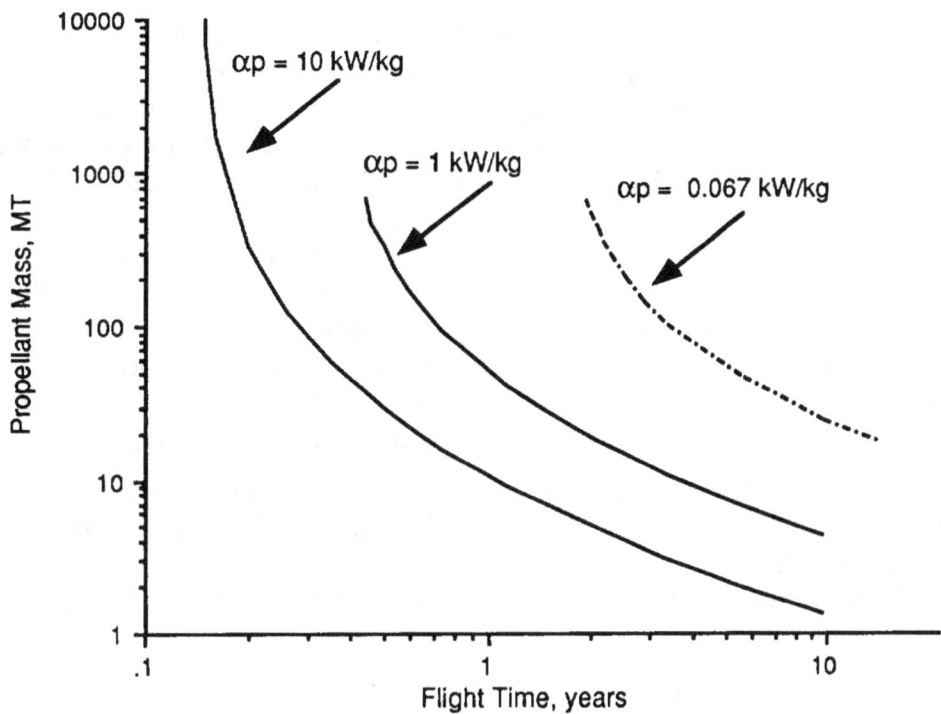

Fig. 2.10. Propellant mass variations with flight duration for a Manned Mars Mission.

A reduction in flight time of only 11 days during a 3-month one-way flight time increases the propellant demands from 335 MT to 681 MT, i.e., we pay a 100 % propellant mass (increase) penalty for a 12% reduction in flight time.

One important point to be made here is that "propellant mass" is defined as the sum of the reacting fuel mass plus the mass of the diluent, assumed to be hydrogen. For reasons of propellant thermalization (mixing efficiencies in the reactor) or for vehicle mass optimization, other elements can be considered. For example, oxygen with its superior density could be a better system trade element. The use of heavier elements may not be compatible with plasma burning. That subject has to be studied further.

2. Cost benefits

Not only does the high specific power propulsion system reduce the flight time, but it also provides substantial cost benefits. A half-year trip for the α_{p10} system consumes 30 MT of propellant whereas the α_{p1} propulsion system uses 335 MT. That would save 10 Shuttle launches, approximately $3B in launch savings just from reducing the propellant mass launch-to-orbit costs.

2.0 High Energy Mission Applications

3. Payload efficiency

Considerable developmental effort to maximize specific power is the best investment in the pursuit of fusion propulsion. As an example, compare the payload efficiency between α_{p1} and α_{p10}. Payload efficiency increases dramatically for the 6-month missions. For a α_{p1} system, the payload comprises **22%** of the initial vehicle mass and **30%** of the return trip mass at the end of the mission. For spacecraft flying a α_{p10} system, the payload comprises **72%** of the outbound vehicle mass and **73%** of the end of mission mass. These two efficiencies bracket that of commercial airlines (~50%), which provides an index for the relative efficiencies. If we are ultimately to become cost effective in the conduct of space operations, the airline standards for payload and operational efficiencies, preferably better, are a must. The commercialization of space depends upon the availability of high payload mass fractions.

In the $\alpha_{p0.067}$ case, the propellant consumed would be 650 MT of krypton, xenon, or argon, and the round trip flight time would require nearly two years. This curve is asymptotic at this time, so shorter flights cannot be anticipated without the achievement of significant reductions in the propulsion system's inert weight or tremendous expenditures in propellant.

4. Higher Δv

For the α_{p10}/130 day mission, the energy input, i.e., the delta velocity, was 130 km/sec, for a 6 month mission – 90 km/sec. The $\alpha_{p0.067}$ mission's Δv was a maximum of almost 43 km/sec. See Fig. 2.11. Clearly the shorter mission elapsed times are made possible by the higher velocities.

2.0 High Energy Mission Applications

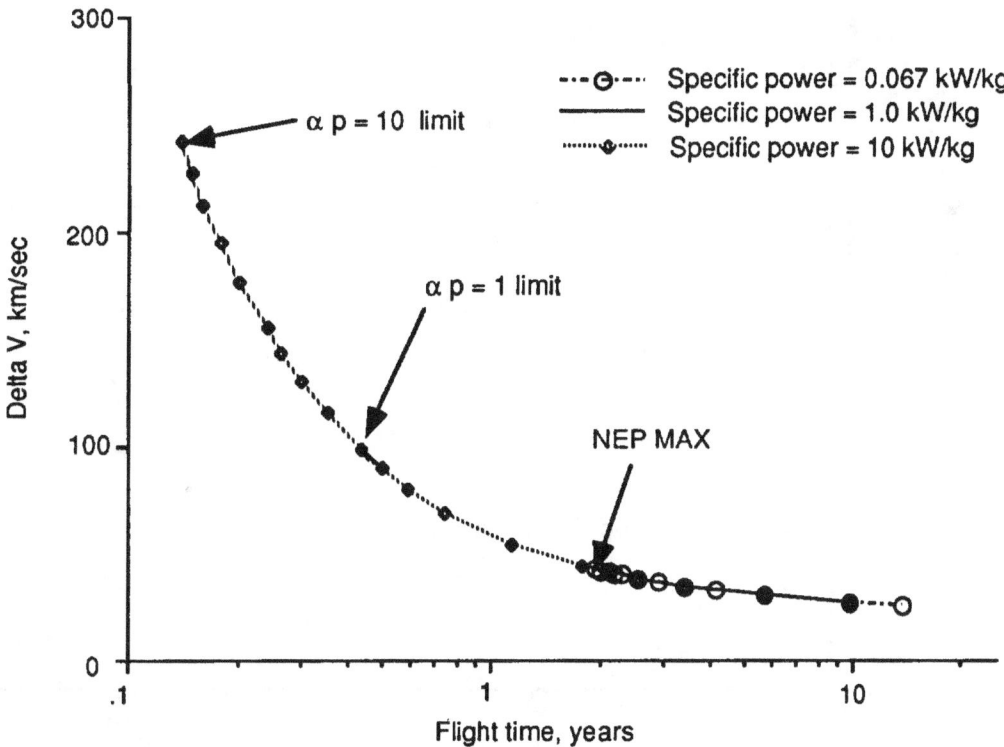

Fig. 2.11. Vehicle velocity variations with flight duration for a Manned Mars Mission.

Performance potential

The average specific impulse requirements to accomplish this Manned Mars Mission (Fig. 2.12) using the fusion propulsion systems should not be difficult to achieve, based upon the theoretical considerations discussed in Section 5.0 "Theoretical Performance Capability."

2.0 High Energy Mission Applications

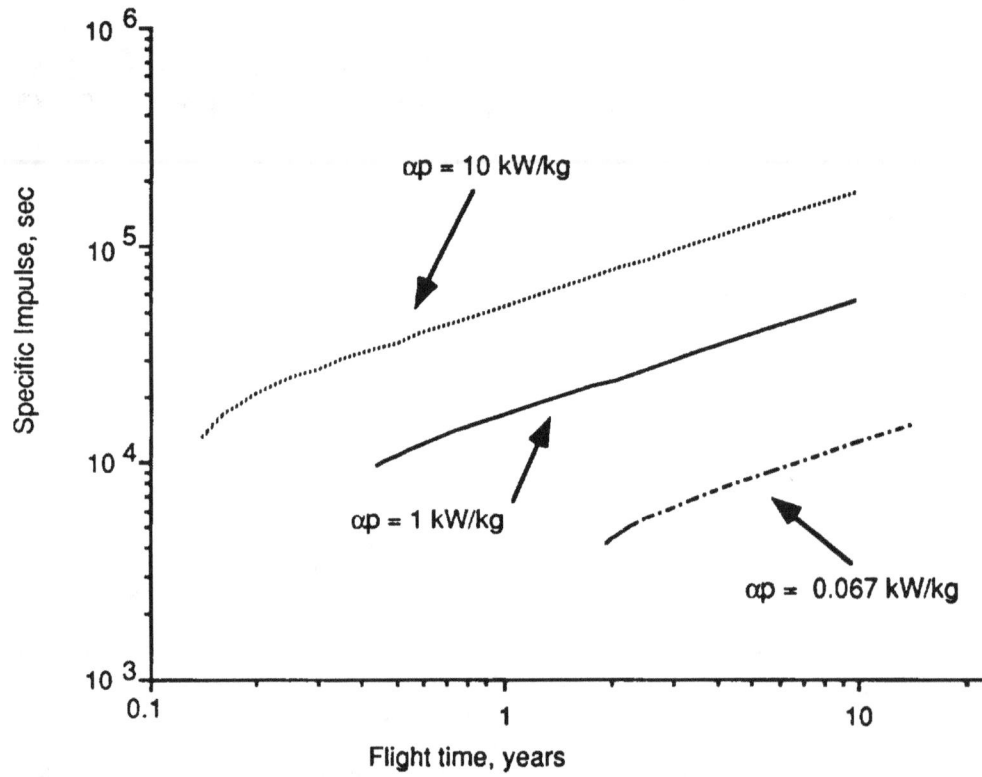

Fig. 2.12. Specific impulse variations with flight duration for a Manned Mars Mission.

The curve shows for a $\alpha_{p1}/0.44$ mission an average Isp of 9,440 seconds (4,450 minimum to 12,900 seconds maximum) is optimal for accomplishing this particular mission. Only an average specific impulse of 10,606 seconds is required for the $\alpha_{p1}/0.5$ mission. This permits a significant mass flow enhancement opportunity for attaining the acceleration rates required. The 35,766 seconds requirement for the α_{p10} mission is attainable. Notice that the specific impulse is higher for the longer flight time-more massive systems. That is a consequence of flying a constant acceleration trajectory. Arrival at the target more quickly requires the expenditure of greater power and more propellant, hence a more massive vehicle, and, therefore, a higher thrust to maintain a constant acceleration. To provide the higher thrust, specific impulse is reduced.

Similarly for $\alpha_{p0.067}$, the average specific impulse of 4,174 seconds value is considered to be possible for xenon and argon. However, the capability of the $\alpha_{p0.067}$ system to deliver a vehicle acceleration of 9×10^{-4} m/s² is doubtful for an initial vehicle mass of 1,004 MT. That requires a 900 N thrust ion engine system design. A 17,000 hour burn duration ion engine would have to be designed. The life limit at the present is considered as less than 5000 hours in the 0.3 to 0.7 N thrust range, that thrust being the average over the burn duration.

2-34

2.0 High Energy Mission Applications

Manned Mars is not a good mission for $\alpha_{p0.067}$, at least as established by the mission mass and energy parameters herein, because the flight time becomes asymptotic at just under two years flight time, a period too long. The fission space reactor power level becomes large. To achieve a 1.93 year flight time, a 650 MW reactor is required. At the present time NASA is developing the 100 kW fission reactor, designated as SP-100.

The fusion jet power necessary to accomplish these missions is 93 MW for the $\alpha_{p1}/0.5$ mission and 178 MW for the $\alpha_{p10}/0.5$ mission. The actual reactor will be larger to account for inefficiencies. The average thrust for the heavier α_{p1} vehicle is 2,400 N, for the α_{p10}, 1,040 N. These are anticipated to be achievable.

An optional power design concept is the Inertial Confinement Fusion (ICF) approach. The VISTA (Vehicle for Interplanetary Space Transport Applications) study referred to earlier used an ICF system for performing a manned flight to Mars (Ort87). That design carried a 100 MT payload to the Red Planet, allowed the crew to stay there for ten days, then returned to Earth for a 100 day total mission duration. The propulsion system is designed to burn D-T and consumes 20 MT of tritium. The study assumed a target gain of 1,500 and pulse repetition rate of 30 Hz. This was a high thrust engine, producing 2×10^5 N. The initial vehicle mass was 6,000 MT; the quantity of propellant carried was 4,400 MT. The propulsion system Isp was 17,000 seconds and total jet power, 2.0×10^4 MW. A view of VISTA is shown in Fig. 2.13.

2.0 High Energy Mission Applications

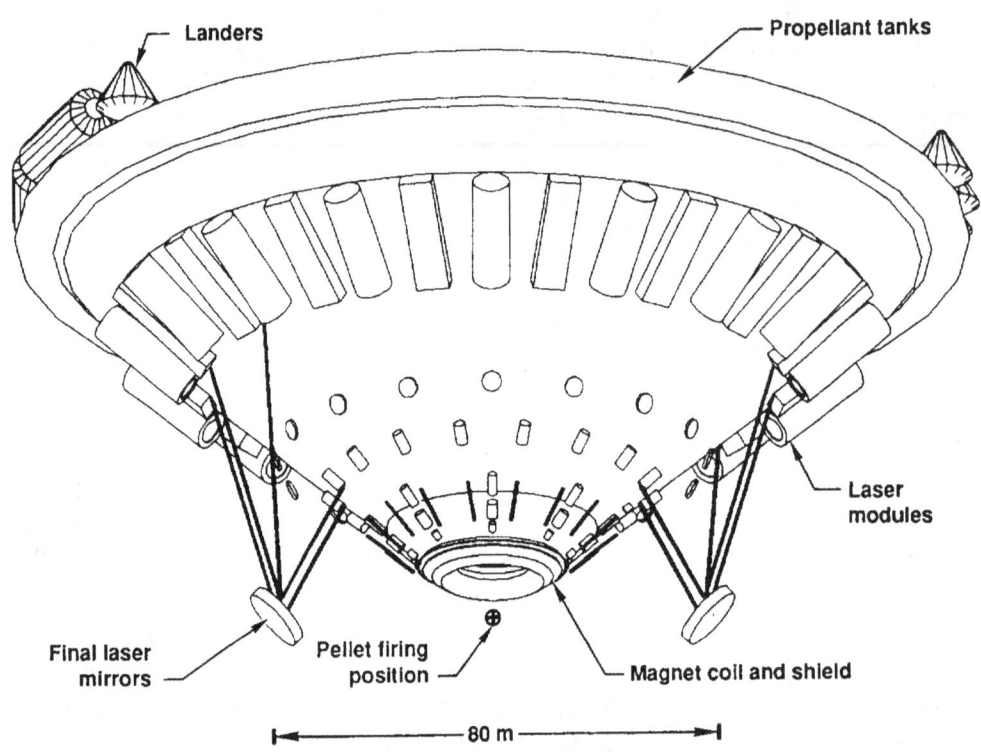

Fig. 2.13. ICF powered manned Mars spacecraft. (VISTA).

The size of the spacecraft attests to earlier statements that fusion is currently intended for large power applications. A performance comparison (100 MT outbound and 100 MT return payload) of the two systems, low thrust MCF and high thrust ICF, was made: assuming an α_{p10}, the initial mass in LEO is 280 MT (100 days) whereas for α_{p1}, the M_o is 6,000 MT for a total flight duration of 125 days.

2.2.1.1 GAINS IN OUTER PLANET EXPLORATION EFFICIENCY

Just as the mining of materials is key for the habitat fabrication for settlement of Mars, so too will the mining of materials for the processing of propellants become an essential technology for efficient space operations. In-situ propellant manufacturing can be anticipated to become an important element in the space mission architecture. Production of propellants on Mars makes use of a planet as a space resource and space operations center. Martian manufactured propellants can be used either for the transport of vehicles back to Earth or toward the outer parts of the solar system, where launch vehicles can serve as a more efficient, time expedient launch platform to the outer planets than vehicles launched from Earth. Mars could become a major scientific outpost for a quicker, cheaper means for exploration of at least the outer planets. If established as a propellant depot as part of the space mission architecture, a Mars depot could provide a gain in the economy of future space activities.

Another approach is the use of lunar volatiles produced as a by-product of the mining of Helium-3 as mentioned earlier. While the cost of the chemical propellants on Earth are by themselves insignificant, when one considers the cascading effects in the launch vehicle cost equation, propellant costs, nevertheless, ultimately become a very significant parameter in terms of sizing vehicles, their support equipment, plus logistics. In the final cost equation, a vehicle's size is a key parameter in the overall program costs. Hence, propellants/fuels, as a large mass component of a launch system, significantly impact the ultimate vehicle cost, the time to accomplish missions, operational costs, ground launch equipment, and even safety of flight, particularly when a greater number of operations is necessary to achieve the same end objective. High specific power and specific impulse fuels reduce the space propellant requirements dramatically.

For fusion fuels, where the fuel requirement is large, incentives exist to explore other extraterrestrial sources for deuterium and ^{3}He. Any bulky, massive material that can be picked up along the way is very cost/performance advantageous. This is preferred over the current approach of requiring the return trip's propellants being transported to LEO and out to the designation.

Looking into the not too distant future for projecting life here on Earth, such as in the time frame of just several decades, one can seriously question whether the relative cost of energy to the cost of consumer products will remain in the current ratio that we now enjoy. Costs are ultimately regulated by the two variables: supply and demand, and two things are certain to happen – a reduced energy supply and an increased demand. It is quite clear that both variables will serve to increase the cost ratio of energy to product value. According to one demographic model recently developed, the Earth's population is mathematically increasing at a rate sufficient to cover the land mass in well under a century. As communications continue to improve and increase, it is a natural assumption that the lesser developed countries will desire the same benefits now enjoyed by the industrialized nations. The US, for example, with 4% of the world's population consumes 25% of the energy. Those two energy demanding factors, population growth and third world development, will combine to rapidly escalate an insatiable thirst for energy, skyrocketing energy prices and creating a corresponding lack of availability. Perhaps if some of those global problems are not resolved by then, energy will have become so precious that it will be consumed only for the very basic functions of life on Earth, namely, food, shelter, and warmth. While utilization of the local planetary resources at Mars and beyond are not being suggested here as a means to resolve this problem, possibly those extraterrestrial energy resources could well extend man's ability to conduct space exploration without

2.0 High Energy Mission Applications

exacerbating the Earth's energy supply situation. The bottom line is that alternative energy sources are needed for space's future.

2.2.2 MANNED MISSIONS BEYOND MARS

Mission Description

Some thought was given to manned outer planetary missions. This mission category is being introduced since it has not yet received consideration, and not surprisingly so, since the Martian trip is already a very ambitious, energy demanding trip. There is currently no transportation means for accomplishing such missions. High energy demands render the chemical systems incapable of performing this mission category, and the NERVA fission propulsion system was shelved almost two decades ago as not being necessary to the space transportation system.

Obviously the first technical thought is whether or not man could safely explore the moons of those planets, taking into consideration the severe radiation environment associated with Jupiter, and to a lesser degree, Saturn. Neither appears likely for missions to the inner moons. At Saturn the radiation belt is weaker than at Jupiter; but since it is still comparable to Earth's Van Allen belt, manned exploration at Saturn is improbable although the outer moons might be considered. The extended exposure of the crew to the natural cosmic radiation environment is additive to those levels. Uranus, Neptune, and Pluto hold some degree of promise for outer planet manned explorations, at least from the standpoint that the radiation present is not known to possess a high degree of risk. Galactic cosmic rays can activate materials used in the construction of the spacecraft due to the long trip times, causing one additional hazard source from radiation. Sporadic radiation due to solar flares comprises another significant radiation hazard to the crew in those missions. The low planetary temperatures also provide an interesting challenge for exploration of those bodies.

Asteroid exploration

Manned exploration of the asteroid belt could also be accomplished and is one of the more interesting and potentially more rewarding missions for exploration. A massive 150 MT payload transported to 3 separate asteroids at a separation distance of 1 AU each could be accomplished in 2.25 years total round trip flight time using an initial vehicle mass of 460 MT, with a mission average specific impulse of 23,600 seconds. That mission design, which assumes a specific power of 1 kW/kg, returns to LEO a 60 MT payload.

Uranus

An exploration trip to Miranda, is probably the first planetary mission beyond the Asteroid belt that man could accomplish without concern over planetary generated radiation. Using the same vehicle for this mission as Mars, i.e., 133 MT outbound payload – 61 MT inbound payload, the mission round trip flight time is 2 years for a specific power propulsion system of 10 kW/kg. The initial vehicle mass is 620 MT; the specific impulse is 65,400 seconds; and the jet power is 1.5 GW. If only a specific power system of 1 kW/kg is available, the trip time will be lengthened to 5 years total round trip flight time.

Neptune

An identical mass used for this flight showed that the time to Triton and return is slightly less than 5 years for 1 kW/kg and ~2.5 years for 10 kW/kg.

Pluto

The outermost planet can be reached by a 133 MT outbound/61 MT return payload in slightly over 5 years with a 1 kW/kg system and ~3.5 years for a 10 kW/kg system.

Comets

Comets present a great interest for manned exploration simply due to the dynamics involved with comets particularly when in the presence of the sun. Man could serve to perform focused in-situ examination of the more attractive geologic features and for the retrieval of specifically selected comet samples.

Mission performance analysis

Mission performance capabilities were not calculated for comet missions which can be studied as part of the recommended activity discussed at the end of Section 2.0. For long manned trips, we can expect the use of hydroponics to raise plants, and the use of lights for plant growth. If that approach is deemed desirable, significant energy demands, perhaps of a fusion scale, can be expected to be required. Transportation energy requirements and flight time can be reduced by launching from Mars.

2.0 High Energy Mission Applications

II. SPACE SCIENCE SOLAR SYSTEM AND INTERSTELLAR MISSIONS

For the far outer planets and interstellar space science missions, a high energy capability is uniquely mission enabling. We can consider two mission classes:

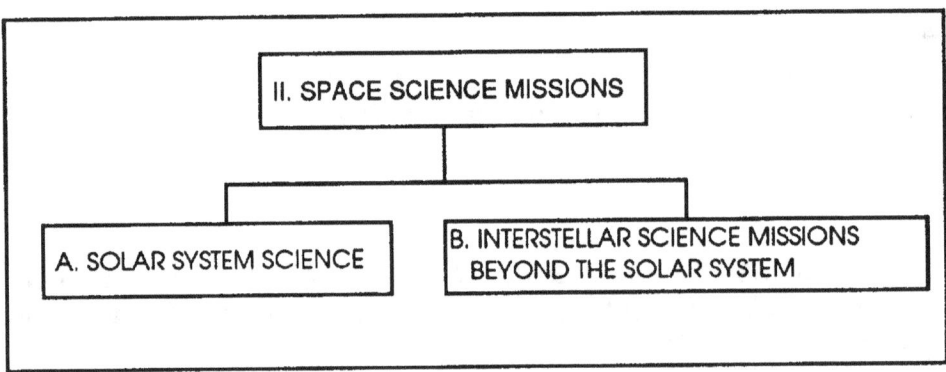

A. SOLAR SYSTEM SCIENCE

Scientific exploration missions of the solar system can be accomplished in a relatively short time span; science experiments can be quickly transported to its destination; and scientific exploration of interstellar space can be performed.

The list below provides a summary of the major solar system science objectives:

I. Space Science Mission Objectives: Solar System

- Winds
- Composition
- Temperatures
- Climate
- Atmospheric dust
- Pressures
- Soil motions
- Structure of the planet's interior
- Search for organics and water
- Geologic history: meteor impacts, flows of surface matter

2.0 High Energy Mission Applications

-Mapping:
- surface of planets
- asteroidal distributions
- Surface elements and compounds
- Soil age
- Magnetic and gravitational fields
- Solar wind plasma
- Density/composition of subsoil
- Imaging of surfaces at altitude and locally directly from the surface
- Science of long term high and low level radiation exposures
- Seasonal changes.

The means by which these objectives can be achieved is by the following Space Science Infrastructure:

A. Scientific Outposts
- Orbiter
- Lander
- Surface rover vehicle
- Atmospheric craft
- Permanent surface laboratory

B. Science Return vehicles
- Soil sample
- Atmospheric samples.

2.2.3 UNMANNED OUTER PLANETS AND MOONS

Mission Description

The propulsion and electrical power technology requirements for accomplishing the high energy science missions of the type envisioned herein to the outer planets, are essentially the same as for Manned Mars Missions. These missions are even greater beneficiaries of high energy since the kinetic energy demands increase with distance from the Earth. Access to high specific energy sources increases substantially the quantity and quality of the space science returned. For example, consider the improvements in spacecraft launch frequency. Rather than waiting for the next available gravity-assisted launch window to open, we could realize launches within expanded launch windows by the use of spacecraft designed with a large reserve of power for Δv. Expanded launch windows will reduce the impact on program costs. Tolerance to launch windows is of greater importance to the high energy missions.

2.0 High Energy Mission Applications

Scientific instruments remaining in orbit or on the surfaces of planetary moons are important for synoptic data gathering purposes such as for monitoring the body's physical characteristics of winds, temperature, pressures, and solar wind plasmas. Remote soil analyses for chemical composition determinations are also of interest.

These science missions can be accommodated in a manner similar to the Manned Mars Mission scenarios discussed in Section 2.2.1. Mars is not included below since the precursor, premanned science exploration missions are presently envisioned as being required too early for the presence of high energy systems. The Manned Mars Mission is assumed to provide for the subsequent Mars science program activities. If the desire exists to use this system for Mars sample return missions, then a specific power system of 1 kW/kg would deliver the 20 MT outbound–10 MT inbound payload in 0.6 years using one Shuttle payload of fuel (27 MT). For a 10 kW/kg system, the same mission could be accomplished in ~3 months. As missions extend further out, however, high energy assumes an ever increasing importance. Depicted below is the scope of solar system science missions considered.

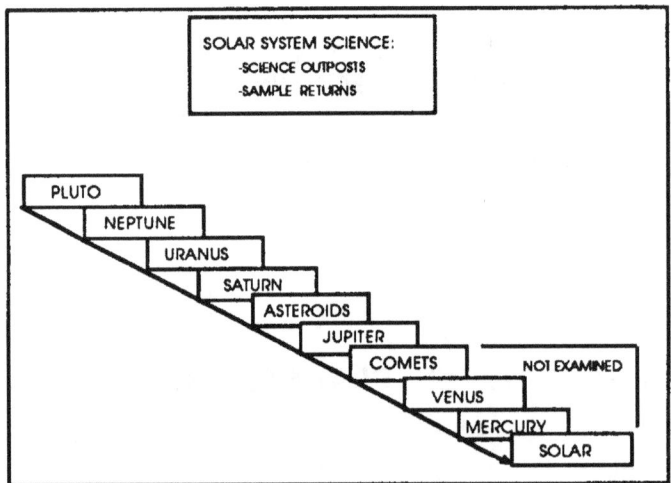

The benefits of higher specific power are: the science instrumentation package arrives at its destination more quickly; the vehicle carries a more comprehensive, increased reliability payload package of science instruments; the on-site telemetry station transmits a higher data bit stream; and the instruments are able to serve for longer mission durations.

Jupiter

For spacecraft that deploy probes, more probes could be carried per launch, lowering their unit cost while providing a greater, quicker return on science. Missions which permit planetary rendezvous, a large consumer of energy, are preferred to the fly-by missions in order to study the changes in the planet as seasons advance and to provide time to study other solar system phenomenon.

With high energy, an opportunity exists to perform some unique physics and materials science. Consider Amalthea, the closest moon of Jupiter. A rover mission there offers the unique opportunity to study the effects of extended radiation exposure to a variety of materials. The proposal is to monitor the emissions from Jupiter close up and to examine samples of soil for establishing the effects of very long term high energy radiation. Samples would be returned to Earth for further analysis. The differential effects of decreases in the intensity level of environmental exposure can be achieved by similar missions to each of the moons.

Saturn

Rendezvous with the rings and a ring sample return mission, a new mission, would hold great interest, as well as rendezvous missions with the rings of Uranus for comparative composition analytical purposes. A comparison of those ring compositions with asteroids would also be of great interest. The asteroid missions are discussed later.

Titan's atmosphere and surface can be characterized and possibly mapped using some type of "aircraft" that would map the atmosphere in all three dimensions.

Uranus

Rendezvous with Uranus's rings and a ring sample return mission, a new mission, would be of great interest, as well as rendezvous missions with the rings of Saturn for comparative composition analytical purposes. Probes through the atmosphere and exploration of the planet's atmosphere by "aircraft" can be studied as a possible future mission. If the mission is feasible, the payload can be anticipated to be massive. Payload mass determinations are not available since this has not been considered earlier. SAIC examined an aircraft for Mars in the ISPP (In-situ Propellant Production) Study. The total payload mass was ~6.5 MT, 0.21 MT of which was allocated for a solar powered aircraft. Solar energy would not be an option at Uranus, requiring the use of the planetary atmosphere for propellant or alternatively the use of on-board flight propellant supplies.

2.0 High Energy Mission Applications

Neptune

This mission has an orbiter and surface rover to conduct surface science exploration missions on Triton.

Pluto

Observations of the spacecraft's trajectory data would provide benefits from improved mass determinations which currently holds a significant uncertainty. Two Plutonian missions, fly-by and rendezvous, can be contemplated. A robotic sample return mission permits detailed soil analysis on Earth. An orbiter will characterize the planet's physical properties, the composition of the atmosphere, temperatures, magnetic fields, overall structural configuration (roundness, mass concentrations, impacts, topography, etc.), presence of new moons, and similar information regarding their characteristics. With the aforementioned data, conclusions can be drawn regarding Pluto's origin, i.e., whether it resulted from a gravitational capture or formed from primordial matter in the solar system.

Mission performance analysis

Mission performance calculations show that science outpost missions during which the spacecraft orbits the target, probes the atmosphere, and provides sample returns are possible on a substantially shortened mission elapsed time in comparison with that provided by current technology. Five round trip sample return missions were examined which included trips to: Jupiter's Europa, Saturn's Titan, Uranus' Miranda, Neptune's Triton, and Pluto's Charon. Plots of the round trip time for carrying an outbound payload of 20 MT and a 10 MT return payload are shown in Figs. 2.14 to 2.33 for each of five moons respectively.

Europa

This is a massive moon which is 1.6 times the distance of Io from the center of Jupiter. A parking orbit of 670,987 km from Jupiter comprised the destination of this spacecraft. For this mission's performance data, the Jupiter mean distance assumed was 4.2 AU from Earth. The payload masses were as previously indicated. Otherwise, the mission input parameters and calculation technique were unchanged from the Martian calculations.

Consider first a α_{p10}-powered spacecraft. The shortest round trip flight time calculated was 295 days. The initial vehicle mass of 92 MT (Fig. 2.14),

2.0 High Energy Mission Applications

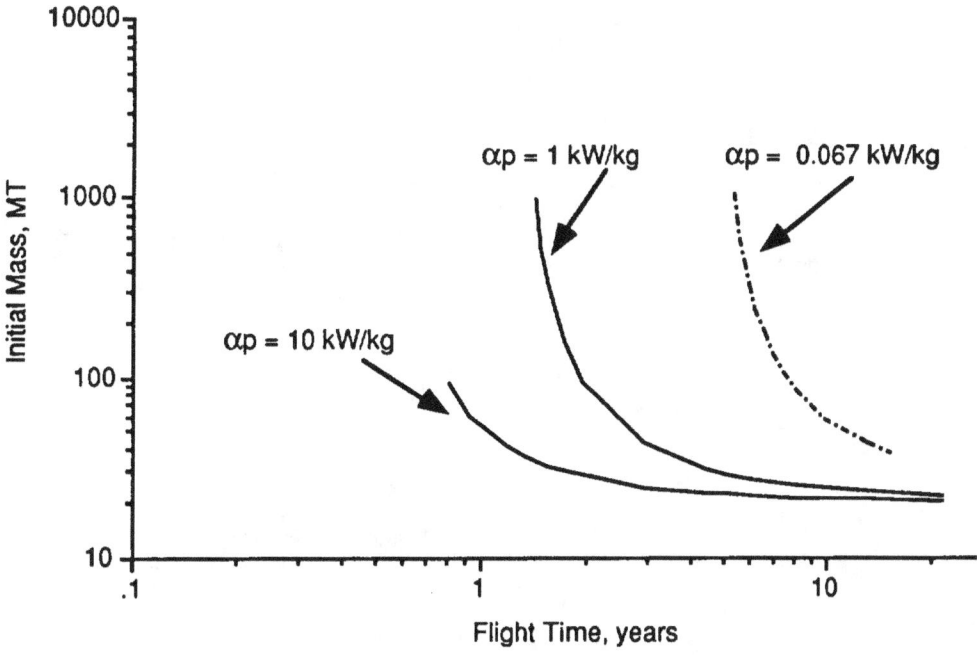

Fig. 2.14. Europa sample return mission, initial vehicle mass variations with flight duration.

carrying a propellant mass of 50 MT (Fig. 2.15),

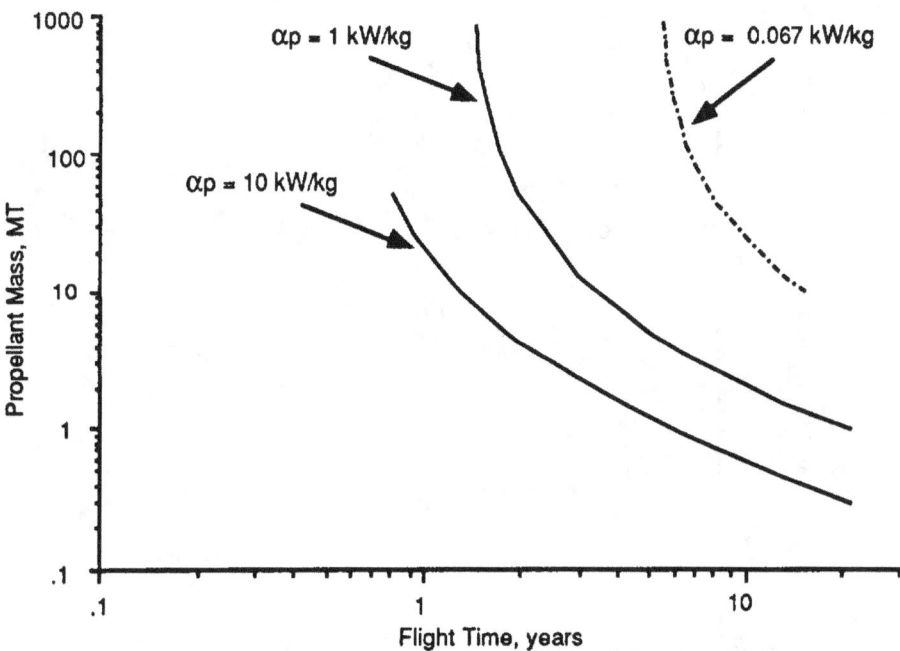

Fig. 2.15. Europa sample return mission, propellant mass variations with flight duration.

2.0 High Energy Mission Applications

provides a Δv of 352 km/s (Fig. 2.16),

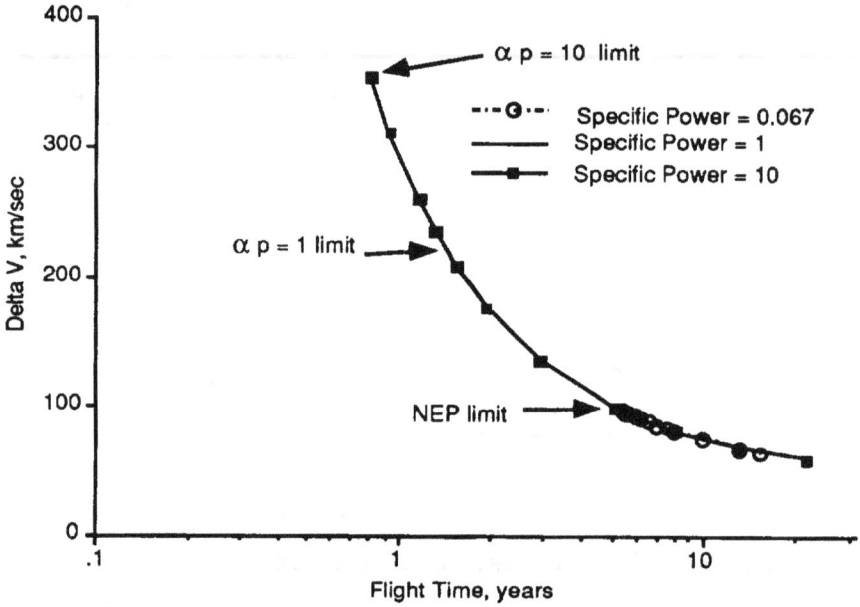

Fig. 2.16. Europa sample return mission, velocity variations with flight duration.

and requires an average specific impulse of 42,205 seconds (Fig. 2.17).

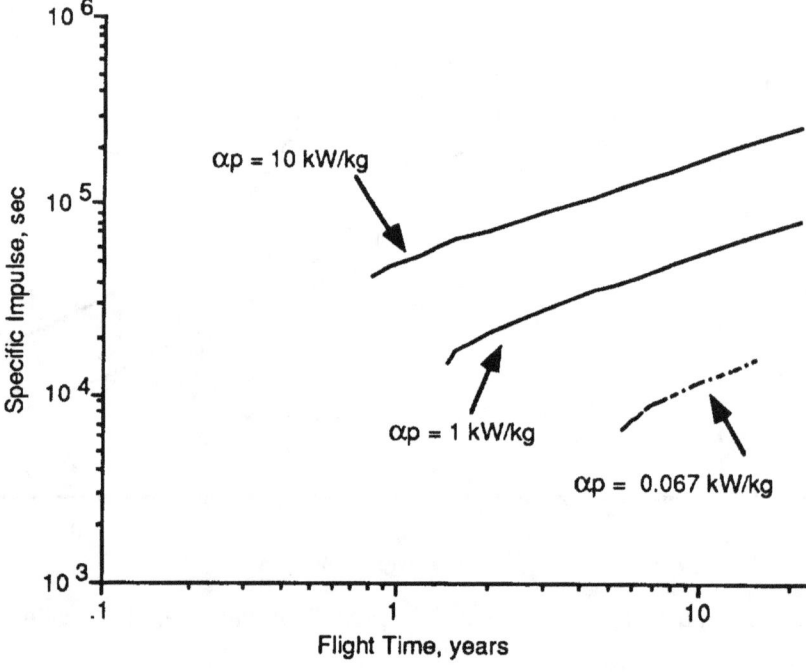

Fig. 2.17. Europa sample return mission, specific impulse variations with flight duration.

For the α_{p1} design, the data show an order of magnitude increase of the initial vehicle mass over the α_{p10} design. The lower specific power system, α_{p1}, requires an increase of M_o to 976 MT and a propellant increase to 843 MT (Fig. 2.16), a factor of 17 greater than the α_{p10} design. The Δv (Fig. 2.17) imparted is 223 km/s, requiring an average specific impulse of 14,724 seconds.

A more likely acceptable mission scenario is to drastically decrease the propellant flow rate sufficiently to accommodate a 420 MT propellant mass, resulting in an increase in total flight time of 22 days. Decreasing it to 240 MT, increases the flight time by only 47 days. The engine thrust is an averaged 1920 N for a $\alpha_{p1}/1.56$ mission; the jet power requirement is 57 MW.

The $\alpha_{p0.067}$ spacecraft's initial mass is asymptotic at ~4.62 years: M_o = 1,037 MT, M_p = 900 MT, Δv = 96 km/s, <Isp> = 6,304 seconds. A six year mission lowers the mass requirements significantly but at a significant increase in specific impulse. The initial vehicle mass for a 6.2 year round trip mission is 240 MT. The propellant mass is 173 MT and average Isp is 7,709 seconds, a value approaching the upper limit of $\alpha_{p0.067}$ systems (at the completion of thruster life). The average thrust ranged from 6 N to 767 N for the shorter trip; the fission reactor power required is 7.8 MW for a 5.4 year round trip mission. While this mission can theoretically be accomplished, the reactor power is 2 orders of magnitude greater than the largest space reactor now being researched. The thrust level is 1 to 2 orders of magnitude higher than current technology.

Amalthea

This, the closest moon of Jupiter, provides a unique opportunity to study the effects of extended radiation exposure to materials. It is a natural laboratory for researching materials which have been exposed to the high and low Jovian radiation levels for millions of years. Amalthea's orbit lies within the most severe part of Jupiter's radiation belt. The purpose of the sample return mission is, therefore, to investigate the physics of materials after exposure to high and low fluxes of particles over very long duration exposures. The prime source of data is from Pioneers 10 and 11 and Voyagers 1 and 2 as obtained from several on-board instruments. At Amalthea's altitude, fluxes of >1 MeV electrons and protons were measured at $10^8/cm^2/s$. Measurements were made at 0.1, 3, and 21 MeV (electrons) and at 1, 20, and 80 MeV (protons) energy levels (Figs. 2-5, Div83). The heavy ion fluxes have not been well characterized at this time.

The 20/10 MT Amalthea payload sample return mission can be performed in less than two years using α_p = 1.0 kW/kg, jet power = 22 MW, M_o = 93 MT, and Isp = 21,140 seconds.

2.0 High Energy Mission Applications

Titan

This unexplored moon of Saturn, having almost twice the mass of Earth's moon, has held great scientific interest for many years, particularly its atmosphere. A target parking orbit of 1,221,855 km from Saturn's surface was used. The results of the calculations for this sample return mission are displayed in Figs. 2.18 to 2.21. Consider the initial vehicle mass for the $\alpha_{p0.067}$, α_{p1}, and α_{p10} spacecraft in Fig. 2.18.

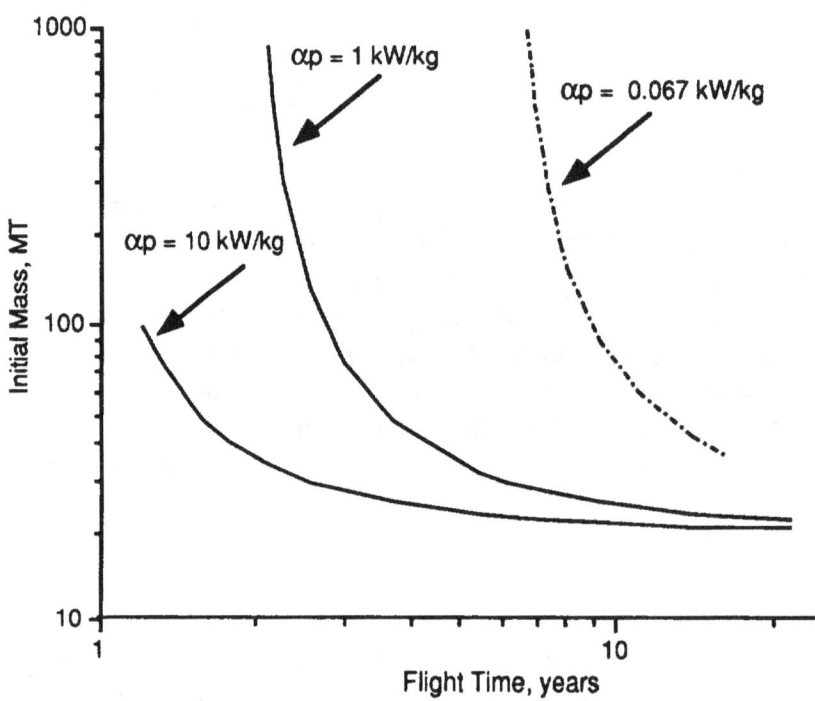

Fig. 2.18. Titan sample return mission, initial vehicle mass variations with flight duration.

The $\alpha_{p0.067}$ spacecraft mass is asymptotic at approximately 8 years. A α_{p10} propulsion system can accomplish the mission in less than a year using a reasonable initial vehicle mass. For a 1.2 year mission, a vehicle having an initial mass of 100 MT, a propellant mass of 56 MT (Fig. 2.19),

2.0 High Energy Mission Applications

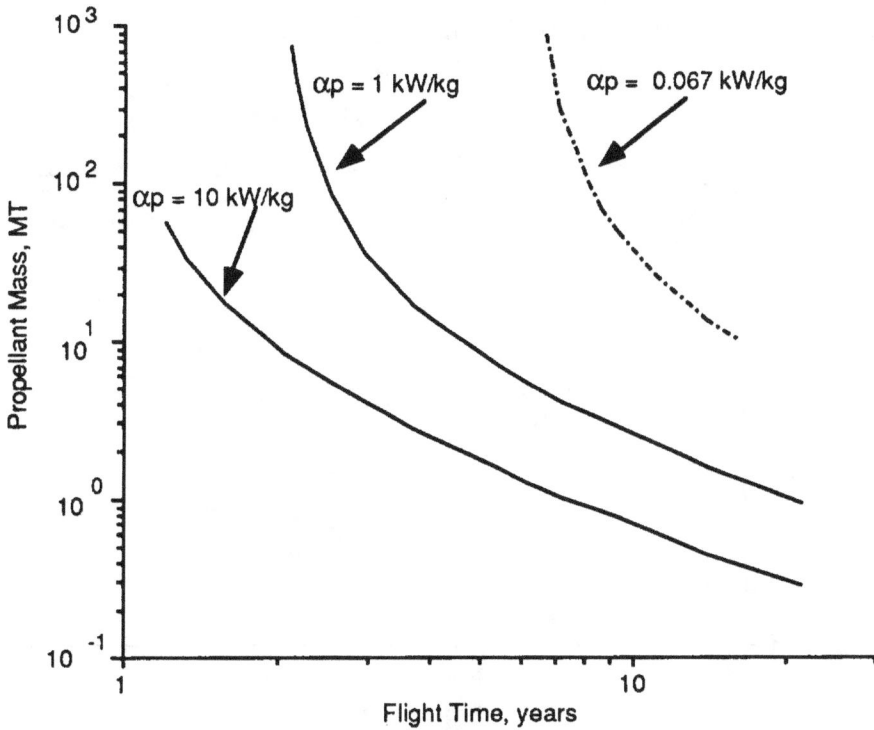

Fig. 2.19. Titan sample return mission, propellant mass variations with mission duration.

and a propulsion system producing an average specific impulse of 50,650 seconds (Fig. 2.20)

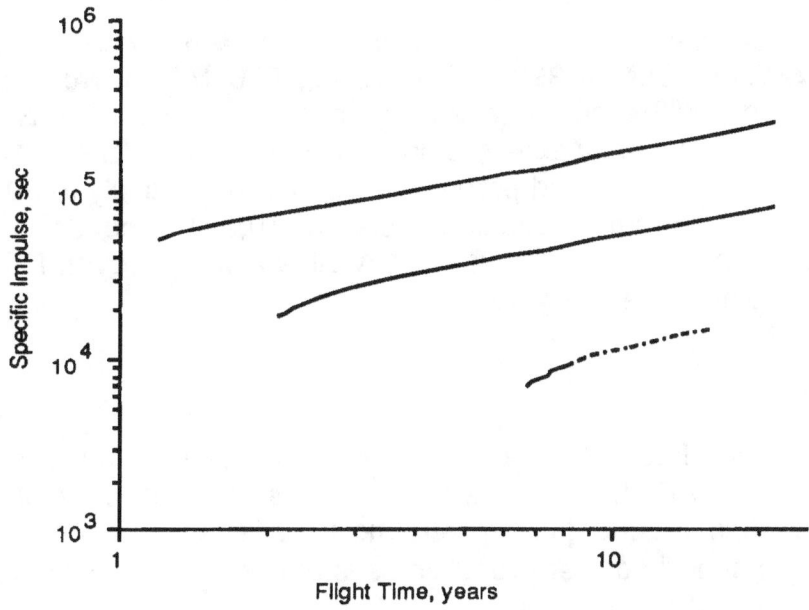

Fig. 2.20. Titan sample return mission, specific impulse variations with flight duration.

2.0 High Energy Mission Applications

will impart to the payload a Δv of 437 km/s (Fig. 2.21).

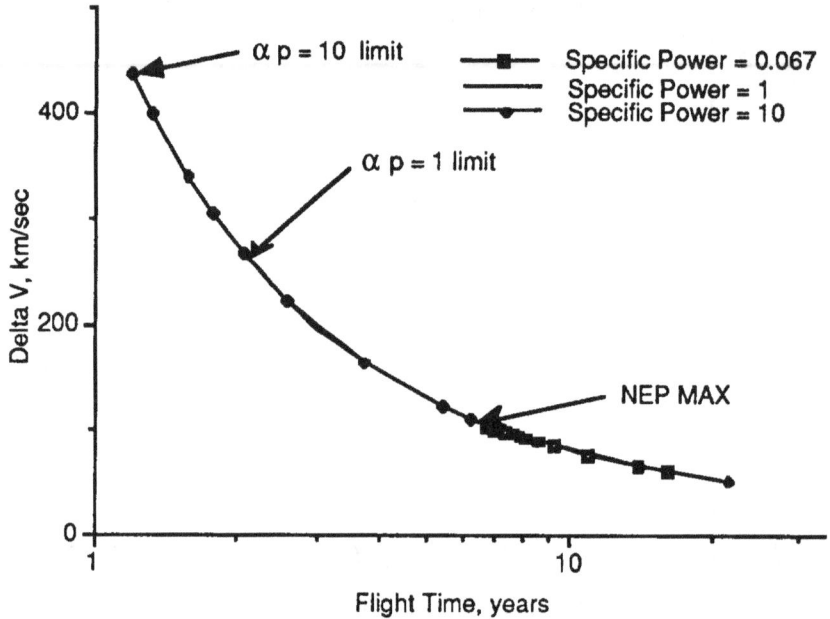

Fig. 2.21. Titan sample return mission, velocity variations with flight duration.

For α_{p1}, a 3-year sample return flight offers a reasonable propellant trade; 36 MT are consumed from a vehicle whose initial mass is 74 MT. The average Isp is 26,202 seconds. The power required is 18 MW. This light mass mission places a low thrust level requirement on the engine -- 220 N.

For the $\alpha_{p0.067}$ specific power propulsion system, a 6.7 year mission requires an initial vehicle mass of 990 MT, including 856 MT of propellant, and an engine yielding 6,889 seconds (average) specific impulse. The Δv imparted is 104 km/s. A more reasonable mission mass-wise would be a 10 year flight which reduces the vehicle and propellant masses respectively to 90 and 49 MT. But the mission averaged specific impulse of 10,221 seconds is beyond the NEP limits as currently conceived, as well as being beyond the currently researched reactor's life of 7 years.

Miranda

This mission was targeted to a parking orbit of 129,886 km at a distance of 18.18 AU from the Earth. Results from the mission performance calculations are shown in Figs. 2.22 to 2.25. The round trip mission is accomplished in 1.93 years, using a specific power propulsion system of α_{p10}, an initial vehicle mass of 112 MT (Fig. 2.22).

2.0 High Energy Mission Applications

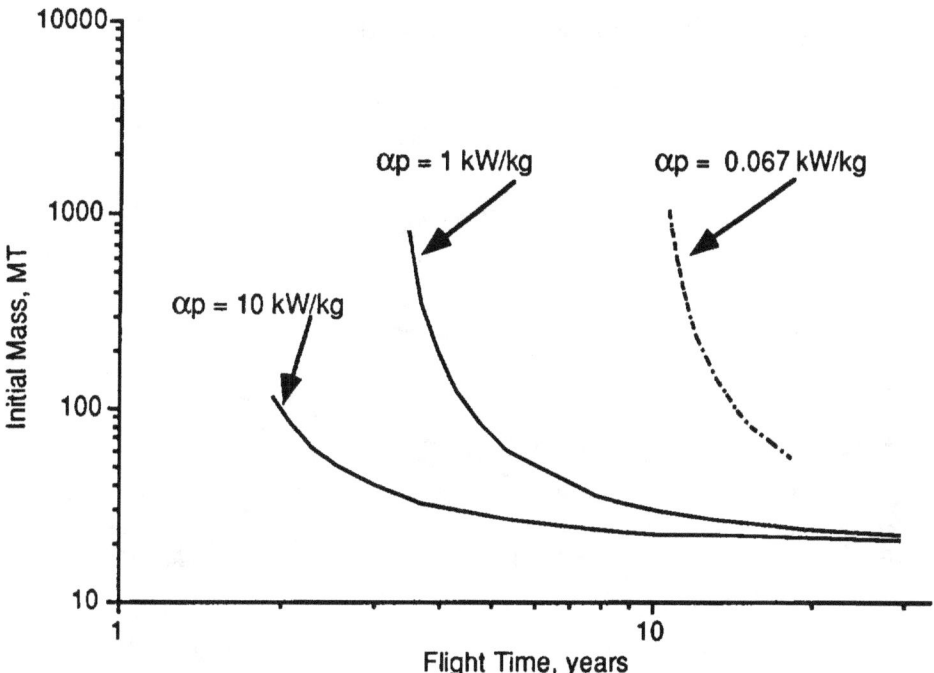

Fig. 2.22. Miranda sample return mission, initial vehicle mass variations with flight duration.

a propellant mass of 66 MT (Fig. 2.23),

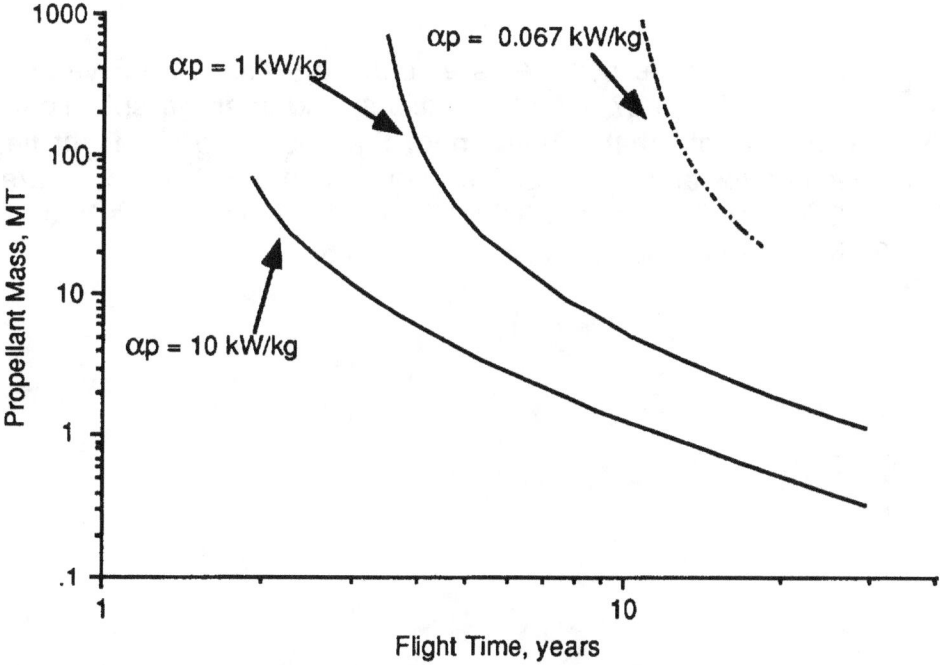

Fig. 2.23. Miranda sample return mission, propellant mass variations with flight duration.

2.0 High Energy Mission Applications

and an average specific impulse of 63,303 seconds (Fig. 2.24).

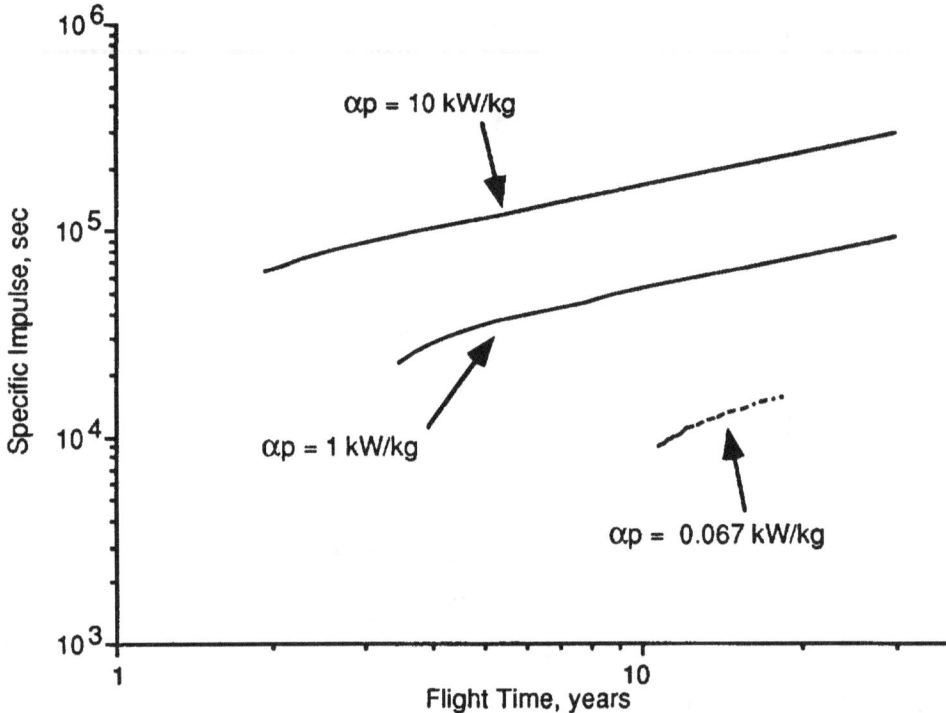

Fig. 2.24. Miranda sample return mission, specific impulse variations with flight duration.

The α_{p1} powered vehicle completes a round trip time in 3.5 years using an initial mass of 809 MT, 687 MT of propellant, and average specific impulse of 22,858 seconds. Note that slightly more than doubling the flight time to 7.8 years results in substantially lower fuel requirements, 8.9 MT compared to 687 MT (Fig. 2.24). The jet power is 6.2 MW versus 101 MW. The thrust level is a modest 35 N, and the Δv is 172 km/sec (Fig. 2.25).

2.0 High Energy Mission Applications

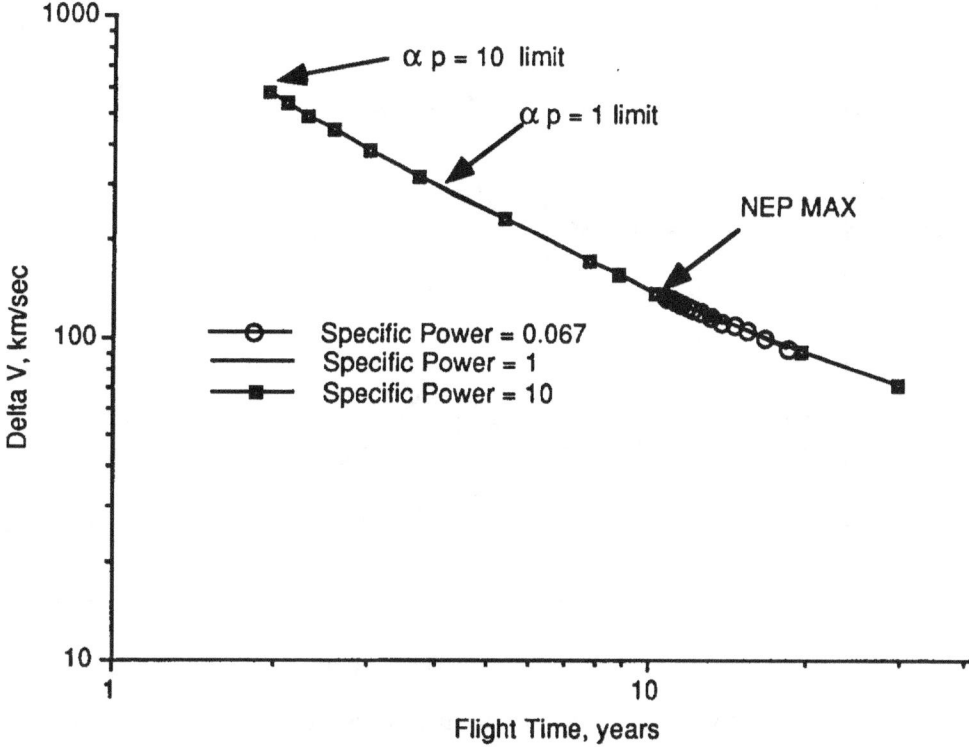

Fig. 2.25. Miranda sample return mission, velocity variations with flight duration.

The $\alpha_{p0.067}$ vehicle is a 10.7 year mission, using a 1,033 MT initial mass and an average specific impulse of 8,836 seconds, exceeding the maximum. The required firing duration is on the order of 10^8 seconds compared to only 10^3 demonstrated for NEP. This length of mission duration exceeds the life expectancy of the SP100 fission reactor by about 59%.

Triton

This massive moon of Neptune contains about twice the mass of Earth's moon. The sample return mission was targeted for a parking orbital altitude of 354,681 km at 29.06 AU from Earth. Performance curves for a science vehicle to this moon are shown in Figs. 2.26 through 2.29. The divergences of initial vehicle mass and flight time due to specific power variations demonstrates the importance of that parameter and the effects on mission requirements even more significantly. The α_{p10} system accomplishes this mission in a reasonable 2.87 years when designed to an initial vehicle mass of 77 MT (Fig. 2.26),

2.0 High Energy Mission Applications

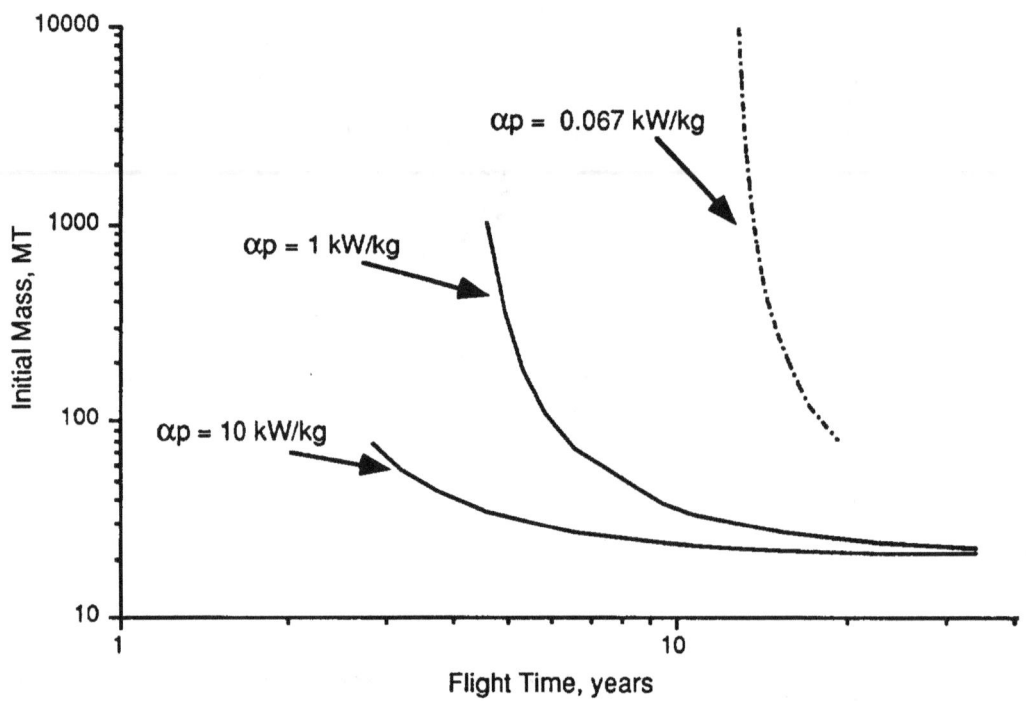

Fig. 2.26. Triton sample return mission, initial vehicle mass variations with flight duration.

propellant mass of 39 MT (Fig. 2.27),

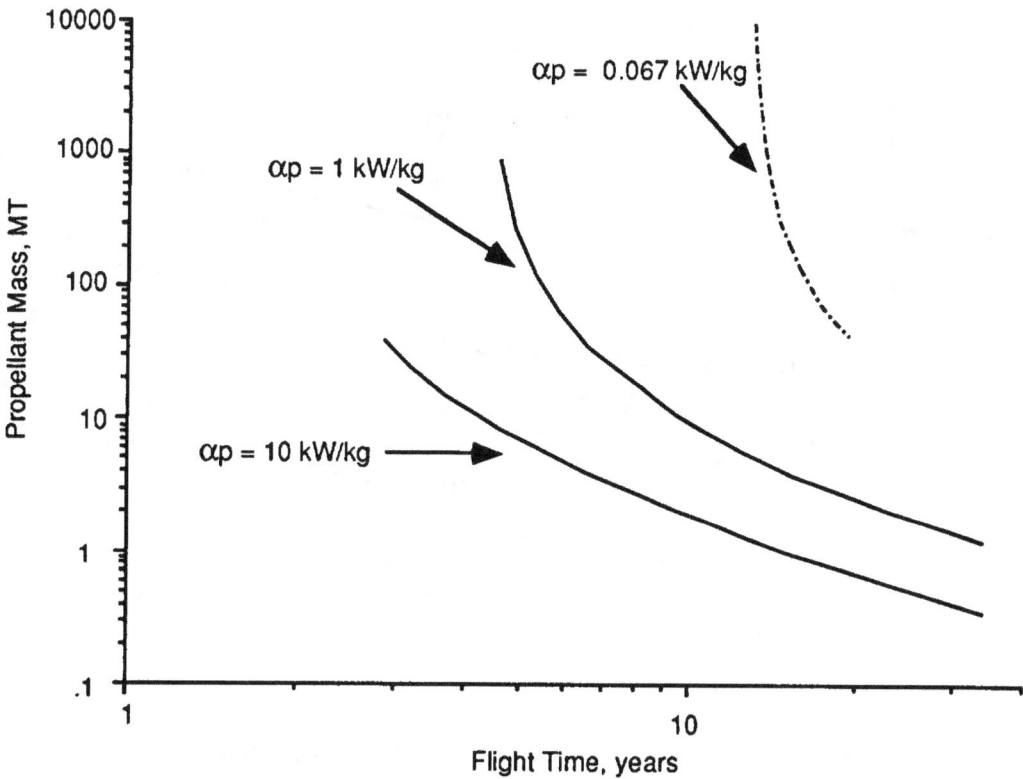

Fig. 2.27. Triton sample return mission, propellant mass variations with flight duration.

and average specific impulse of 79,815 seconds (Fig. 2.28).

2.0 High Energy Mission Applications

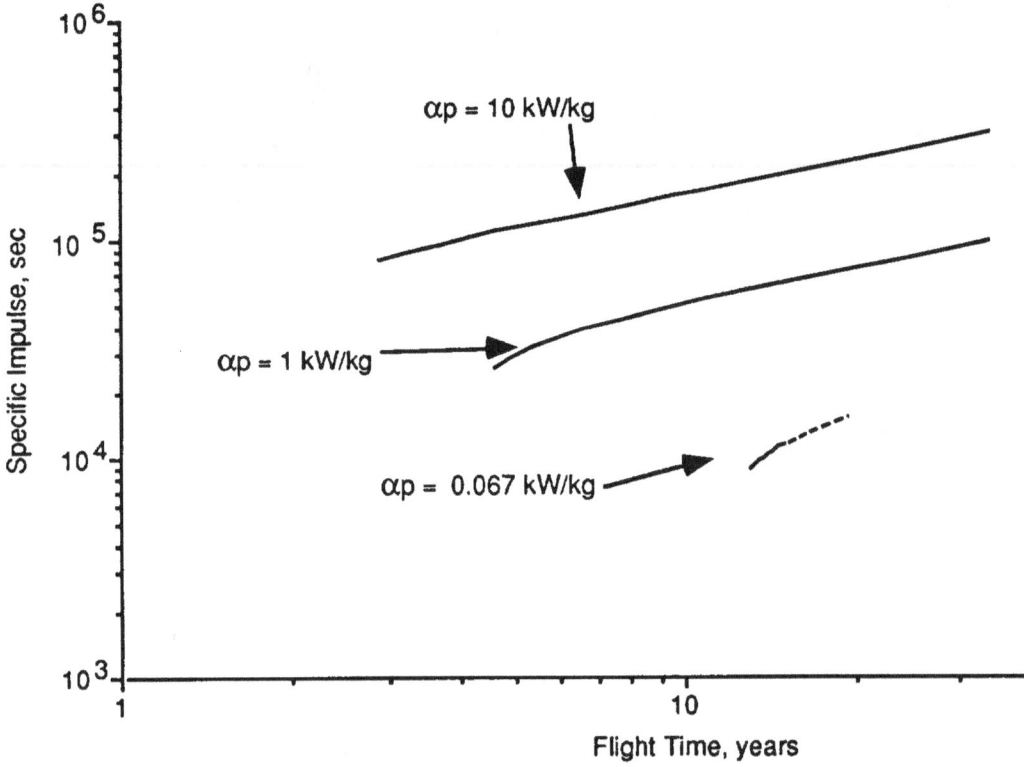

Fig. 2.28. Triton sample return mission, specific impulse variations with flight duration.

The flight is less initial mass sensitive, and the time could be reduced further without impact from the high rate of mass rise shown for the α_{p1} design or as shown by the even greater rapid rise rate of the $\alpha_{p0.067}$ specific power value. If the propulsion system design is limited to α_{p1}, the propellant penalty is very severe. The vehicle kinetic energy requirements increased to 393 km/s for the 4.6 year round trip mission, Fig. 2.29.

2.0 High Energy Mission Applications

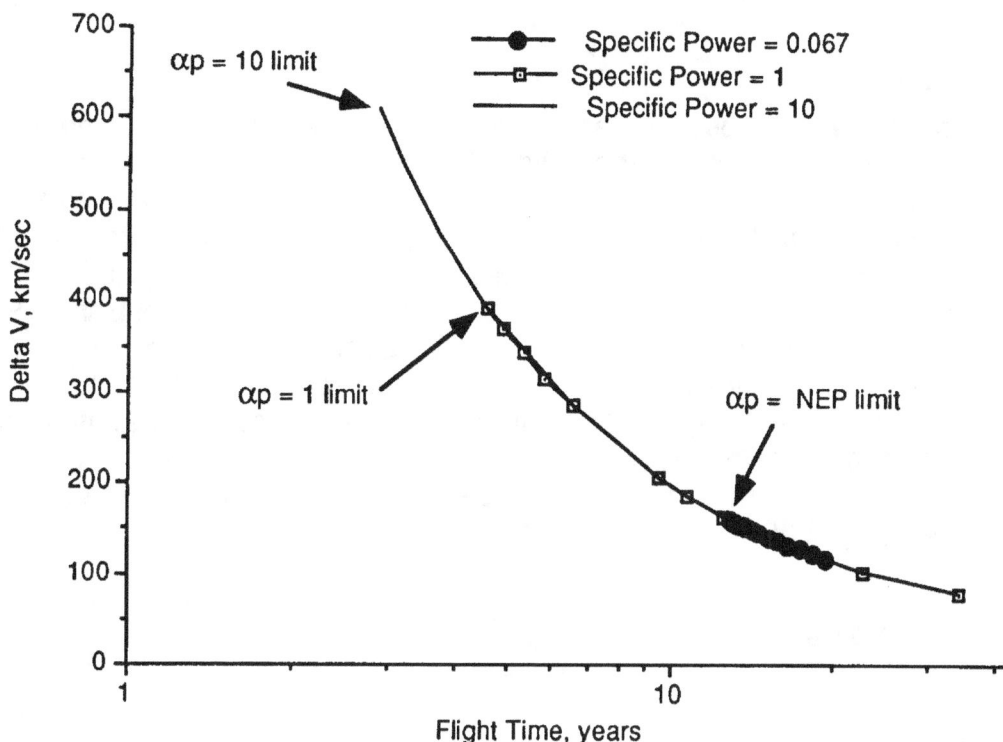

Fig. 2.29. Triton sample return mission, velocity variations with flight duration.

Table 2-1 below compares **two** equal flight time trips.

TABLE 2-1. Comparison of specific power performances for a round trip Triton sample return, 4.6-year mission.

α_p, kW/kg	M_p, MT	M_o, MT	F, N x 10^3	P_j, MW	Δv, km/s	Minimum Isp, seconds	Maximum Isp, seconds
0.067		∞ at ~12.7 years					
1.000	895	1,031	40	117	393	5,770	43,750
10.00	8	34	140	58	393	87,400	114,730

For $\alpha_{p0.067}$ to accomplish a 13-year mission, an initial vehicle mass of 9,800 MT, propellant mass of 9,383 MT, and averaged specific impulse of 8,818 seconds over the mission duration are required. Based upon specific impulse limitations, a limit of 14 years occurs at the 10,000 second specific impulse value. It should be noted here, too, that these long flight times exceed the current 7-year fission reactor life by a factor of 2.

2.0 High Energy Mission Applications

Charon

Pluto has recently been determined to have an atmosphere. Characterization of that atmosphere and of the physical properties of Charon are of great interest with regard to understanding the planet and its relation to the solar system's development and evolution. There will also be great interest in planetary capture data. In this mission Charon was targeted at a parking altitude of 17,233 km from Pluto's surface at a distance of 38.44 AU from Earth. Figs. 2.30 through 2.33 illustrate the propulsion capabilities. The mission's target requirements are similar to those for Triton, the distance being greater; but with the target mass being smaller, this mission offers some relaxation in energy requirements. With a α_{p10} propulsion system design, the round trip mission can be accomplished in only 2.76 years. To do so, the initial vehicle mass is 237 MT (Fig. 2.30);

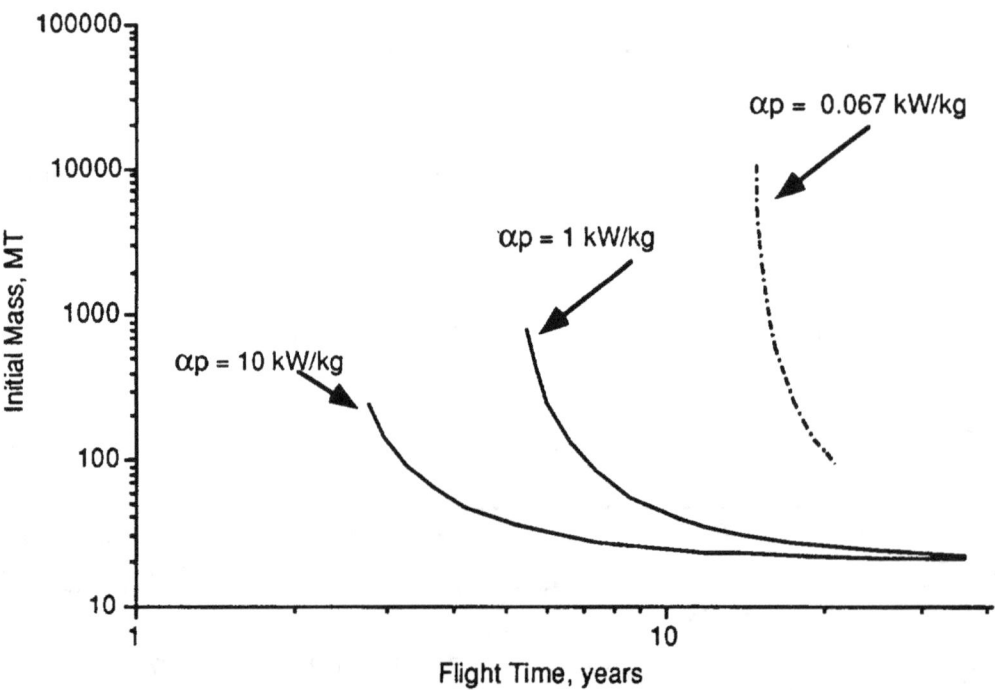

Fig. 2.30. Charon sample return mission, initial vehicle mass variations with flight duration.

the propellant mass is 171 MT (Fig. 2.31);

2.0 High Energy Mission Applications

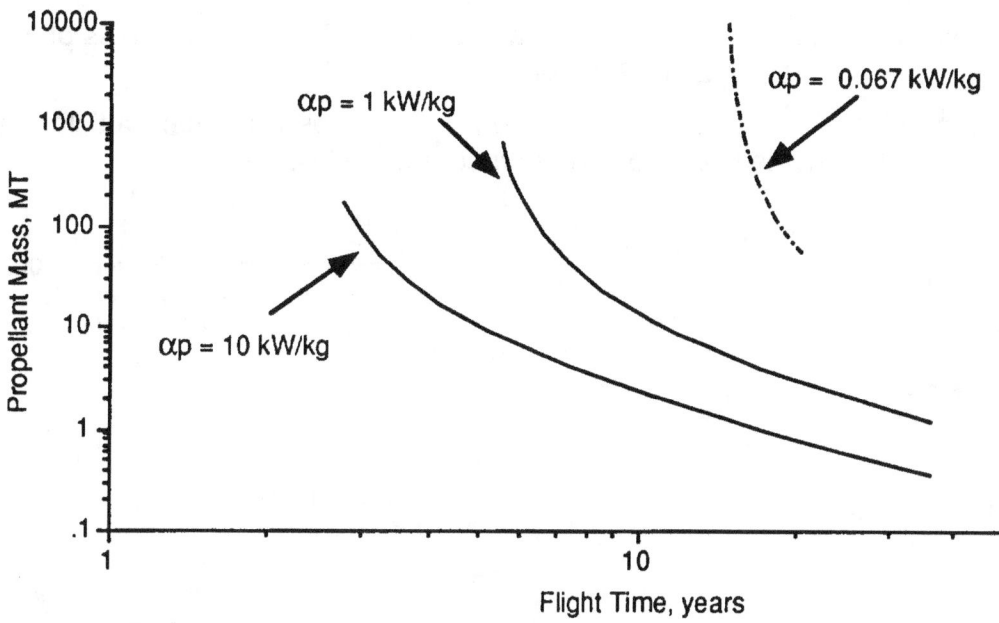

Fig. 2.31. Charon sample return mission, propellant mass variations with flight duration.

and the average specific impulse is 70,134 seconds (Fig. 2.32).

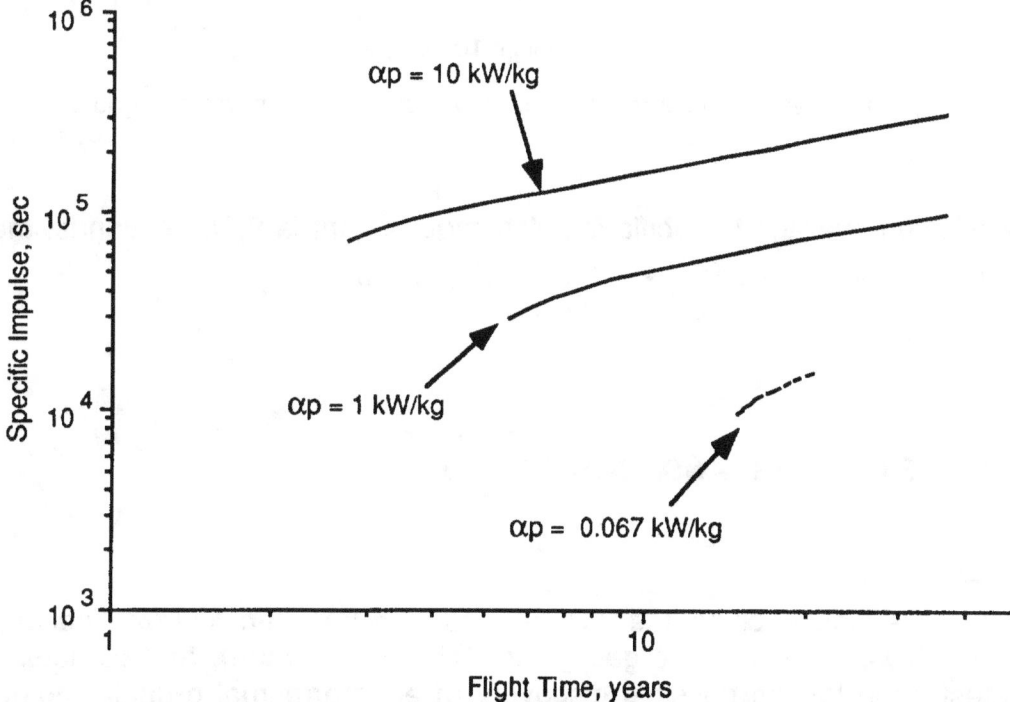

Fig. 2.32. Charon sample return mission, specific impulse variations with flight duration.

The power requirement is 464 MW, and the mission averaged thrust level is 3,320 N. A α_{p1} propulsion system will require 5.5 years to complete the round

2-59

trip mission, using an initial vehicle mass of 797 MT, propellant mass of 676 MT, and specific impulse of 28,568 seconds.

By comparison, a system flown with an $\alpha_{p0.067}$ has an initial vehicle mass of 10,247 MT to provide the Δv of 170 km/sec (Fig. 2.33).

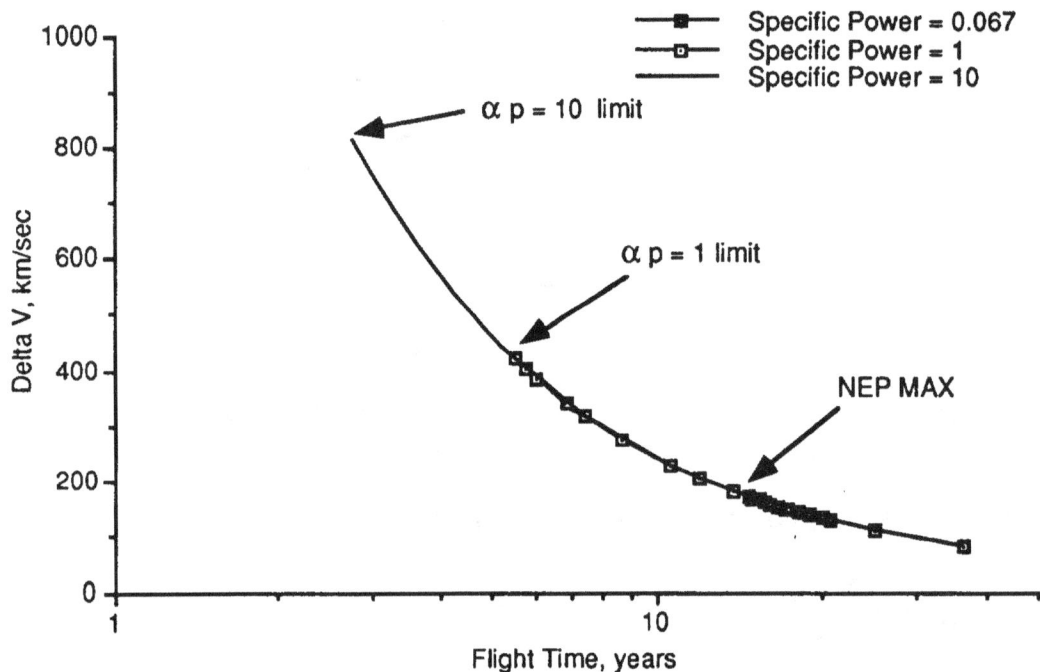

Fig. 2.33. Charon sample return mission, velocity variations with flight duration.

The mission averaged specific impulse requirement is 9,385 seconds; the trip time is 14.81 years, obviously not an NEP mission.

2.2.4 ASTEROIDS AND COMETS

Mission Description

Three of the most exciting and fruitful areas for pure scientific investigations are the asteroids, moons of the gas giants (Titan and Triton), and comets. The greatest hope for the possible discovery of extraterrestrial organic molecules inside the solar system resides in some of these bodies assuming that no scientific subtleties remain to surprise us on Mars. The solar system has been shown to naturally contain amino acids, the basic building constituents of life. What is the amino acid's origin?

Asteroids

Explorations of the asteroid belt will be particularly aided by the payoff of high performance propulsion. High energy technology will enable the mission planner to perform asteroid hopping missions, instead of targeting singular Earth launched vehicles in serial (or parallel) flights using low performance propulsion. Mission enabling advancements in space transportation of a large magnitude of the type contemplated here will produce tremendous gains in the amount of science data returned, data which is provided more quickly by flying multiple asteroid visits on one mission.

The science of the origin and development of elements and compounds as extracted from the planets is expected to play a major role in understanding the solar system. Because the asteroids have been in a dormant geologic condition, they offer the possibility of even greater knowledge regarding the origin of the solar system or planetary formation. Are the Asteroids remnants of a once existing planet, or are they masses which never accreted into a planetary structure? If the former is proven, determinations of planetary composition and structure will be used to compare with existing theories of internal planetary constituents and the origin of planets. If the latter, accretion theory will be better understood. To gain a valid statistical sampling of the chemical constituents could require a very energy intensive series of missions, requiring many trips to a large number of Asteroids. Multiple Asteroid hopping will be a much more efficient approach provided that the propulsion systems with sufficiently high energy are available.

As stated in the recent National Research Council Space Science Board report,

> ...intensive study and exploration of the wide diversity of asteroids will remain. In this area we will want to know the following: the overall structure of the asteroid belt and its radial variations of composition and physical characteristics, which are expected to reveal clues about the structure of the protoplanetary nebula; the mechanisms that powered the evolution of differentiated asteroids; and the chemical composition and physical character of comet nuclei, in order to determine under what conditions these most primitive planetesimals formed. (Don88 *Planetary and Lunar Exploration,* p. 24)

A more analytically penetrating and rapidly conducted scientific exploration program could be made possible by the availability of large power sources. Large, massive instruments can be taken to targets to accomplish in-situ analysis and to perform measurements that address the above science issues and others. For a single launch from Earth orbit, additional samples covering a larger number of asteroid visits encompassing a larger returned payload sample mass, can be accomplished without the time losses and costs associated with multiple Earth launched probes. By performing such missions, NASA achieves a tremendous improvement in science productivity using high energy. An observatory spacecraft, parked in the Asteroid belt in a retrograde orbit, could more quickly and accurately map Asteroids.

Comets

The comets are considered to comprise primordial matter. They have experienced less exposure to the solar environment than the asteroids. A significant contrast can be anticipated between the asteroids and the comets by nature of the differences in solar environmental exposure, i.e., the solar winds, plasma radiation, etc. The change and variations in chemical composition experienced over the life of the solar system will be a key piece of information in assembling the solar system history and, we can assume, in projecting its future. Perhaps the theory on the origin of life may be written within the comets. Do they contain the basic building block of life, deposited as they passed through Earth's atmosphere. How did they originate? Did they supply Earth's water? Maybe, as man becomes more knowledgeable about asteroids and comets, we will learn more about the extinction of species on Earth and become more aware of the function of meteor activity and their effects on Earth and its environment. The study of comets should include Oort Cloud science.

<u>Mission performance analysis</u>

Multiple asteroid targets were analyzed for sample return missions of multiple destinations. The mission performance computational techniques used were the same as for the other planetary missions. The only variation was to calculate mission performance data for multiple targets ranging from 3 to 6 Asteroid visits. The Asteroids were considered to be massless bodies with which to rendezvous at a distance of 1.5 AU. The assumed Asteroid separation between each target was one AU.

Based upon the same propulsion system designs used for the moons of the outer planets to perform sample return missions, that is, 0.067 kW/kg, 1.0 kW/kg, and 10 kW/kg specific powers, the calculations show that the high energy missions could be performed in times varying from less than one year to less than three years depending upon the specific power and the number of targets selected. The curves are shown in Figs. 2.34 to 2.37. The calculations assume that a 20 MT payload was transported to the first Asteroid where 10 MT was expended in acquiring the sample and/or providing scientific outpost data.

Fig. 2.34 shows the effect of specific power on the initial vehicle mass as a function of the number of asteroids visited, ranging between 3 and 6.

2.0 High Energy Mission Applications

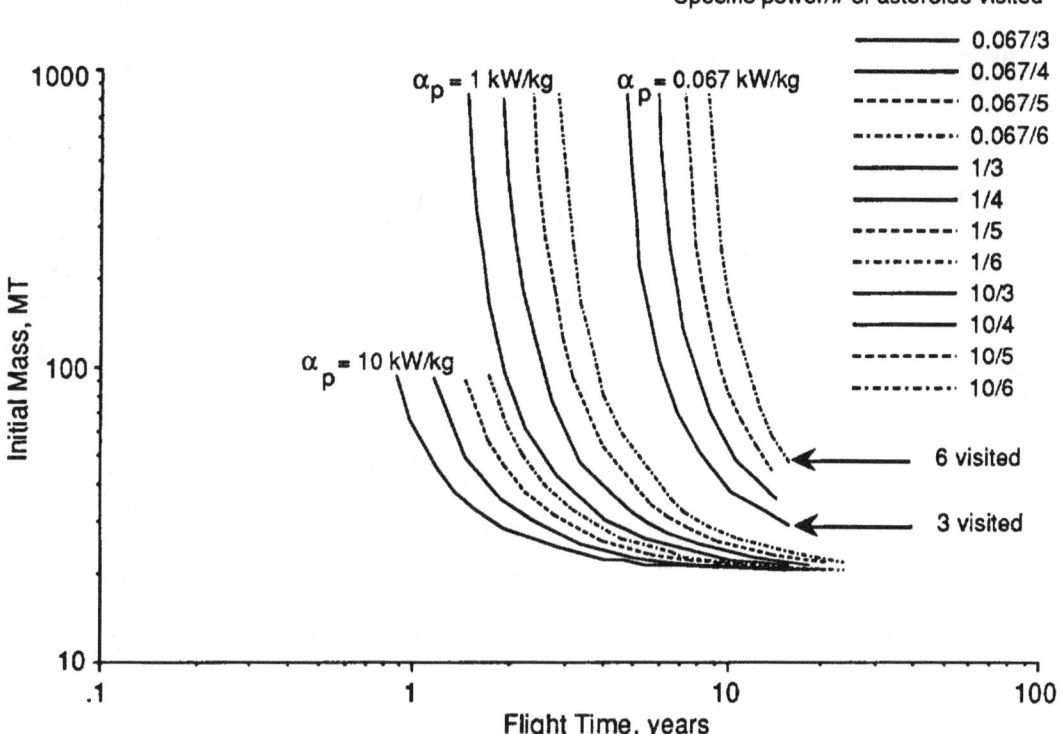

Fig. 2.34 Asteroid sample return mission, initial vehicle mass variations with flight duration for 3 through 6 asteroids visited.

To consider the most optimistic specific power system first, 10 kW/kg, one will note that for a modest propellant investment of 50 MT, Fig. 2.35, 3 asteroids can be visited and the samples returned, using a 10 MT in-bound payload, in only 0.9 year.

2-63

2.0 High Energy Mission Applications

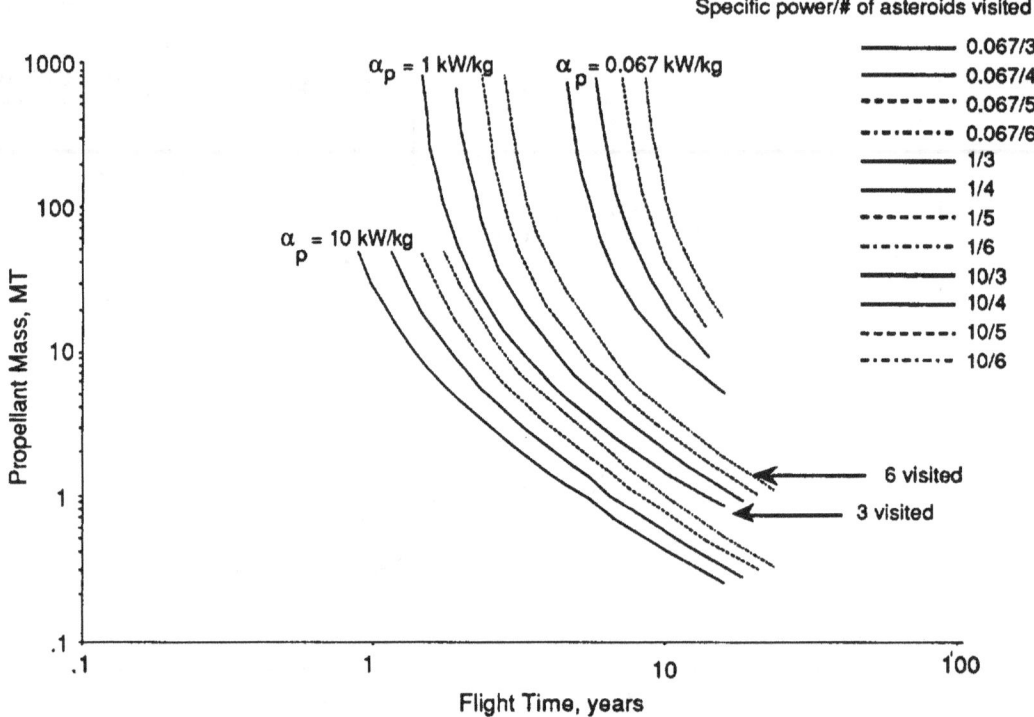

Fig. 2.35. Asteroid sample return mission, propellant mass variations with flight duration for 3 through 6 visits.

Alternatively, if the flight time is extended to 1.7 years, 6 visits with samples returned can be accomplished using the same quantity of propellants.

In the case of a system designed to meet a specific power of 1 kW/kg, there is a drastic increase in propellant consumption from 50 MT to a 820 MT consumption in order to reduce the trip time only by one half year from 1.9 to 1.4 years, an investment not likely to be made during the course of a 3 visit mission. The same 50 MT of propellants expended over slightly less that 4 years allows a visit to 6 asteroids.

The averaged Isp requirements for the 3 and 6 visits are respectively 20,900 seconds and 30,070 seconds, Fig. 2.36.

2.0 High Energy Mission Applications

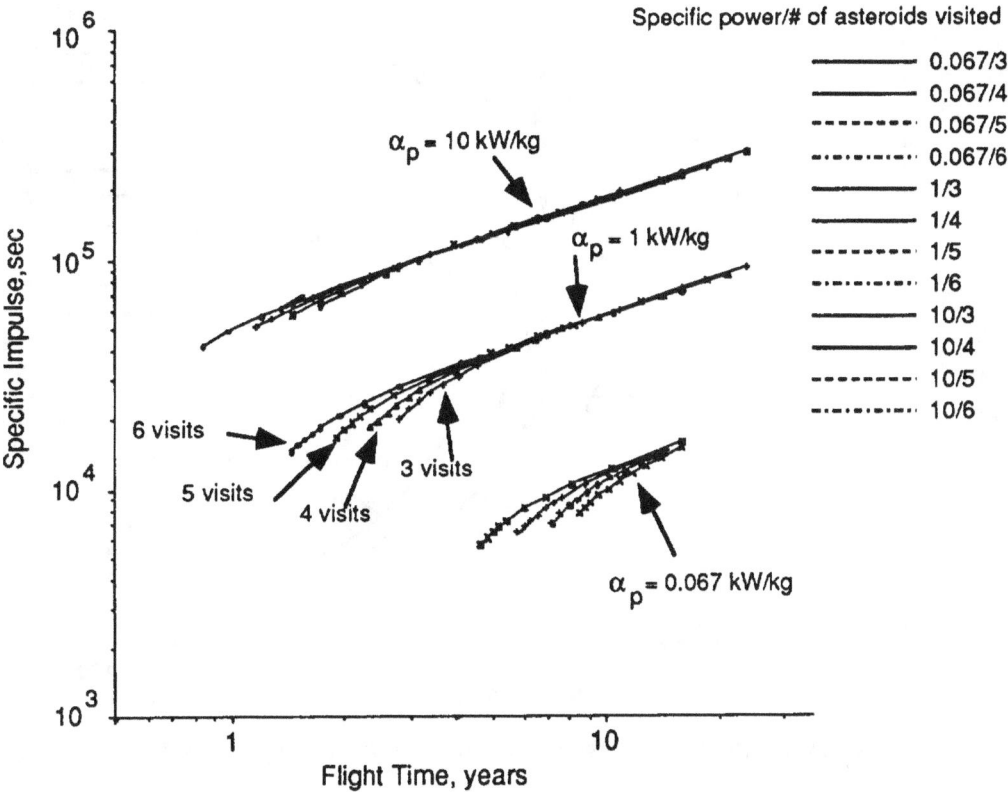

Fig. 2.36. Asteroid sample return mission, specific impulse variations with flight duration for 3 through 6 visits.

The average specific impulse for the α_{p10} missions is substantially higher at slightly over 50,000 seconds for 3 visits and nearly 60,000 seconds for the 6 visit mission. These all appear feasible based upon current understandings of fusion's performance potential.

The α_{p1} system's Δv requirement (Fig. 2.37) for the 3 site visit is 167 km/s, a 1.9 year trip time, and 225 km/s for the 6 site visit.

2.0 High Energy Mission Applications

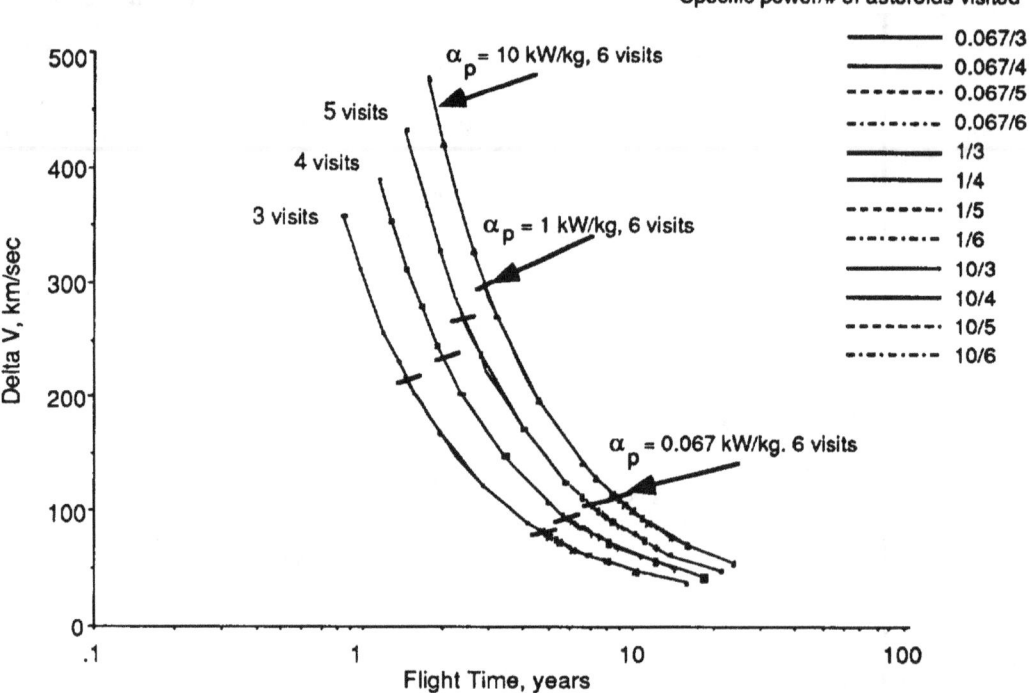

Fig. 2.37. Asteroid sample return mission, velocity variations with flight duration for 3 through 6 visits

A $\alpha_{p0.067}$ vehicle, to accomplish this 3-site mission, requires a minimum of approximately 7.7 years for the same mission and to remain under 10,000 seconds of averaged specific impulse. That mission consumes nearly 25 MT of propellant. A gross increase of the quantity of propellant by 705 MT will lower the flight time from 7.7 to 4.6 years. To meet the $\alpha_{p0.067}$ vehicle performance capability for a 6 visit mission, a flight time of 10 years results. That mission consumes 112 MT of propellant but exceeds the 7-year reactor life. If the reactor's life can be extended, then, in order to provide the reactor power necessary to accomplish these missions, at least an order of magnitude increase to one megawatt for a 7-year, 3-site visit is needed. Now if we wish to accomplish the 6-asteroid visit and remain within the upper specific impulse limit of ion engines, then a reactor size of 3 to 4 megawatts is required. Another technology gap is the production of higher thrust levels, 200 N, to accomplish the 10-year mission.

2.2.5 INNER PLANETS

Mission performance calculations were not performed for Venus and Mercury. These, too, are targets which could benefit by a high energy transportation system which performs sample return missions. The low thrust approach was not considered to be as accurate where the effect of gravity is greater. Hence, mission flight performance parameters were not computed.

B. Unmanned Science Missions Beyond the Solar System

Discussed previously have been examples of the anticipated future flight missions for the conduct of science within the solar system. In this section we examine a lesser defined set of mission categories, namely, those involving missions and science beyond the solar system – the next step in space science missions. These missions comprise a set that have not been given much attention since the energy means for propulsion and power to conduct them is not available. Two subcategories are considered, one which examines the space science of interstellar medium and the other concerning stellar missions. Included in the latter category are capabilities to enhance and conduct science where the targets are astrophysical related, but the science instrumentation lies within the solar system.

The chart below presents the type of missions termed as being beyond the solar system.

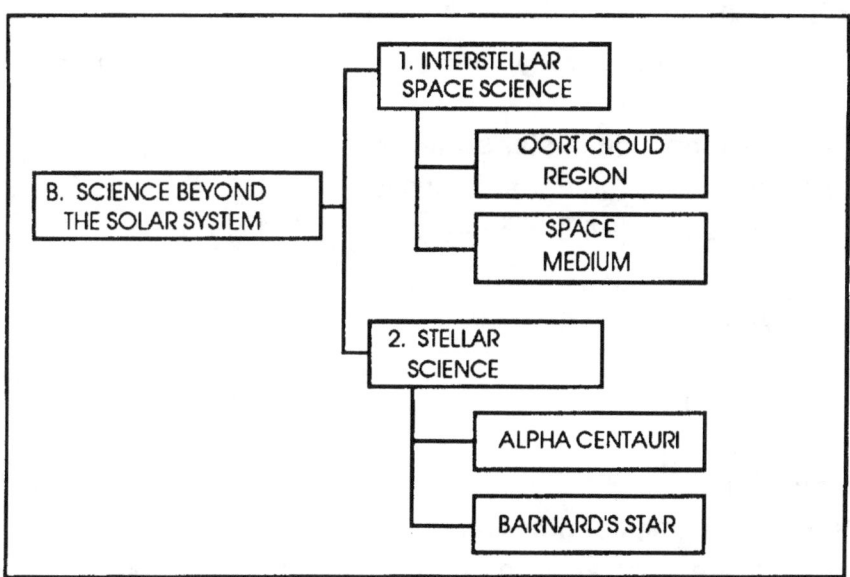

2.0 High Energy Mission Applications

A list of Interstellar Space Science Missions, the second category of Space Science Missions considered, is provided below:

1. Interstellar Space Science
    - Interstellar astronomy
    - Astrometry
    - Plasma
    - Heliosphere limit
    - Element and compound compositions
    - Fields and particles
    - Cosmic ray measurements
        • high energy
        • low energy
    - Dust particles
    - Gravitational waves (potential)
    - Gravitational lens
    - Oort Cloud
        • Physical properties: temperatures, reflectivity
        • Size, mass
        • Distribution
        • Composition [probe/lander]: compounds (water, organics), elements
        • Plasma/chemical analysis

2. Stellar Science
    - Presence of extrasolar system planets, also Oort Cloud and comet-type formations
    - Planetary formation in star systems
    - In-situ analysis of star characteristics
        • sun's age (~4.5 billion years)
        • old
        • young
        • sun size
        • smaller
        • larger
    - Search for life
    - Heliosphere limits of other stars
    - Comparative solar astronomy (relative to Earth-based)
    - Dynamics of star systems

            -Imaging
            -Formation of star versus planet
            -Presence of water and organics beyond the solar system
            -Fields: gravitational and magnetic
            -Electromagnetic radiation spectrum
            -Solar System Based Instrumentation
                    •Solar system based telescopes (remote from Earth)
                    •Solar observatories
                    •Polar solar characteristics

2.2.6 OORT CLOUD

Mission Description

The Oort cloud resides in the region of 20,000 AU and beyond. The mass within it can not be explored by optical telescopes due to the low reflectivity of the mass present. The temperature is only several degrees Kelvin making infrared observations from Earth difficult. No other means of energy exists for direct exploration of those bodies other than fusion. The science to be gained by a rendezvous, in-situ station analysis, is not achievable without the large magnitude of power increase provided by fusion. A Large Space Based Science Laboratory (LSBSL) could autonomously conduct experiments to determine the presence of organic molecules, their composition, and if present, their concentration. Physical property data for the Oort planetesimals will be acquired similar to that for the planets and moons. The imaging data from the Oort Cloud planetesimals will hold great interest since the data will not ever likely be obtained from Earth based telescopes. Comparisons of the Oort cloud's physical and chemical characterizations with comets and asteroids may yield very meaningful information on the origin and dynamics of the solar system, particularly since the Oort cloud mass is considered to comprise primordial solar system matter. Better knowledge of solar system dynamics and changes should result from the above to benefit the construction of solar system models. Unique astronomy from instruments located outside the solar system will also be achieved in this mission.

Mission performance analysis

As the result of the long distances to be traversed, a large energy source is essential if the science data are to be obtained within a reasonable time. Very little is known regarding the Oort Cloud. Another Oort Cloud mission suggestion is to fly stellar spacecraft through the region with other stars targeted. This would serve as a multiple purpose mission.

2.0 High Energy Mission Applications

A series of mission performance values was calculated for propulsion systems having specific powers of 1 kW/kg, 10 kW/kg, and 100 kW/kg. The value of 100 kW/kg was added for comparative purposes rather than to imply that such values are currently considered achievable. A 10 MT payload was flown on both fly-by and rendezvous missions to 20,000 AU. The $\alpha_{p0.067}$ vehicle system was not considered applicable to missions beyond the solar system because of the enormity of the distances and mission duration far exceeding the reactor's life. Vehicles using 1 to 3 stages were evaluated. The results from the mission performance calculations for the fly-by and rendezvous missions are presented in Figs. 2.40 to 2.47.

The mass impact of staging the vehicle for the trajectories shown is negative; for the design selected and flown, the multiple stage vehicles actually penalize the overall performance due to the high payload to initial mass vehicle design considered here. The advantage of more than one stage is the additional reliability gained. This results from including an added stage for pure redundancy. That additional stage further enhances system reliability since the long engine thrusting duration inherent with optimal low thrust missions is reduced. These mission performance calculations used the optimal flight approach to minimize the propellant and propulsion system mass; therefore, propulsion systems must endure a firing period lasting for 2/3 of the flight time, a non-trivial matter. Two mission categories are considered – fly-by and rendezvous.

FLY-BY MISSIONS

The fly-by mission flight time is approximately 35 to 70 years in length (Fig. 2.40) depending upon the performance of the reactor design, those flight times being for specific powers of 10 kW/kg to 1 kW/kg respectively. If a system could be designed to a specific power of 100 kW/kg, the time would decrease to about 13 years. There is a substantial vehicle mass penalty for the high speed trips. A mass increase of nearly two orders of magnitude for a 3-stage α_{p10} vehicle mass is required just to reduce 8 years of flight time in a 38-year flight. That is, the initial vehicle mass is increased from 1,000 MT to 80,000 MT for a 30-year mission.

A more reasonable solution from a mass perspective is to fly a 40 to 45-year mission on a 2-stage α_{p10} vehicle with a gross weight of approximately 200 MT. The flight time for either, one or two-stage vehicles, is nearly identical at 43 years, and hence no mass-time penalty is incurred by adding the second stage, a good approach to enhance mission reliability with no performance penalty. That intersection is referred in this text as the "stage invariant" mass. The addition of another stage adds 4 more years to the flight time. Refer to Fig. 2.38.

2.0 High Energy Mission Applications

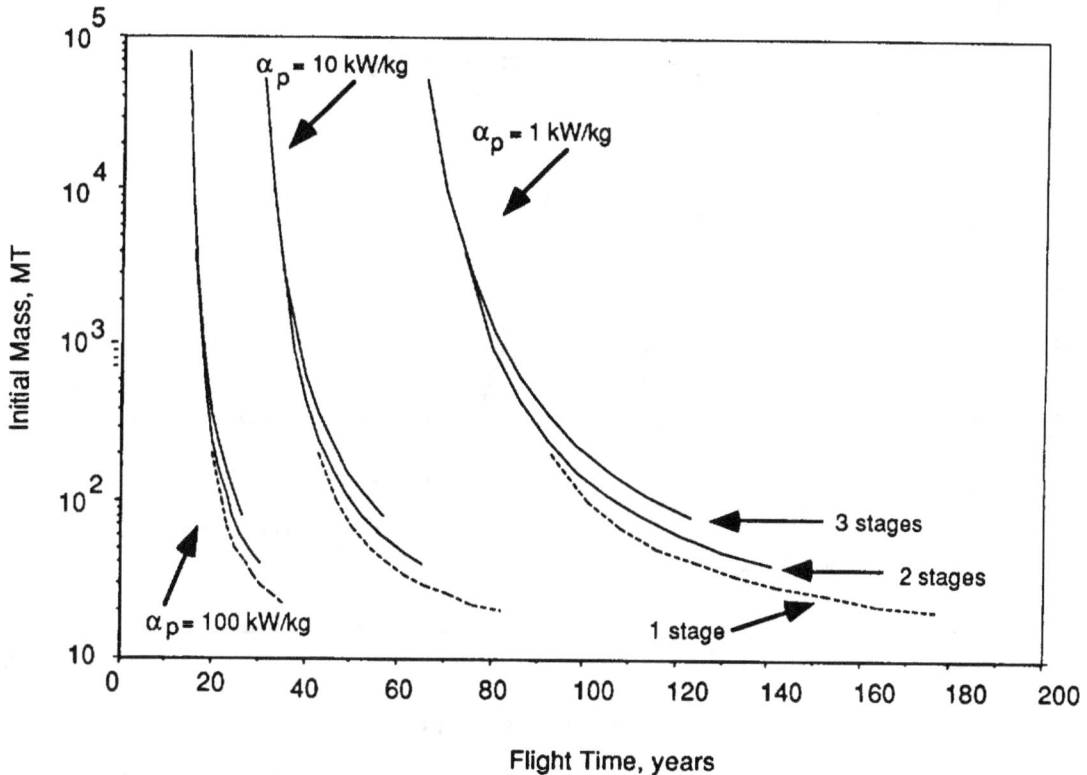

Fig. 2.38. Oort Cloud fly-by mission, initial vehicle mass variations with flight duration for 1, 2, and 3 stage configurations.

The quantity of propellant consumed is approximately 140 MT to 150 MT for the α_{p10} vehicle during that 40 to 45-year period. The propellant masses are shown in Fig. 2.39.

2.0 High Energy Mission Applications

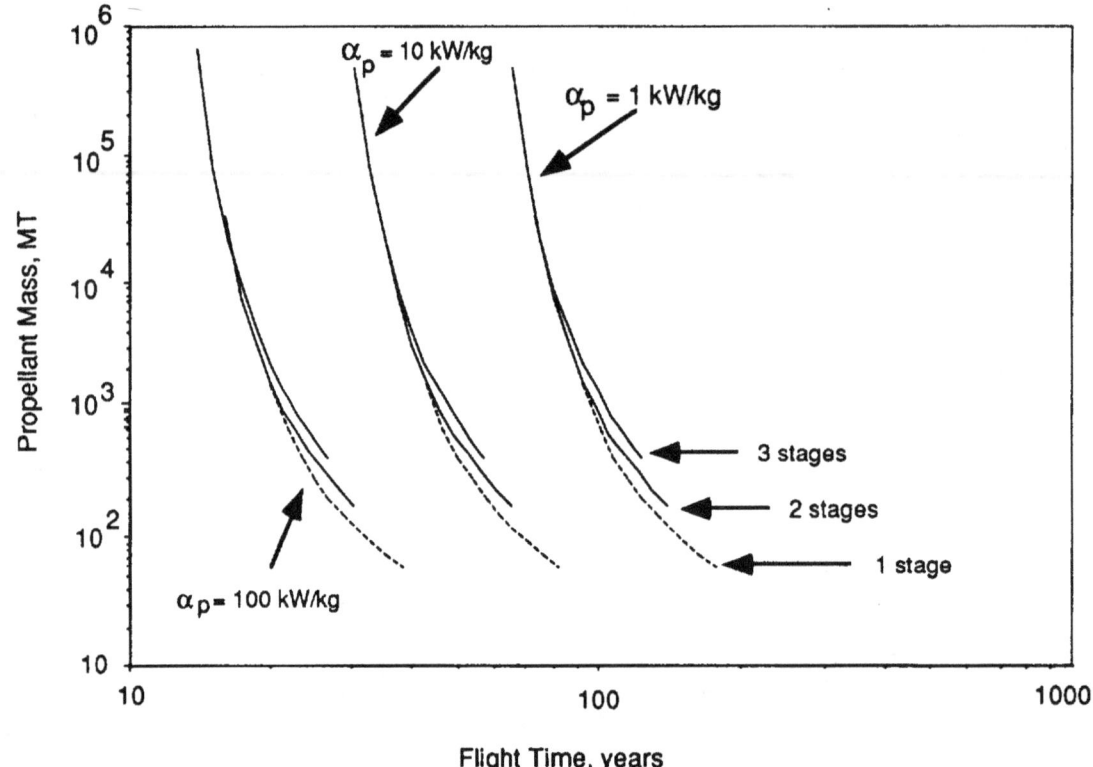

Fig. 2.39. Oort Cloud fly-by mission, propellant mass variations with flight duration for 1, 2, and 3 stage configurations.

The average specific impulse (Fig. 2.40) is between 210,000 and 220,000 seconds.

2.0 High Energy Mission Applications

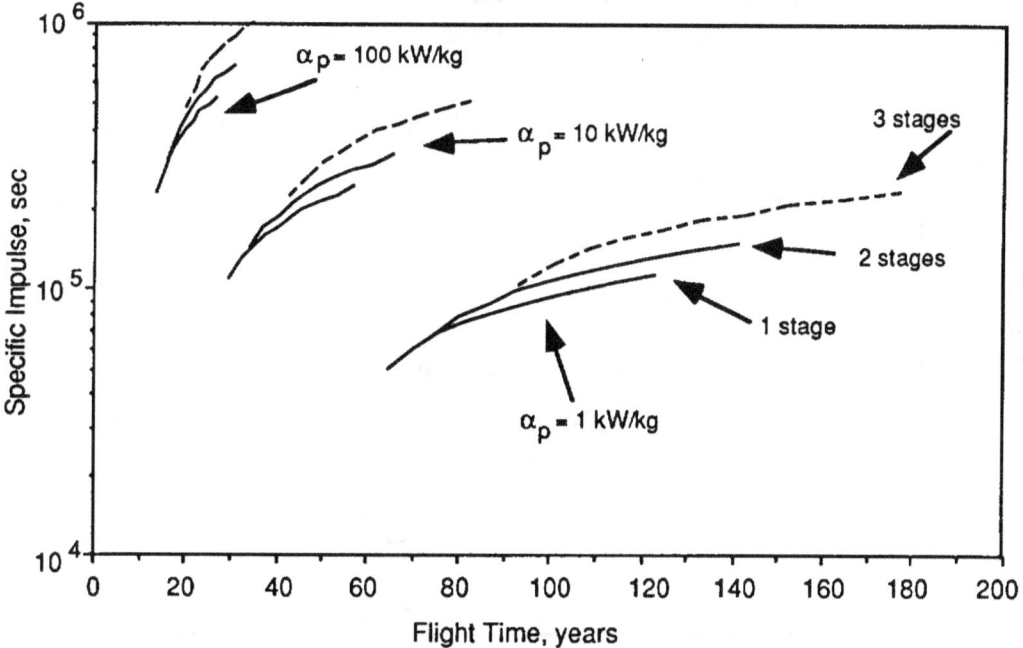

Fig. 2.40. Oort Cloud fly-by mission, propellant mass variations with flight duration for 1, 2, 3-stage vehicle configurations.

That range is considered achievable. The Δv (Fig. 2.41 a, b, and c) is approximately 3,000 km/s.

2.0 High Energy Mission Applications

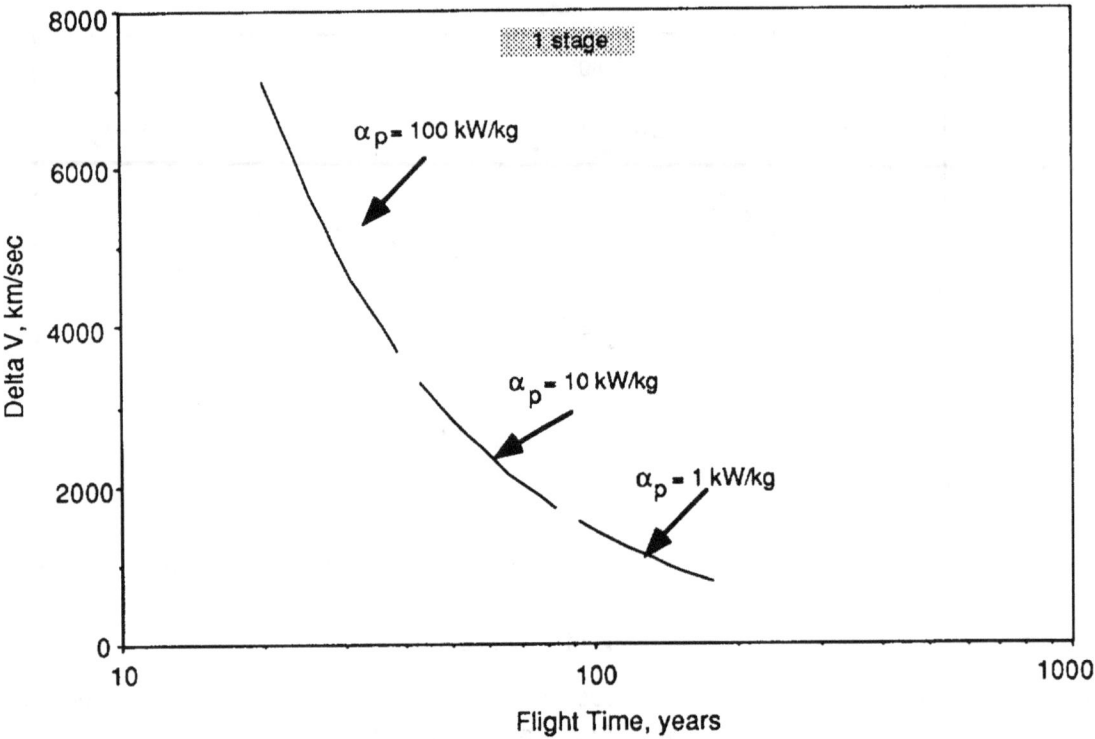

Fig. 2.41a. Oort cloud fly-by mission, velocity variations for a single-stage vehicle.

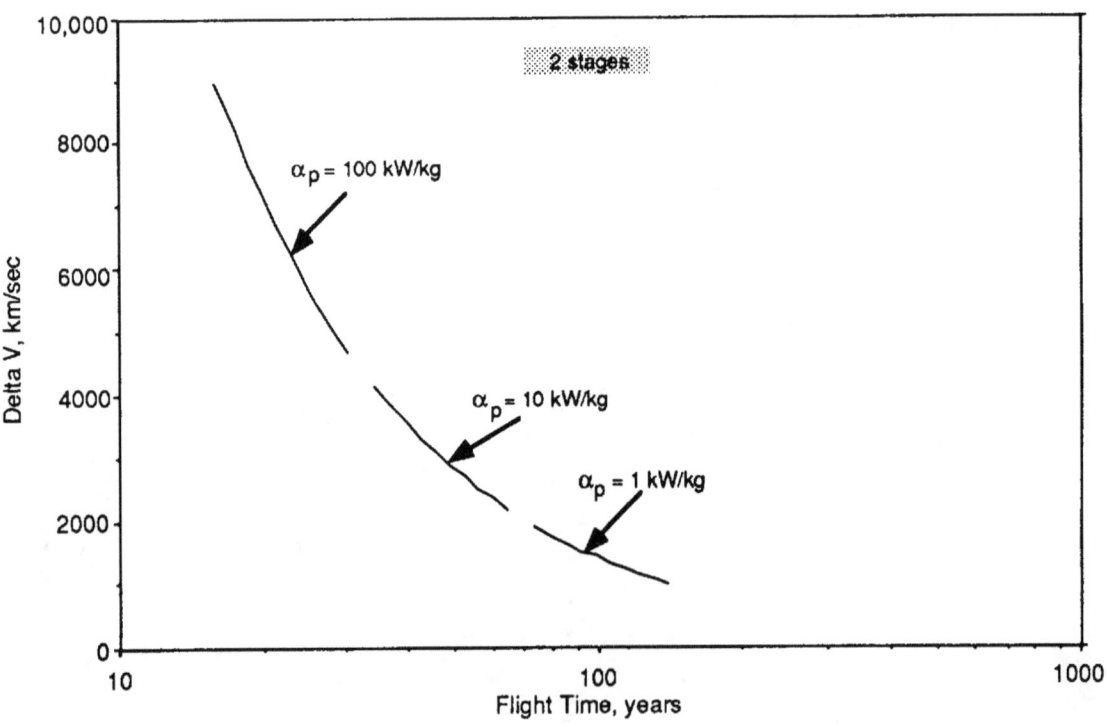

Fig. 2.41b. Oort cloud fly-by mission, velocity variations for a 2-stage vehicle.

2.0 High Energy Mission Applications

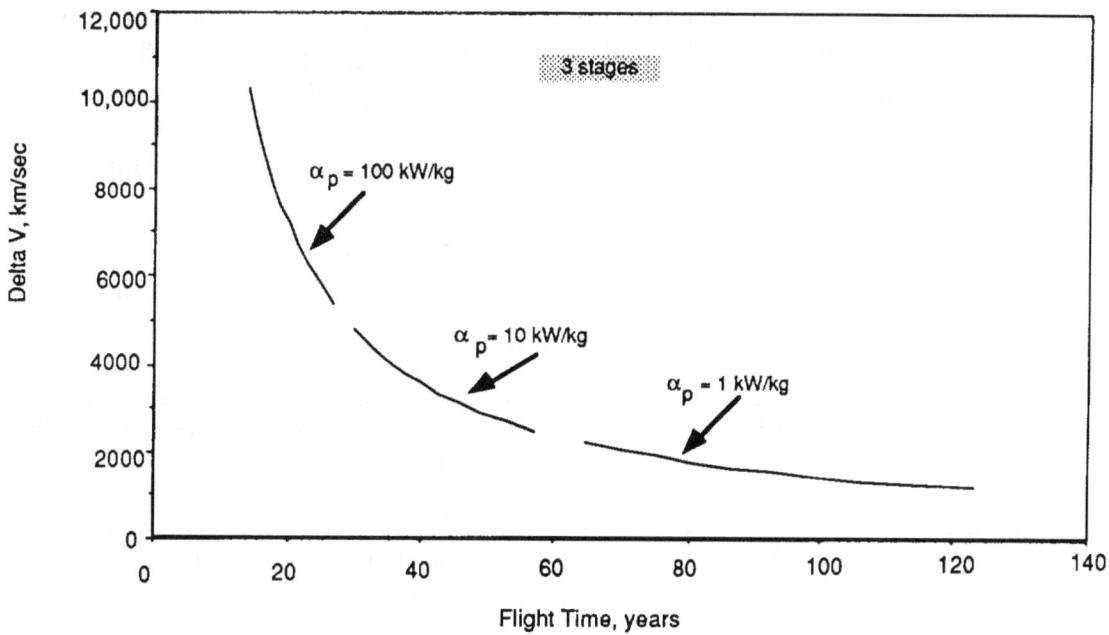

Fig. 2.41c. Oort cloud fly-by mission, velocity variations for a 3-stage vehicle.

The α_{p1} system suffers a severe time problem because a comparable 2-stage vehicle mass of 200 MT takes 95 years of flight time in comparison with the 40-50 year interval for the α_{p10} vehicle. The reactor jet power output requirement for a α_{p10} system, 45.8-year flight duration mission is 400 MW for the first stage and 100 MW for the second. The single stage jet power requirement is 347 MW to accomplish a 43-year mission. The key mission figures of merit are compared in Table 2-2 a. and b. for both the stage invariant flight times and for a fast mission. Because of the vast size of the Oort Cloud, a 20,000 AU mission will only commence to penetrate the region. Another approach, as mentioned, is to conduct a joint Oort Cloud - Stellar mission. Stellar missions are discussed below.

2-75

2.0 High Energy Mission Applications

TABLE 2-2a. Figures of merit for the Oort Cloud fly-by mission using fusion energy, 2 and 3 stage vehicles, approximate stage invariant initial mass, 10 MT payload.

α_p, kW/kg	Stages	t, years	M_o, MT	M_p, MT	P_j, MW		$<Isp>$, seconds	Δv, km/s
α_{p1}	2	99	160	100	1st stage:	40	106,000	1,441
					2nd stage:	10		
					3rd stage:	0		
α_{p1}	3	106	156	90	1st stage:	36	99,690	1,344
					2nd stage:	15		
					3rd stage:	6		
α_{p10}	2	49	111	65	1st stage:	275	245,940	2,904
					2nd stage:	83		
					3rd stage:	0		
α_{p10}	3	52.9	110	60	1st stage:	242	228,985	2,690
					2nd stage:	109		
					3rd stage:	49		
α_{p100}	2	21.3	160	100	1st stage:	4,000	492,000	6,689
					2nd stage:	1,000		
					3rd stage:	0		
α_{p100}	3	23	156	90	1st stage:	3,632	462,700	6,236
					2nd stage:	1,453		
					3rd stage:	581		

TABLE 2-2b. Figures of merit for the Oort Cloud fly-by mission using fusion energy, 3-stage vehicle, fast flight time, 10 MT payload, 80,000 MT initial vehicle mass, 65,375 MT propellant mass.

α_p, kW/kg	t, years	P_j, MW, Stage 1	P_j, MW Stage 2	P_j, MW Stage 3	$<Isp>$, seconds	Δv, km/s
1	64	13,890	694	35	50,190	2,212
10	30	138,900	6,943	347	108,130	4,766
100	13.9	1,389,000	69,430	3,470	232,960	10,267

RENDEZVOUS MISSION

The Oort Cloud mission presents the mission planner with somewhat of a paradox. Flight through the Oort Cloud at 3000 km/s does not afford the opportunity to conduct extended science measurements on science targets of interest. Yet at 20,000 AU the spacecraft is only beginning to penetrate that region, and certainly a strong desire to explore the region for the first time will undoubtedly exist. That is a point taken into consideration in the discussion below.

The results of the rendezvous mission performance calculations are collectively presented in Figs. 2.42 to 2.45. The minimum flight time calculated in this study varied from 102 years for a 3-stage α_{p1} vehicle, to 47 years for a α_{p10} vehicle, to 22 years for a α_{p100} vehicle.

Initial vehicle masses are shown in Fig. 2.42.

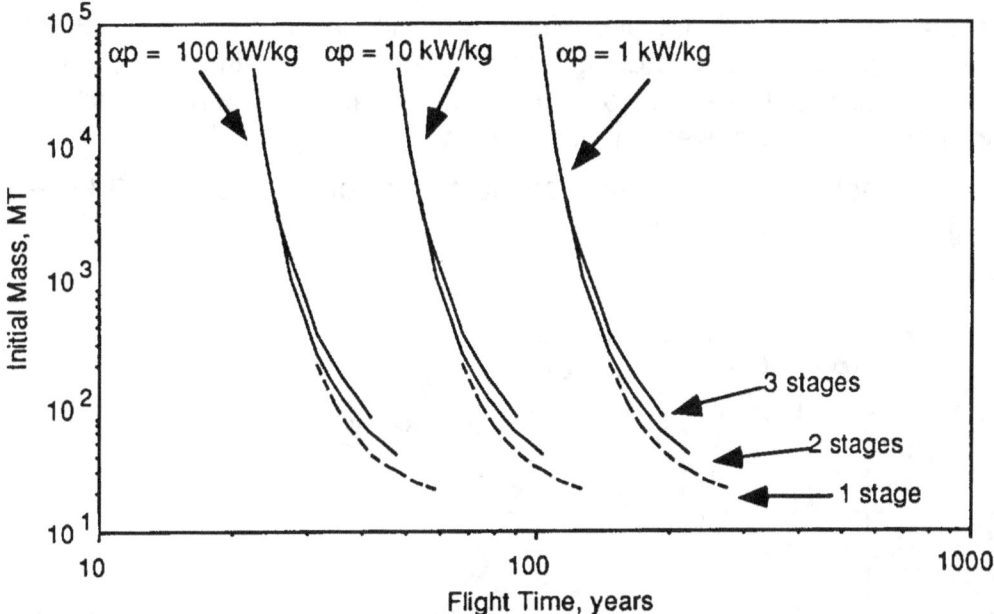

Fig. 2.42. Oort Cloud rendezvous mission, initial vehicle mass variations with flight duration for 1, 2, and 3-stage configurations.

The change in performance between propulsion systems using specific powers of 1 kW/kg, 10 kW/kg, and 100 kW/kg is illustrated, as is the massive size of the vehicle – up to 10^5 MT. If only the lower specific power reactor designs are achievable, there will be a science payoff to more heavily loading the initial vehicle design. That is, if for example the system can be designed no higher than 1 kW/kg, a more massive vehicle will be flown to shorten the flight time as compared to a 10 kW/kg system.

Use of the stage invariant mass approach offers advantages as previously discussed, but this time a difference in mission structure is suggested. One stage with one payload conducts braking maneuvers for rendezvous while another stage continues with an Oort Cloud penetrating mission from the mother spaceship to effect a joint fly-by and rendezvous mission from a single

2.0 High Energy Mission Applications

launch from LEO. That divides the 10 MT payload into two separate 5 MT payloads or some mass combination thereof. This 50-50 split is a conceptual, simplified solution to the goal of securing the maximum science benefit from one visit to the Oort Cloud. Further analysis would refine the split, but it points out a technique that can be further examined to optimize a joint fly-by/rendezvous mission that extracts the maximum science. This study did not explore whether the 5 MT payload size would adequately accomplish the desired science objectives and still provide sufficient spacecraft electrical power. The spacecraft's electrical power is a subject discussed in greater depth in the next section. More sophisticated mission performance calculations will have to be performed to determine the appropriate Δv splits for rendezvous and for fly-by.

A rendezvous mission's initial vehicle mass will be on the order of 3,000 MT. If that value is chosen as the design point, then the flight time can be expected to be somewhere between 55 to 120 years, depending upon man's ingenuity to build a high specific power unit, i.e., 10 kW/kg versus 1 kW/kg.

The computations for propellant consumption for vehicles at 10 kW/kg and 1 kW/kg are shown in Fig. 2.43.

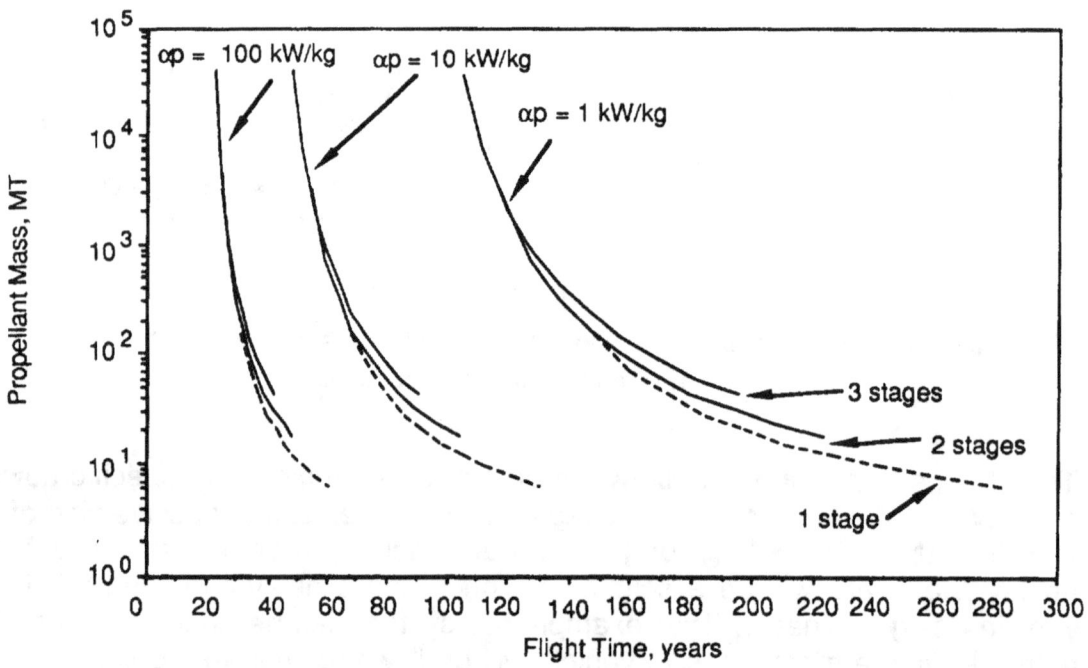

Fig. 2.43. Oort rendezvous mission, propellant mass variations with mission duration for 1 to 3 stages.

Both spacecraft vehicles – the α_{p10} and the α_{p1} – would use 2,100 MT of propellant. The specific impulse variations are presented in Fig. 2.44.

2.0 High Energy Mission Applications

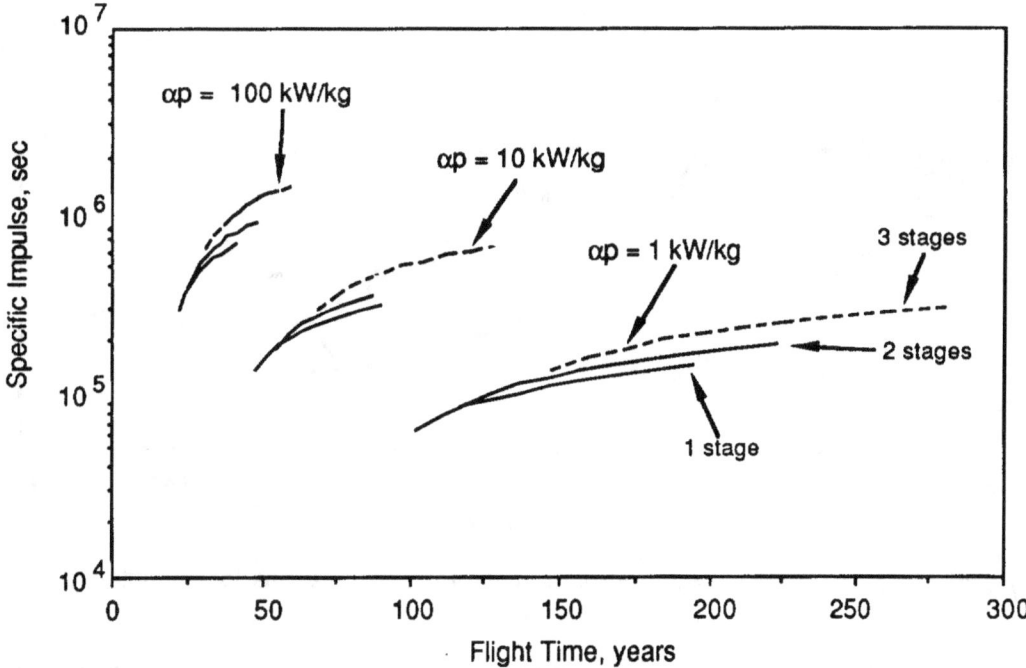

Fig. 2.44. Oort Cloud rendezvous mission, specific impulse variations with flight duration.

The upper limit for fusion systems is considered to be 10^6 seconds. The specific impulse requirements for the single stage system for an α_{p100} propulsion system exceeds that limit during some of the longer flight times. The vehicle Δv's are given in Fig. 2.45 a, b, and c.

2.0 High Energy Mission Applications

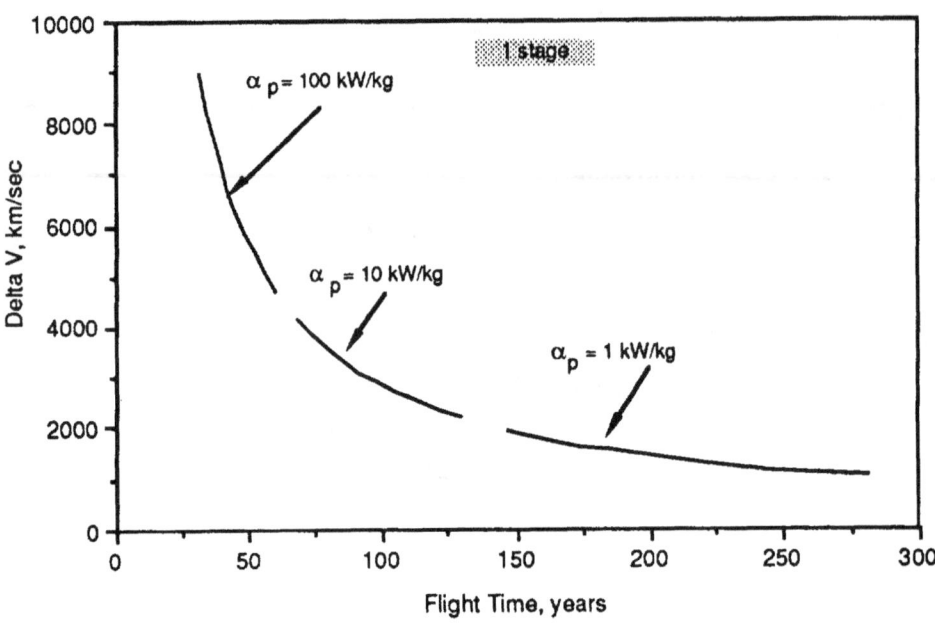

Fig. 2.45a. Oort cloud rendezvous mission, velocity variations for a single-stage vehicle.

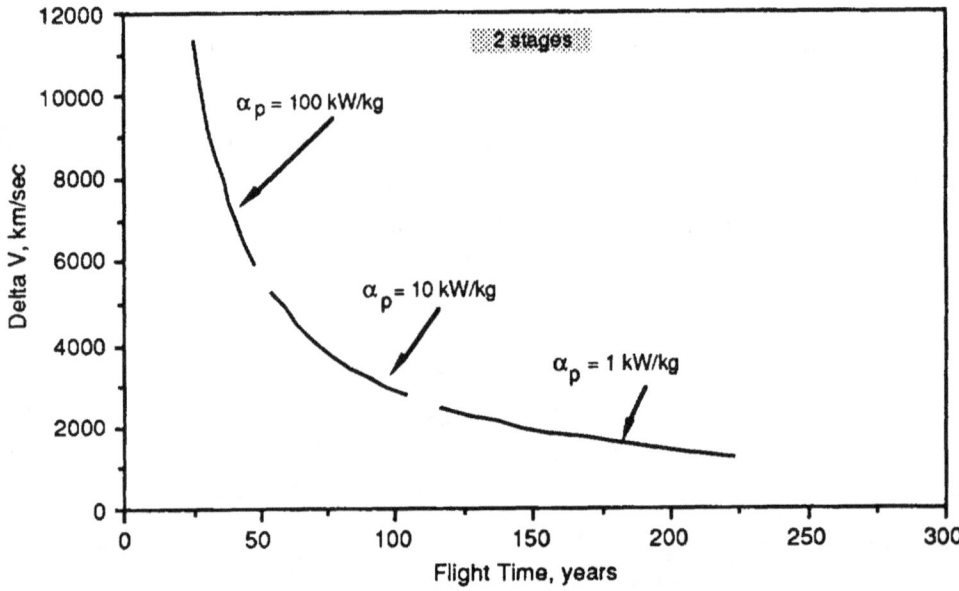

Fig. 2.45b. Oort cloud rendezvous mission, velocity variations for a 2-stage vehicle.

2.0 High Energy Mission Applications

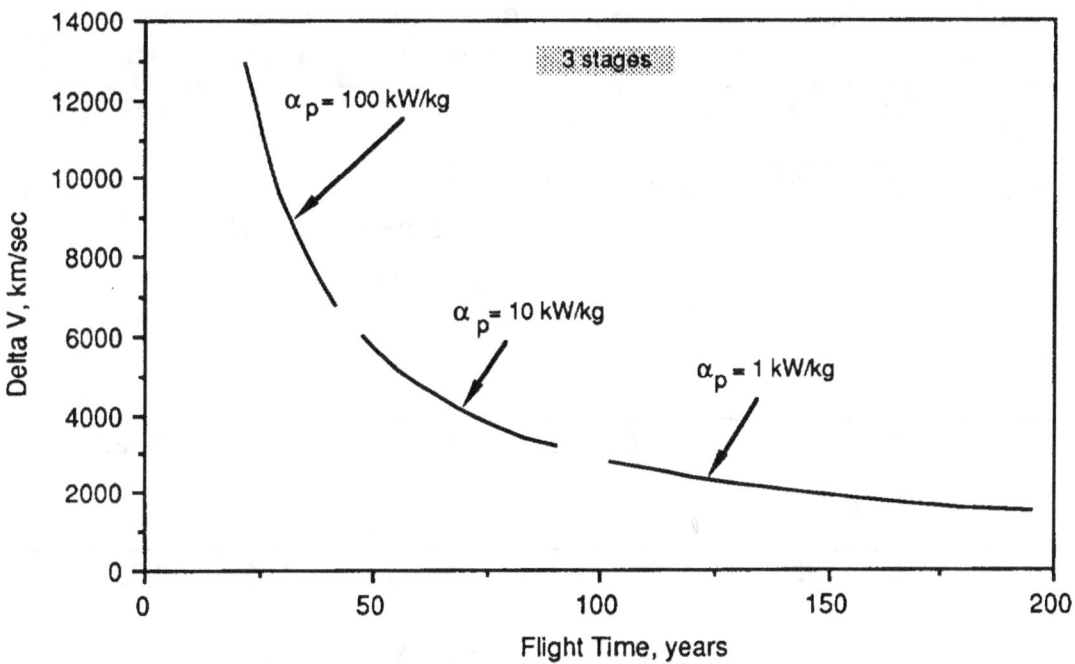

Fig. 2.45c. Oort cloud rendezvous mission, velocity variations for a 3-stage vehicle.

The 3-stage α_{p10} vehicle attains a maximum velocity increment of 6,000 km/s. These high energy missions demand high reactor power levels. To perform the 55-year mission, the initial vehicle mass is approximately 3,000 MT which requires 140 GW first stage power, 7 GW second stage, and 350 MW third stage jet power. Table 2-3 a. and b. compare in summary format the stage invariant and fast mission requirements.

2.0 High Energy Mission Applications

TABLE 2-3a. Figures of merit for the Oort Cloud rendezvous mission using fusion energy, 2- and 3-stage vehicles, approximate stage invariant initial mass, 10 MT payload.

α_p, kW/kg	Stages	t, years	M_o, MT	M_p, MT	P_j, MW		<Isp>, seconds	Δv, km/s
α_{p1}	2	117	4,000	3,261	1st stage:	694	82,860	2,435
					2nd stage:	35		
					3rd stage:	0		
α_{p1}	3	111	2,963	2,129	1st stage:	703	85,280	2,380
					2nd stage:	105		
					3rd stage:	16		
α_{p10}	2	54	4,000	3,261	1st stage:	6,940	178,520	5,245
					2nd stage:	350		
					3rd stage:	0		
α_{p10}	3	55.5	2,963	2,129	1st stage:	7,030	183,730	5,127
					2nd stage:	1,055		
					3rd stage:	158		
α_{p100}	2	25	4,000	3,261	1st stage:	29,430	384,600	11,300
					2nd stage:	3,470		
					3rd stage:	0		
α_{p100}	3	26	2,963	2,129	1st stage:	70,300	395,840	11,050
					2nd stage:	10,550		
					3rd stage:	1,580		

TABLE 2-3b. Figures of merit for the Oort Cloud rendezvous mission using fusion energy, 3-stage vehicle, fast flight time, 10 MT payload, 80,000 MT initial vehicle mass, 65,375 MT propellant mass.

α_p, kW/kg	t, years	P_j, MW, Stage 1	P_j, MW Stage 2	P_j, MW Stage 3	<Isp>, seconds	Δv, km/s
1	102	13,890	694	35	63,240	2,787
10	47	138,900	6,943	347	136,240	6,004
100	22	1,389,000	69,430	3,470	293,500	12,936

2.0 High Energy Mission Applications

B. Stellar

2.2.7 NEAR EARTH STELLAR MISSIONS

Mission Description

The star nearest to Earth is Alpha Centauri at 4.3 light years. The next is 1.7 light years further as shown by Table 2-4 below which presents the distances of the 6 closest stars. This class of science missions is clearly the most difficult to accomplish, requiring a lengthy period of time.

TABLE 2-4. Stars nearest Earth

Star	Distance
Alpha Centauri	4.3 light years
Barnard's star	6.0 " "
Wolf 359	7.7 " "
Lalande	8.2 " "
Sirius System	8.7 " "
Luyten 726-8	9.0 " "

Measurements which characterize the interstellar physics, plasma, fields, and particles and which provide mass spectrographic data, solar and stellar wind relationships are expected to comprise major mission objectives. These missions will provide the first plasma and cosmic ray in-situ data from far out of the solar system's ecliptic plane.

A prime mission objective concerns answers to the question of whether planetary systems, or the remnants of one, exists under such multiple star conditions. Multiple star systems comprise a very significant proportion of all known star systems. Abt and Levy concluded that 54% of all solar type stars are multiple star systems, having periods of less than 100 years (Abt76) There is great interest in learning more about the dynamics of multiple star coupled systems. An on-site observation and data gathering approach offers a granularity not otherwise possible. What are the dynamics and evolutionary processes that transition from pure multiple star systems to planetary solar systems? At what point does solar science converge with planetary science? Understanding of multiple star systems and the interrelated dynamics and the impact of that dynamics on the star system development could be beneficial in understanding the formation of the solar systems in general since a large percent of "stars" are in reality star systems. Work accomplished by Harrington shows that orbits of planets in some orbital zones will be stable for billions of years for a binary star system (Har77). No information was found concerning

2.0 High Energy Mission Applications

the comparable analysis of 3 star systems. Interest in the question of planetary presence is illustrated by Harrington's summary (Harrington:82:151):

> Observations have not yet provided any persuasive evidence for the existence of extrasolar planetary systems. The observed angular momenta of binary stars, coupled with current ideas of how the solar system formed, suggest that most of the 40 percent or so of stars without close stellar companions may be accompanied by planets. It is doubtful whether planets could also exist in binary star systems. Such indirect arguments, however, are weak because the origin of our own planetary system is poorly understood. We still cannot dismiss the proposition that planetary systems are extremely rare. (Har82)

If we are to look for life, young stars are eliminated on the basis of inadequate time for evolution to have occurred. Old stars are eliminated because the necessary chemical elements, carbon, oxygen, nitrogen and other elements required for life as we know it, are believed deficient in these stars and therefore in their planetary system. The heavier elements were created by the older stars in the process of fusion fuel burning. At the time of the birth of the sun some 4.5 billion years ago, there apparently was sufficient time for the necessary heavier elements to have formed, so slightly older and younger stars than the sun are candidates to examine. The other important parameter is size. A large sized sun will burn out more quickly, and a small one will have a furnace that is less well lit. As mentioned, we may also not be able to consider (multiple) star systems due to planetary instabilities, although there seems to be debate on this subject.

What are the candidates? Lets examine them. Alpha Centauri is the closest. Barnard's Star, an old star, is the second closest. The next several stars are considered too small or too old. The first beyond Barnard's Star is Epsilon Eridani at 11 light years away. Beyond Epsilon Eridani several more stars are not considered as candidates; the next candidate is Tau Ceti at 12 light years away. That is a total of twenty stars within 12 light years distance of Earth. The known odds for encountering stars having life potential properties, based on these limited statistics, is 20%. Thus, in the search for extrasolar systems, the question of planetary formation in multiple star systems becomes very important.

Another topic which may be addressed by interstellar observatories is the formation of planetary systems. Young stars can, at least potentially, be visited where disc formations are in the process of giving birth to planetary systems.

Of interest to the ultimate fate of the universe is the missing matter which is estimated would have to comprise 90% of the mass of the universe. One possibility is that the matter may reside in the presence of brown dwarf stars, those having considerably less mass than the sun, approximately 6% – an amount insufficient to ignite a fusion energy release. Perhaps brown dwarf stars, if they do exist, can be better detected by an outpost astronomy observatory, and, if found, the quantity of matter and statistical distribution must be such that they could well be within the range of a fusion powered spacecraft. Thus, they could be studied at close ranges by the use of remote sensing laboratories. Interstellar dust can be accumulated for analysis during the trip out to the stars as an additional benefit of the mission. The ultimate fate of suns which have burned out can be studied by in-situ spacecraft. Also of interest are any events which are rapidly changing, such as cloud motions, eruptions, explosive events, expansions, contractions, etc., which will benefit from the sensors of remote in-situ spacecraft.

One subject of interest in the formation of solar systems is the presence of water. That leads to the question of comets and whether some type of Oort Cloud surrounds another star. Is our solar system unique in that regard? That is a significant question as we examine the question of life elsewhere and the importance of water for the existence of life.

As another mission objective, it could be very informative to compare our Earth-based perspective of the star with close-up in-situ measurements. Also, this mission provides a larger baseline for astrometry. Further refinements could be achieved by targeting a second spacecraft to a destination at approximately the opposite direction to some other star system on a separate mission. Some very accurate triangulation determinations would result as the relative distances increase. As another mission objective, additional data on the Oort Cloud region could also be gathered as mentioned in the previous section.

This mission category has received only limited interest by the science community. The mission capability does not exist. The NRC report addressed an "Interstellar Probe" mission using megawatt nuclear electric propulsion. However, at the suggested velocity of 100 AU per 10 years, the 270,577 AU trip time to Alpha Centauri will be in transit for 27,058 years carrying only a 500 to 1000 kg mass payload. For imaging, that is an inadequate mass for a mission of this magnitude.

> The objectives of the mission are to determine the characteristics of the heliopause, interstellar medium, low-energy cosmic rays excluded from the heliosphere, and global interplanetary gas and mass distribution of the solar system, and possibly, a much more precise determination of the stellar and galactic distance scale through parallax measurements of the distance to nearby stars. (Don88, *Solar and Space Physics*, pp. 41-42)

Consideration was given in this study to the definition of an appropriate stellar payload size. A large, light weight telescope is essential. The resolution should

2.0 High Energy Mission Applications

provide technical improvements over future performance expectations in the field. This implies a size of no less than 3 meters. A light weight telescope on the order of 4 MT is considered to be a design target. Since the loads imparted by vehicle dynamics are low in a fusion powered vehicle, orbital assembly by the Space Station Freedom facility would aid by permitting a minimal mass structure, less than that which would be possible by fabrication on Earth and transported assembled into orbit.

Why should there be interest in a mission where the results are so far into the future?

One mission design objective is to provide for a mission such that data are available to a continuum of space scientists, rather than to only produce data at its conclusion. Otherwise, it would only serve as an abstraction from an earlier generation being passed along for some future generation, by-passing those in between. The proposed plan, then, is to collect data, store it, and transmit it to Earth on an annual or biannual basis. As discussed, there are fundamental science objectives to be met by the conduct of interstellar science during the trip out to Alpha Centauri, a pioneering trip which defines the space medium in uncharted territory. There is also the pragmatic goal that all science would not be lost if a major malfunction, one causing loss of mission, occurred on the trip out. If an efficient, high power joint propulsion-electrical power reactor can be designed, then a continuum of science data could be transmitted to Earth, provided that the energy consumption requirements can be maintained within reasonable limits.

If a means could be made available to "leave" an instrument package along the way at one or two "astronomical non-imaging outposts," a continuum of data flow would be possible, providing a research tool for the earlier astronomers on the mission. Near term astronomy from outside of the solar system, where the density of hydrogen is less, may benefit from the unique advantage of permitting observations unimpeded by any spectral absorptions experienced in solar system conducted astronomy.

The flight operational plan establishes important life cycle and restart considerations from the aspect of high energy system design requirements. The other significant variable to consider is whether this mission should be a fly-by or a rendezvous mission. The preference is to conduct a rendezvous mission to continue the transmittal of science data from the star system since this is such a rare opportunity. In a fly-by mission, one would not anticipate much additional new information to be obtained on the outbound leg of the trip as it proceeds beyond Alpha Centauri, but data on the star system can be gathered by the spacecraft residing at Alpha Centauri, e.g., the presence of planets – their composition, atmospheres, mass, seasons, etc. The rendezvous mission's system requirements elevate system demands, however, as shown by Figs. 2-50 through 2-53.

2.0 High Energy Mission Applications

There are possibly some science gains to be made in gravitational physics. The NRC *Astronomy and Astrophysics* report (Don88) discusses tests for a restricted range of gravitational waves using spacecraft equipped with dual frequency transponders traveling on different trajectories in the solar system. Current tests are too weak to detect such waves by orders of magnitude. The masses, velocities, and distances may be of a sufficient magnitude to permit the resolution of gravitational waves. Also, gravitational lens experiments might be possible using the sun as the lens.

An out-of-plane galactic spacecraft offers unique opportunities to observe intergalactic matter and radiation, now inhibited by the galactic mass. Is there water in this region? Are there organic compounds? Earth based telescopes have been unable to optically penetrate to the center of the Milky Way galaxy where a black hole is surmised. This mission would contribute to a better understanding of our galaxy.

Alpha Centauri

Alpha Centauri is a three star system, a fortuitous situation since it is also the nearest. One of the stars is Earth's sun size and age and hence a good candidate except for the planetary stability and formation question. Are the multiple stars indicative of a prolific area and conditions for multiple masses to have been bred? Or is the situation such that the planets will have been seized by the competing gravitational fields? At Alpha Centauri the individual stars appear to be at a sufficiently great distance separation that planetary structures may have formed and remained stable. Alpha Centauri α is approximately the same mass and age as the sun and emits similar light. One star is smaller than the sun and emits very little light, particularly in the ultraviolet frequencies which appears necessary for life. The other in this star system is comparable in age to the sun. Thus, one main objective in this mission will be to establish the presence of planetary masses in a star system. How does accretion occur such that star systems form versus solar planetary systems, or do they?

Mission performance analysis

This mission is extraordinarily energy demanding. To establish a reactor duty cycle, the electrical power required to transmit data was examined. A data rate of 100 bits per second was considered as representative of the minimum for imaging. A back of the envelope calculation indicates that approximately 10 MW of directed microwave power is necessary at the Alpha Centauri distance. Another quick independent calculation indicated that the power level requirement is approximately 1 MW. The mass of the transmitter is substantial. None of that size are known to exist. The transmitter mass is expected to be 4.0

2.0 High Energy Mission Applications

MT to 4.5 MT. The telescope mass is expected to be held within 4 MT. That leaves 2 MT for other science instruments and payload structure.

If the energy allocated for electrical power is very limited, it would be considered appropriate to transmit data annually or biannually for 2-week durations to provide a continuum of interstellar physics data including the opportunity to conduct astronomy from outside of the solar system. Designed to provide that operational capability, the mission offers new and comparative astronomy, that is, using departure space science data as reference data, we possess the capability to examine and update that data as the new transmissions are periodically received from distances further and further out into space. Thus, we can observe possible changes in the same parameters to provide a proximity calibration as Alpha Centauri is approached. It should be noted that with regard to the sun the opposite applies.

Mission performance calculations showed that while the mission times are long, they are not intolerably so. Much depends upon the specific power of the fusion reactor. Propulsion system specific powers of 1 kW/kg, 10 kW/kg, and 100 kW/kg were considered in mission performance calculations for two mission classes: fly-by and rendezvous. Gravity losses were not considered due to the magnitude of the distances involved, resulting in small performance losses. Because these are low acceleration missions, there is an additional quantity of energy consumed in the escape from the solar system's gravitational field above that required for a more rapid departure. There will be a significant amount of inertia to overcome due to the vehicle's immense size. For the mission performance calculations that follow, a 10 MT payload mass was considered, the bulk of which is for the telescope and the transmitter, the same mass as assumed for the Oort Cloud mission. A mass optimized technique was used in the mission performance calculations. Vehicles flying 1, 2, 3, and 4 stages were each examined. A high specific impulse of 10^5 to 10^6 seconds was required.

If a planet should be detected, it would be important to have the capability to concentrate on defining its physical properties, including water and oxygen detection instrumentation. Hence, the desirable goal would be to maintain sufficient maneuvering capability at the completion of the interstellar journey to visit the planet up close.

2.0 High Energy Mission Applications

FLY-BY MISSION

Results from the study's mission performance calculations, showing the mission parameters of greatest interest, are presented in Table 2-5 a. to d. for an Alpha Centauri Fly-by Mission. These calculations were selected to encompass a wide range of parameters for evaluating stellar program interest and determining mission practicality.

TABLE 2-5a. Alpha Centauri fly-by mission performance capabilities and requirements for variations in specific power for: single-stage vehicle, 10 MT payload, 4,000 MT propellant mass.

α_p, kW/kg	1.0	10.0	100.0
t, years	460	213	99
Δv, km/s	4,184	9,015	19,422
M_o, MT	4,205	4,205	4,205
P_j, MW	195	1,950	19,500
F, N	1,800	8,440	39,200
<Isp>, seconds	141,180	304,167	655,307

TABLE 2-5b. Alpha Centauri fly-by mission performance capabilities and requirements for variations in specific power for: 2-stage vehicle, 10 MT payload, 4,000 MT propellant mass.

α_p, kW/kg	1.0	10.0	100.0
t, years	414	192	98
Δv, km/s	4,646	10,011	21,569
M_o, MT	4,873	4,873	4,873
P_j, MW, Stage 1	816	8,162	81,620
P_j, MW Stage 2	37	370	3,500
T, N	2,600	12,100	56,000
<Isp>, seconds	153,110	329,870	710,675

2.0 High Energy Mission Applications

TABLE 2-5c. Alpha Centauri fly-by mission performance capabilities and requirements for variations in specific power for: 3-stage vehicle, 10 MT payload, 65,400 MT propellant mass.

α_p, kW/kg	1.0	10.0	100.0
t, years	365	170	79
Δv, km/s	5,270	11,360	24,460
M_o, MT	80,000	80,000	80,000
P_j, MW, Stage 1	13,900	139,000	1,390,000
P_j, MW Stage 2	694	6,940	69,400
P_j, MW Stage 3	35	350	3,500
F, N	54,900	255,000	1,180,000
<Isp>, seconds	120,000	258,000	555,102

TABLE 2-5d. Alpha Centauri fly-by mission performance capabilities and requirements for variations in specific power for: 4-stage vehicle, 10 MT payload, 1,308,000 MT propellant mass.

α_p, kW/kg	1.0	10.0	100.0
t, years	332	154	71
Δv, km/s	5,800	12,500	26,900
M_o, MT	1,600,000	1,600,000	1,600,000
P_j, MW, Stage 1	277,700	2,770,000	27,770,000
P_j, MW Stage 2	13,900	139,000	1,390,000
P_j, MW Stage 3	694	6,940	69,400
P_j, MW Stage 4	35	350	3,500
F, N	1,330,000	6,170,000	28,700,000
<Isp>, seconds	98,700	212,700	458,200

Note the increase in jet power requirements as the specific power increases. Nearly 30 TW are required for specific power systems of 100 kW/kg systems. The complete data set for the Alpha Centauri Fly-by Mission is shown in Figs. 2.46 to 2.49. The above tabulated data provide single specific design points for short flight times. From Fig. 2.46, the initial mass variations, one can observe that a change in the performance of the propulsion system's specific power from 1 kW/kg to 10 kW/kg will reduce flight time by 193 years.

2.0 High Energy Mission Applications

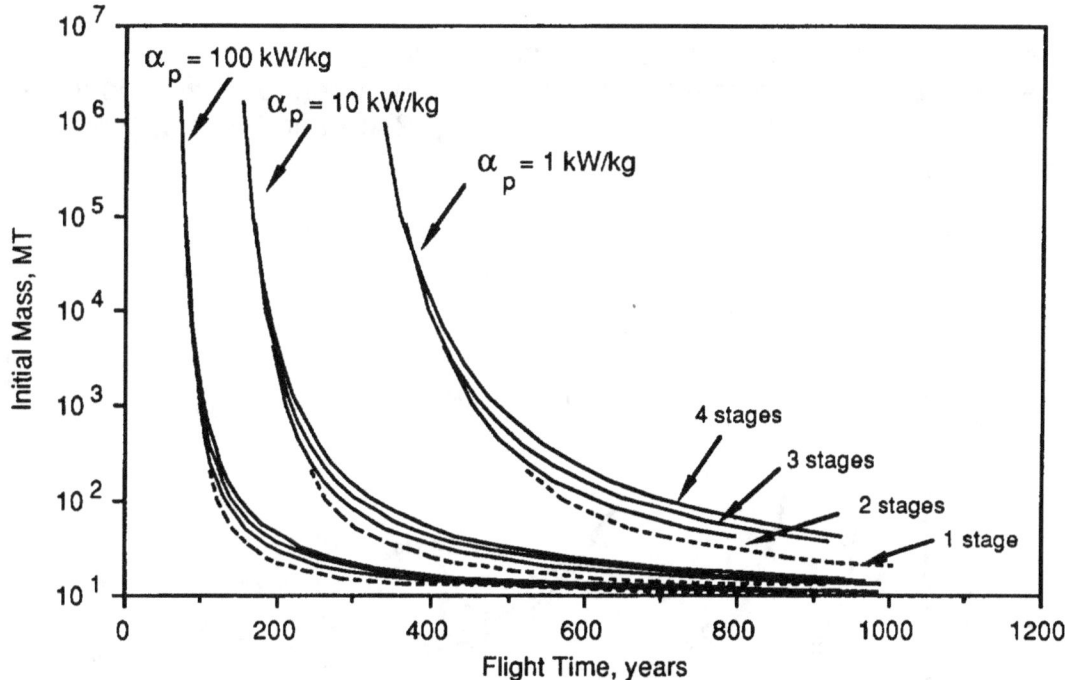

Fig. 2.46. Alpha Centuri fly-by mission, initial mass variations with flight duration for 1 to 4 stages.

Increasing specific power further to 100 kW/kg saves an additional 90 years. The dominance of specific power for this extremely high energy mission is quite evident.

The effect of changes in payload mass were briefly examined. Due to the large mass of these vehicles, a small penalty in flight time results with the addition of payload mass. For example, consider a 3-stage vehicle having an initial mass in Earth orbit of approximately 150,000 MT for a fly-by mission. The change in flight time for a payload mass increase of 1 MT to 15 MT increased only by 22 years, i.e., from 340 to 362 years.

One will note that for large payload mass fractions, where flight times are long, the multiple stages penalize the vehicle's mission performance. The stage mass penalty is attributed to the extensive inert mass of the fusion propulsion systems and the relatively large payload mass fractions permitted by fusion's efficiency. The four curves intercept, at which time vehicle performance is invariant with the number of stages as pointed earlier out in the Oort Cloud discussion. If the data had been computed for still higher initial vehicle masses, the reverse situation will result, and a performance advantage would be shown for the multiple stage designs.

The period of stage invariance occurs at approximately 180 years for the α_{p10} propulsion systems and 85 years for α_{p100}. In the case of α_{p10} that is an increase of 26 years in flight time beyond the quickest times shown; and for an α_{p100} system, the flight time added is 14 years over the shortest times. Use of

2.0 High Energy Mission Applications

the four stage invariance mass approach would appear to be a very worthwhile system trade to implement. Using that approach one will observe that differences in the initial vehicle masses (Fig. 2.46) and the propellant masses (Fig. 2.47) are very astounding.

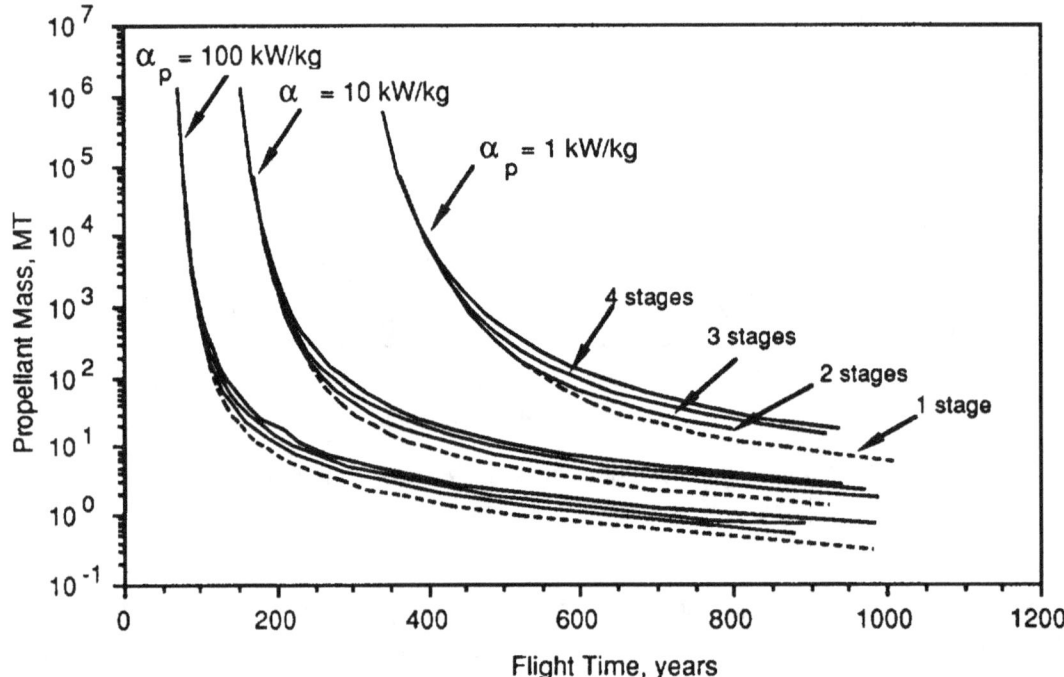

Fig. 2.47. Alpha Centuri flyby mission, propellant mass variations with mission duration for 1 to 4 stages.

The initial vehicle mass decreased from 1,600,000 MT to 10,000 MT, α_{p10}/4-stage configuration, and the propellant decreased from 1,308,000 MT to approximately 8,000 MT. The average specific impulse (Fig. 2.48) will be approximately 290,000 seconds.

2.0 High Energy Mission Applications

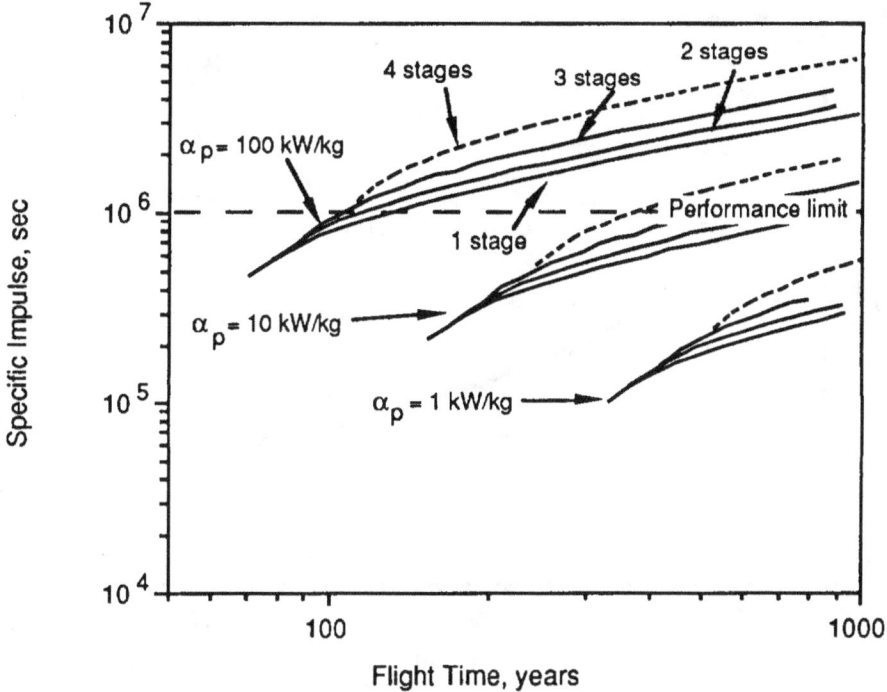

Fig. 2.48. Alpha Centuri flyby mission, specific impulse variations for 1 to 4 stages.

One significant technology to address is the means to attain a sufficiently high fusion thrust (~50 kN) to accelerate the massive vehicle. For this mission it is 6×10^6 N at an averaged specific impulse of 2×10^5 seconds. It is not known that thrust levels this high can be produced, but not inconceivable, since the reactor will be large. Multiple engines operating in parallel offer a potential solution. The first stage power requirement is approximately 24,000 MW for α_{p10}. The Δv imparted is approximately 10,000 km/s (Fig. 2.49).

2.0 High Energy Mission Applications

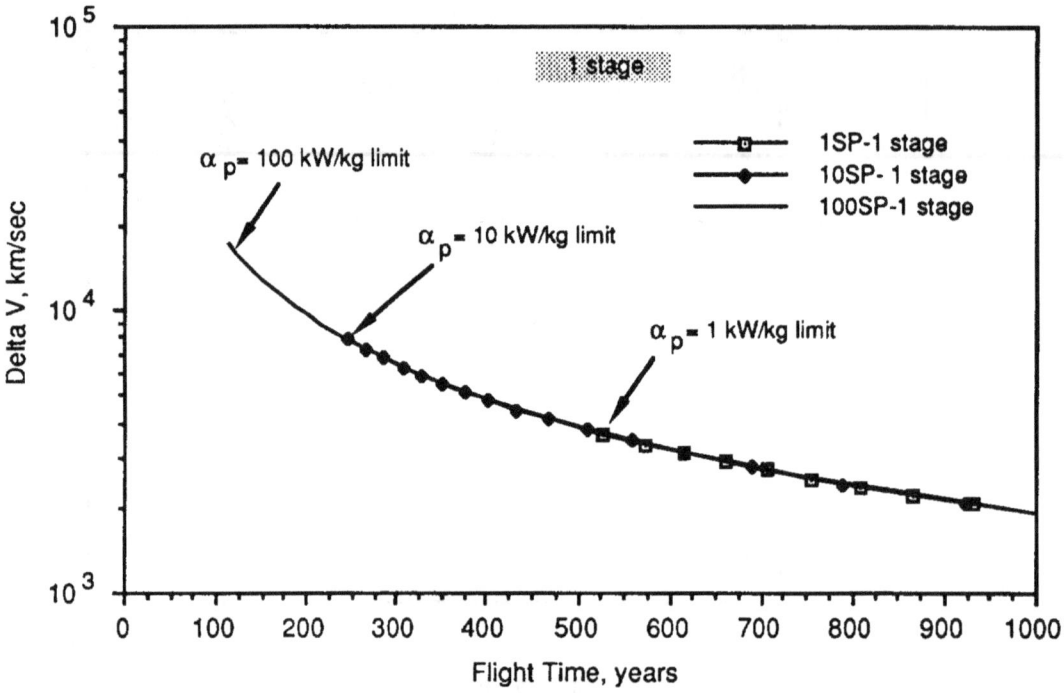

Fig. 2.49a. Alpha Centuri flyby mission, velocity variations for a single-stage vehicle.

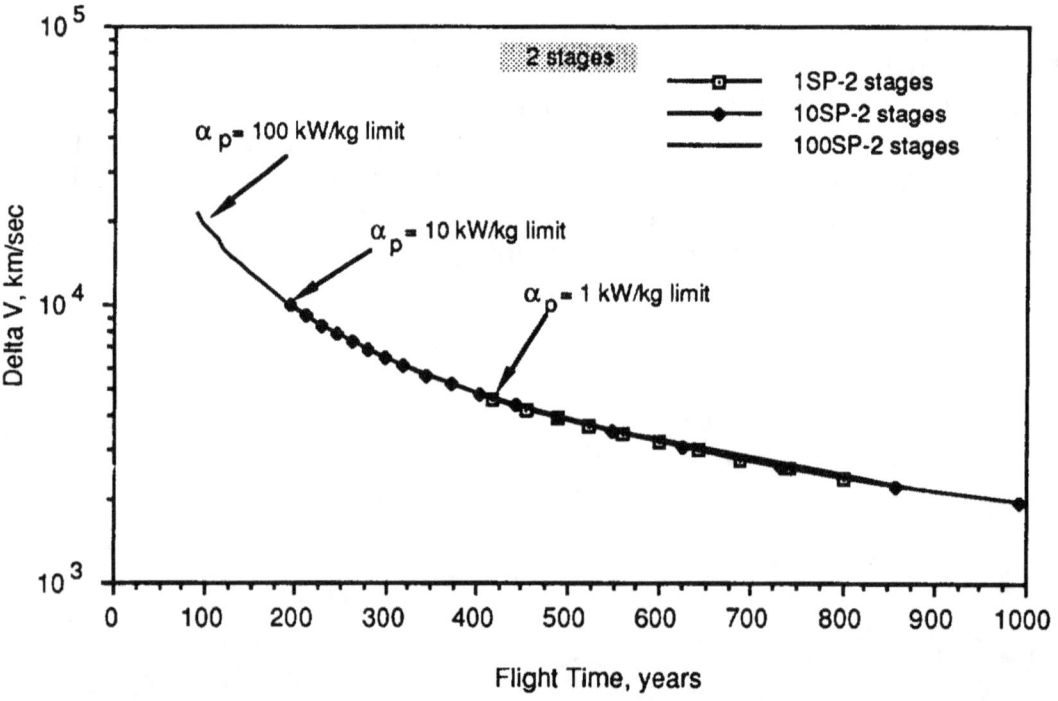

Fig. 2.49b. Alpha Centuri flyby mission, velocity variations for a 2-stage vehicle.

2.0 High Energy Mission Applications

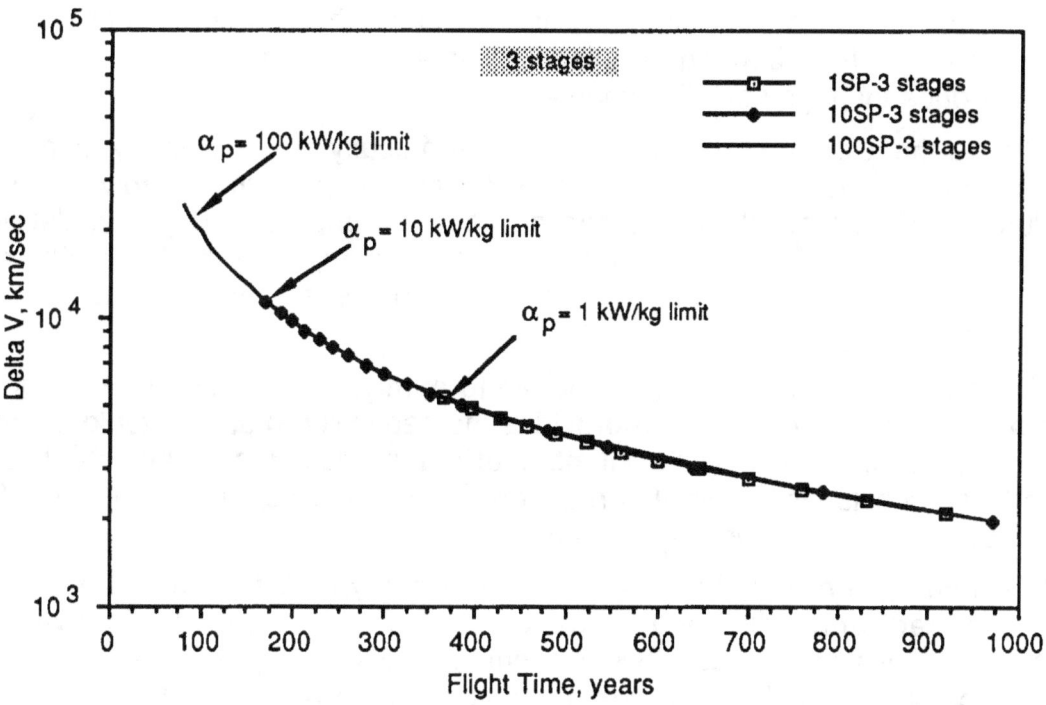

Fig. 2.49c. Alpha Centuri flyby mission, velocity variations for a 3-stage vehicle.

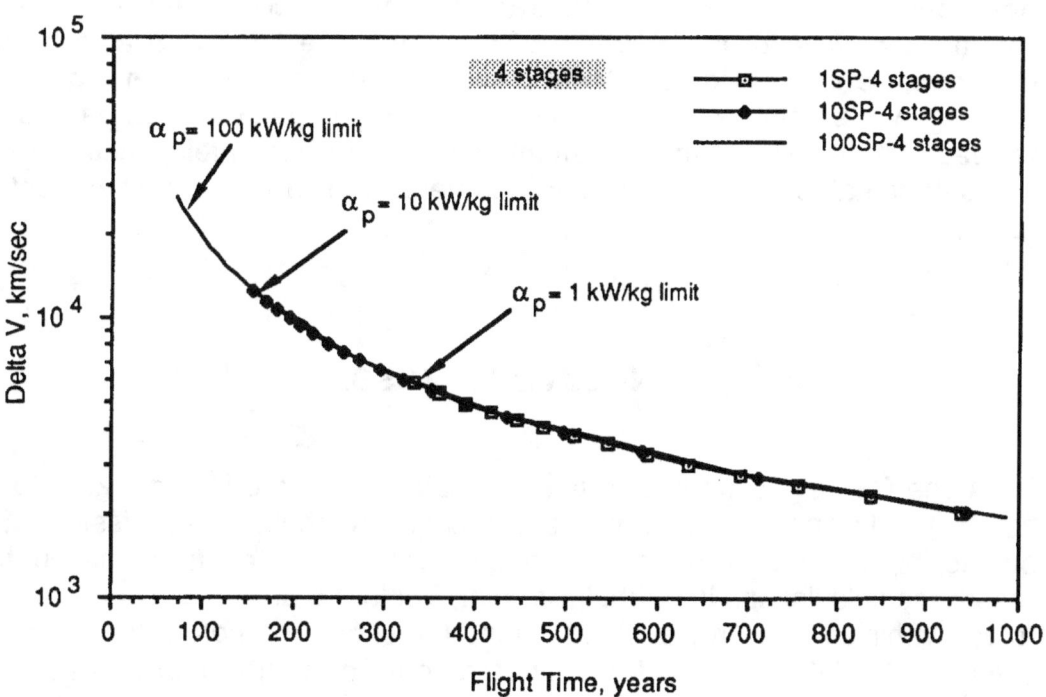

Fig. 2.49d. Alpha Centuri flyby mission, velocity variations for a 4-stage vehicle.

2.0 High Energy Mission Applications

This will be a fusion fuel intensive mission since the source of nearly all of the thrust will be from burning the plasma without much added propellant, in order to produce the high specific impulse.

The Alpha Centauri mission will be technically an extremely challenging engineering feat since the vehicle will have to perform reliably for a flight duration of at least 85 years and perhaps as long as 180 years, depending upon our ability to design a fusion reactor having very high specific power characteristics and with very high output power levels, on the order of 24 GW for the first stage.

The propulsion system will be required to burn for two thirds of the stage flight time. A vehicle design which provides the capability to stage would decreases the life requirements by the number of stages selected. The thrust level is reduced substantially from that required for the massive fast trip to 3×10^4 N and the specific impulse reduced to 3×10^5 seconds.

In summary, a new transmitter design technology (~10 MW output power level) is needed and will require development. To conserve power, data transmissions can be sent twice annually to Earth to provide a continuum of science data, including imaging, in transit. That power level will serve as well to meet the transmission power requirements at Alpha Centauri distances. No theoretical insurmountable obstacles are known. The proposed flight operational plan to provide data transmissions twice annually defines the system duty cycle for the duration of nonpropulsive flight and also a reactor being capable of diverting electrical power during thrusting. The reactor start-up energy will somehow have to be stored as it is expected to be generated by the reactor. New standards for automation and artificial intelligence will be set for a self contained science mission and vehicle for a mission which lasts 100 years.

RENDEZVOUS MISSION

The Alpha Centauri rendezvous mission data, presented in Figs. 2.50 to 2.53, show a highly energy demanding and technically challenging mission. Some specific, "quick" trip duration data points were extracted from the curves to present the calculations in a tabular format, Table 2-6 a. to 2-6 d. for 1, 2, 3, and 4-stage vehicles. As previously mentioned, it should be noted that the specific power of 100 kW/kg is provided only as a figure of merit rather that as a value that we currently consider as achievable.

2.0 High Energy Mission Applications

TABLE 2-6a. Alpha Centauri rendezvous mission performance capabilities and requirements for variations in specific power for: 1-stage vehicle, 10 MT payload, 4,000 MT propellant mass.

α_p, kW/kg	1.0	10.0	100.0
t, years	730	388	180
Δv, km/s	5270	11,360	24,470
M_o, MT	4205	4205	4205
P_j, MW	195	1,950	19,500
F, N	1,440	6,700	31,100
<Isp>, seconds	177,880	383,200	825,600

TABLE 2-6b. Alpha Centauri rendezvous mission performance capabilities and requirements for variations in specific power for: 2-stage vehicle, 10 MT payload, 4,000 MT propellant mass.

α_p, kW/kg	1.0	10.0	100.0
t, years	657	305	89
Δv, km/s	5,855	12,614	27,175
M_o, MT	4,873	4,873	4,873
P_j, MW, Stage 1	816	8,160	81,600
P_j, MW Stage 2	37	370	3700
F, N	2,060	9,580	44,500
<Isp>, seconds	192,900	415,600	895,400

TABLE 2-6c. Alpha Centauri rendezvous mission performance capabilities and requirements for variations in specific power for: 3-stage vehicle, 10 MT payload, 65,400 MT propellant mass.

α_p, kW/kg	1.0	10.0	100.0
t, years	580	270	125
Δv, km/s	6,640	14,310	30,820
M_o, MT	80,000	80,000	80,000
P_j, MW, Stage 1	13,900	139,000	1,390,000
P_j, MW Stage 2	694	6,940	69,400
P_j, MW Stage 3	35	350	3500
F, N	43,600	202,000	939,000
<Isp>, seconds	150,700	202,000	699,400

2.0 High Energy Mission Applications

TABLE 2-6d. Alpha Centauri rendezvous mission performance capabilities and requirements for variations in specific power for: 4-stage vehicle, 10 MT payload, 1,308,000 MT propellant mass.

α_p, kW/kg	1.0	10.0	100.0
t, years	527	244	113
Δv, km/s	7,310	15,750	33,930
M_o, MT	1,600,000	1,600,000	1,600,000
P_j, MW, Stage 1	277,700	2,777,000	27,700,000
P_j, MW Stage 2	13,900	139,000	1,390,000
P_j, MW Stage 3	694	6,940	69,400
P_j, MW Stage 4	35	350	3,500
F, N	1,060,000	4,900,000	22,700,000
<Isp>, seconds	124,380	268,000	577,330

The variation of initial vehicle mass as a function of flight time is shown in Fig. 2.50.

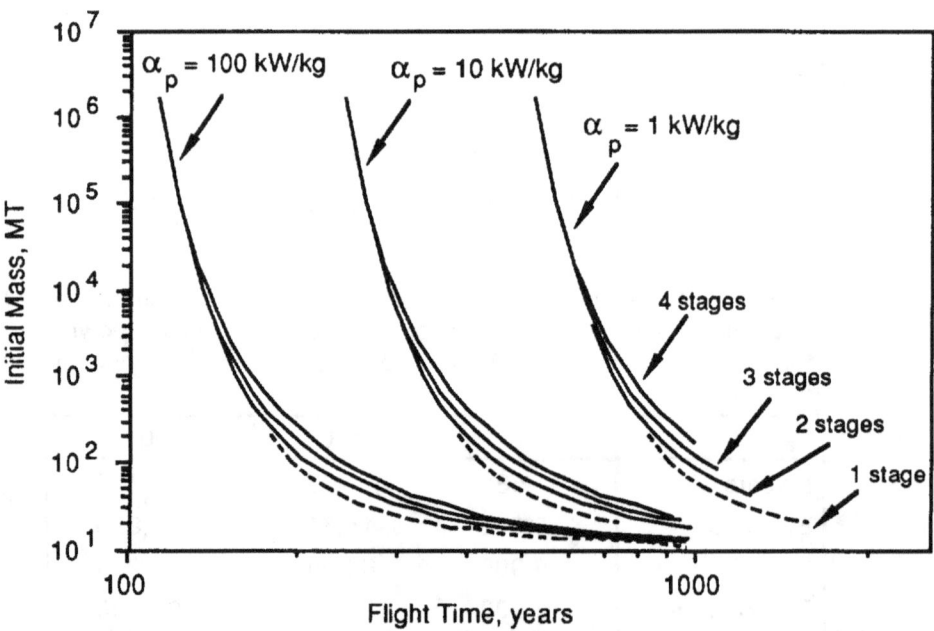

Fig. 2.50. Alpha Centuri rendezvous mission, initial vehicle mass variations with flight duration for 1 to 4 stages.

The order of magnitude improvement for specific power from 1 kW/kg to 10 kW/kg decreases the vehicle's flight time by 300 years. The importance of achieving the highest specific power is particularly noteworthy. The flight time only is shown and does not include the in-situ time for the conduct of the science mission, which in the case of a mission of this nature should be lengthy, such as, at least 20 years or more.

The flight time for the stage invariant design a (α_{p10} system) is at approximately 290 years, 100 years longer than the fly-by mission. To reduce the flight time by 50 years requires an increase of the initial vehicle mass to ~1,590,000 MT, i.e., to 1,600,000 MT from 10,000 MT for a 4-stage vehicle. The propellant mass for the fast flight time is nearly equivalent to the lunar supply of ^{3}He, and clearly some new fuel source such as Jupiter will have to be acquired for a mission demanding such high power levels, lacking a man-made means to manufacture helium-3. The propellant mass, mainly fusion fuel, increases from 8,000 MT to 1,310,000 MT, Fig. 2.51.

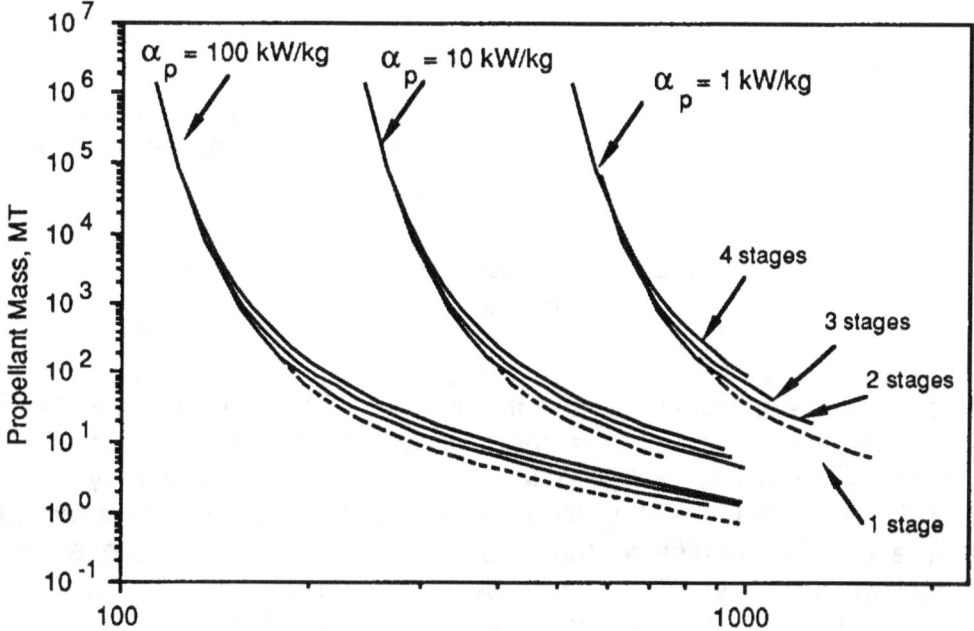

Fig. 2.51. Alpha Centuri rendezvous mission, propellant mass variations with flight duration for 1 to 4 stages.

Specific impulse requirements, Fig. 2.52, become quite high, an average of 380,000 seconds for the 290-year mission.

2.0 High Energy Mission Applications

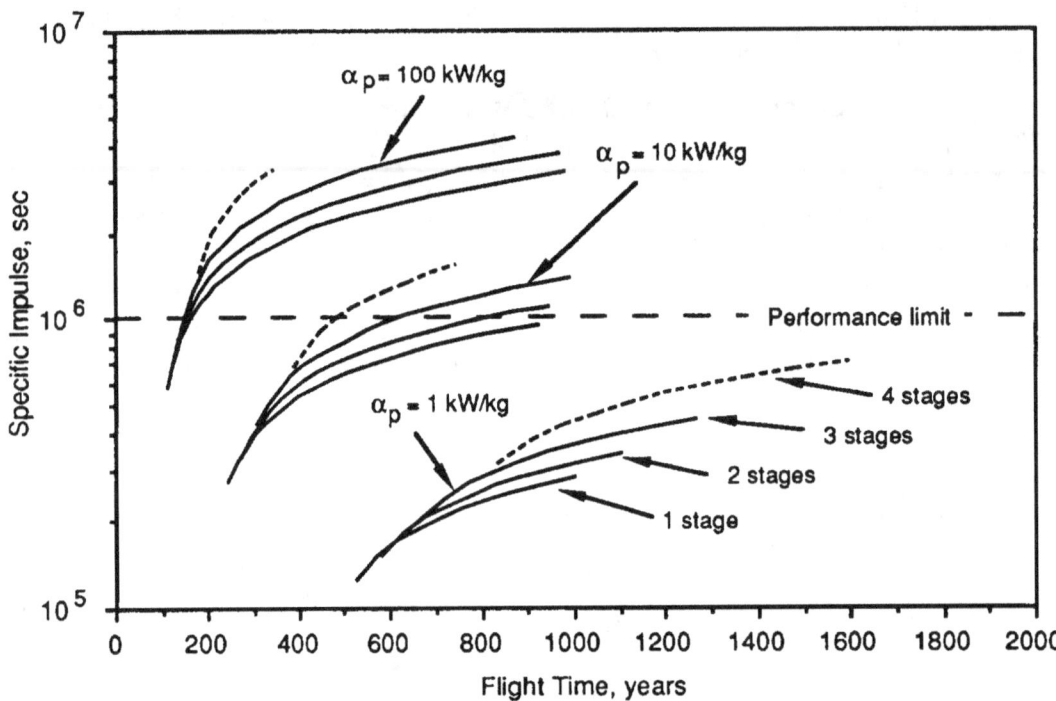

Fig. 2.52. Alpha Centuri rendezvous mission, specific impulse variations with flight duration for 1 to 4 stages.

The jet power requirement for the first stage is 28,000 MW. The Δv, Fig. 2.53 a-d, for the 290-year mission is approximately 13,100 km/s; to shorten it by 50 years increases the Δv by 2,700 km/s; for a 113-year mission, by an additional 20,800 km/s. It is interesting to note that as the specific power increases, so does the specific impulse requirement. For longer flight durations the specific impulse will be beyond the 10^6 seconds performance limit. Thus, any additional vehicle velocity increases will not be optimized.

2.0 High Energy Mission Applications

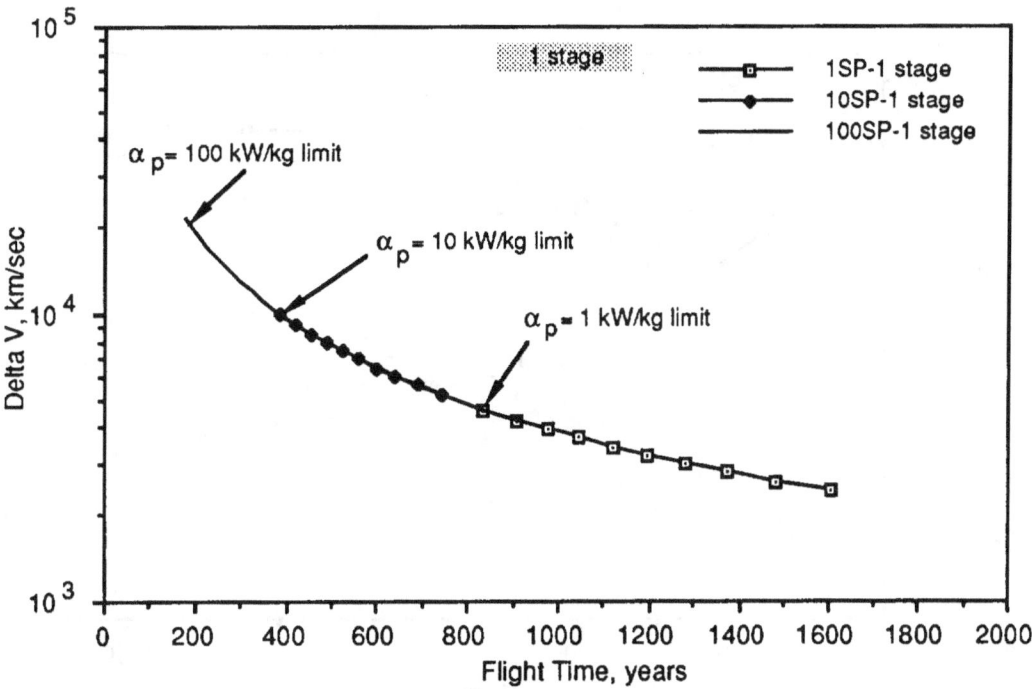

Fig. 2.53a. Alpha Centuri rendezvous mission, vehicle velocity variations with mission duration for 1-stage.

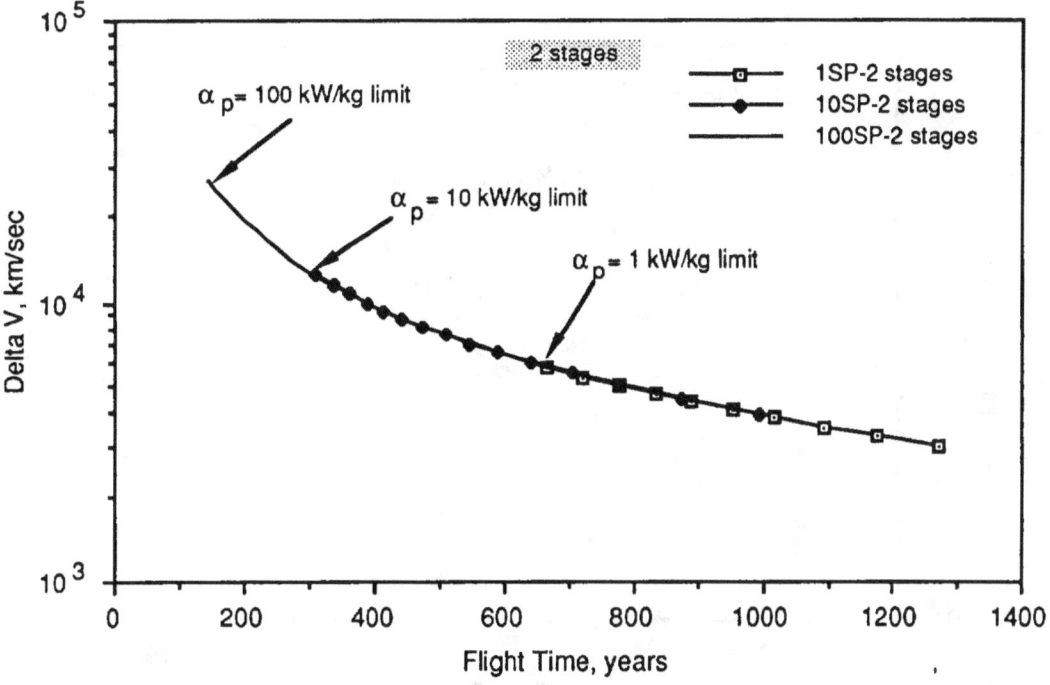

Fig. 2.53b. Alpha Centuri rendezvous mission, vehicle velocity variations with mission duration for 2 stages.

2.0 High Energy Mission Applications

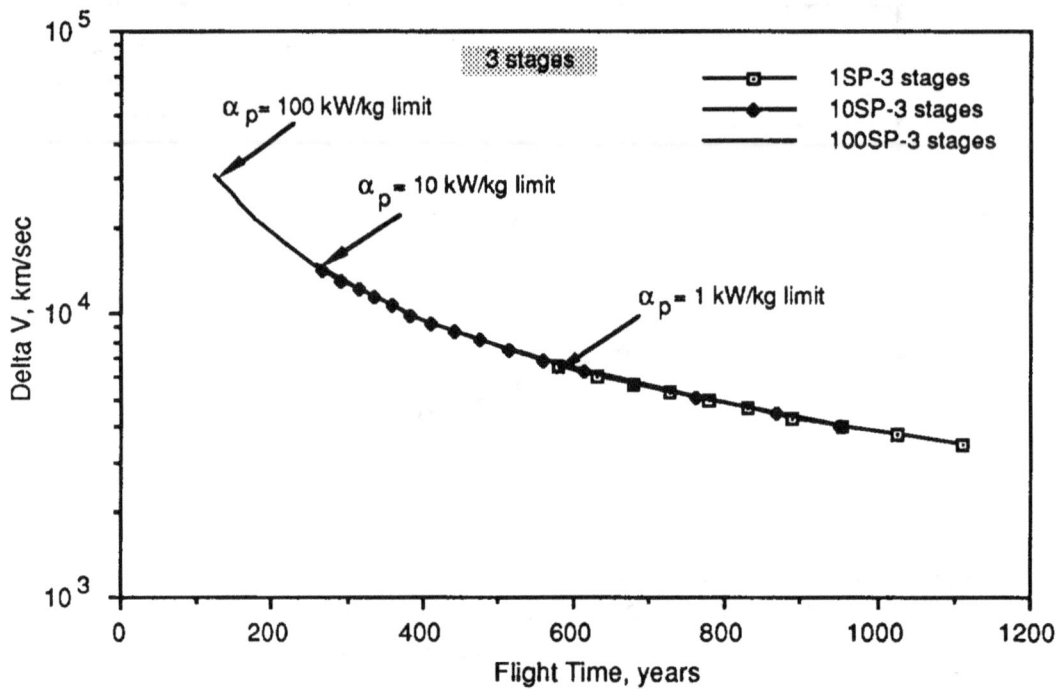

Fig. 2.53c. Alpha Centuri rendezvous mission, vehicle velocity variations with flight duration for 3 stages.

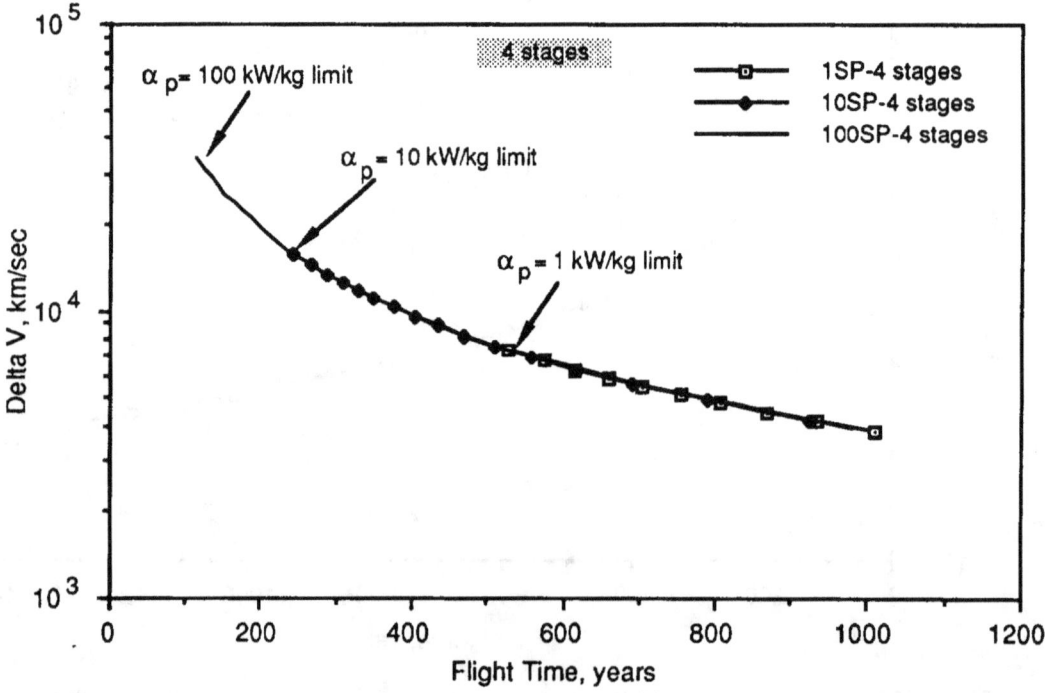

Fig. 2.53d. Alpha Centuri rendezvous mission, vehicle velocity variations with flight duration for 4 stages.

2.0 High Energy Mission Applications

If the 290-year stage invariant design is the one selected, each reactor will have to endure a firing time of nearly 50 years, based upon a 4-stage vehicle. The fourth stage power will be adequate to power a transmitter at a data bit rate more than 3 times greater than the minimum (100 bits per second), so the reactor could serve a dual function, propulsion and power.

The mission strategy, whether a fly-by or rendezvous, is likely to be a subject for considerable discussion. An additional 100 years will be required to accomplish a rendezvous mission over that for a fly-by. On the other hand, the length of time to fly these stellar missions is not expected to decrease significantly without some major new breakthrough in theoretical physics.

The mission's flight time, instead, is best reduced by a technology breakthrough in the development of high specific power. For the same investment in initial vehicle mass, namely 10,000 MT, the α_{p100} powered system arrives having flown for 140 years. That is very attractive. As discussed earlier, it should be recalled that these calculations assume no interstellar drag loads and no relativistic corrections in establishing the energy requirements.

Because of the time and expense, a rendezvous mission would be favored. The search for planets orbiting around Alpha Centauri will be a time consuming task. In a fly-by mission the data may be inconclusive due to the short observing time as a result of the high velocity. The scientist's philosophy regarding rendezvous for other space missions is stated in the Space Science Board report, (Don88, *Planetary and Lunar Exploration*, p. 16),

> Rendezvous missions are important for the study of small bodies, comets, and asteroids. The behavior of comets, with their changing distance from the sun, is of special interest. Although such studies may have begun by 1995, they should be continued in order to explore the diversity of comets and asteroids.

Extrapolated to the conduct of research of another "solar system," the same rationale would be expected to apply even more to their exploration.

During deceleration on the approach to Alpha Centauri, the rendezvous mission allows a more slowly paced mission to take place at Alpha Centauri when the spacecraft is traversing the equivalent of Earth's Oort Cloud distances. That would permit greater time for observations for Oort Cloud-like masses. Also, a rendezvous mission provides greater time for the vehicle's artificial intelligence (AI) system to perform any desired reformatting of science missions and for mission contingency planning in-situ while the spacecraft is resident at Alpha Centauri. As mentioned earlier, the AI design, along with the necessary mission hardware, for a trip to Alpha Centauri will pose a challenge.

Barnard's Star

In addition to Alpha Centauri, a fly-by mission was examined for Barnard's star at 6.0 light years. The 4-stage fusion powered vehicle operating at 10 kW/kg can deliver a 10 MT payload in 260 years, approximately 50 years longer than the trip to Alpha Centauri. The initial vehicle mass is 2,500 MT. The first stage jet power requirement is 6 GW.

Barnard's star has been observed closely for dynamical motions to reveal the presence of planetary bodies. The results of these observations lack sufficient granularity to have been conclusive. It would certainly be a good target for a second interstellar mission. It is a very old star, however, and therefore considered lacking in the important life supporting elements.

Beyond

Earlier it was mentioned that Tau Ceti, at 12 light years distance, would be a candidate star for supporting life. A 4-stage vehicle, transporting a 10 MT payload, can fly by Tau Ceti in approximately 400 years, assuming an initial vehicle mass of 2,500 MT and a 10 kW/kg propulsion system. The first stage power output requirement is 6.4 GW, and the average specific impulse is ~483,000 seconds. A rendezvous mission time increases to ~650 years; the Δv requirement is 16,500 km/sec.

2.2.8 SOLAR

Mission Description

Solar and heliospheric physics require high propulsive energy capabilities for large plane changes in the conduct of solar polar missions and in performing in-situ physical measurements from close to the solar surface. (Don88 Solar and Space Physics, p. 33,ff) In addition, the composition of a poloidal magnetic field and plasma wind composition of the solar system from a spacecraft flying normal to the plane of the ecliptic would provide a characterization of the solar system never before contemplated due to the large energy demands for propulsion systems. This is another high power consuming mission for which only the Ulysses Program has been designed to fly, the first spacecraft to explore the solar poles. Three mission requirements beyond the capability of conventional propulsion systems are:

    (a) an elliptical orbit for the solar probe with a 1-year period;

    (b) a circular or near circular orbit for the heliosynchronous satellite at 30 solar radii;

2.0 High Energy Mission Applications

(c) at least a 40 km/s velocity for a spacecraft leaving the heliosphere. (Don88 *Solar and Space Physics*, p. 54)

While solar electric propulsion or other higher performance systems fulfill these particular mission requirements, the growth trend toward greater energy requirements for the conduct of space science and exploration missions is evident.

Mission performance analysis

No mission performance studies were made in this study to characterize solar mission performance sensitivities.

2.2.9 REMOTE SPACE BASED TELESCOPES

Very large, and massive, near-Earth observatories could benefit from fusion power. A long multi-mirror and/or sensor space based telescope could be transported into Earth orbit and assembled for transport to a more optimally located position away from Earth's electromagnetic interferences. This type telescope has been considered by Buyakis as discussed in Zuckerman's paper on "Searches for Electromagnetic Signals" (Zuc82, p11). As an example, Fig. 2.54 depicts a giant space interferometeric telescope. "The telescopes are 10 km in diameter, and they are separated by 10 astronomical units. The angular resolution of such a system (where the entire universe is in the near field) is ~10^{-10} second of an arc."

2.0 High Energy Mission Applications

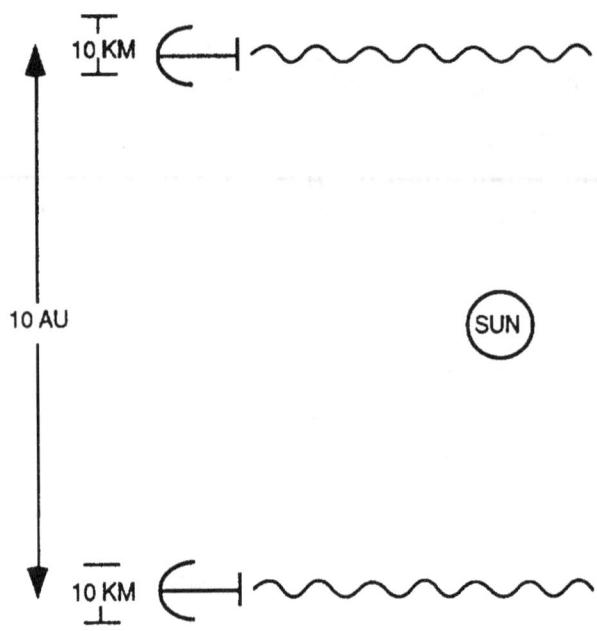

Fig. 2.54. A giant space interferometer having a resolution of 10 second of arc. (BUY79).

The placement of massive devices at large distances is a mission that large energy sources such as fusion can provide.

High energy vehicles will provide the power for transporting scientific observatories that are currently based in low Earth orbit to a point far from the Earth for improved observations. Observatories placed at remote locations from the Earth will permit the conduct of more frequent astronomical observations by avoiding eclipses. Ultimately, such astronomy could be necessary to escape orbital debris and Earth's man-made radiation and to accommodate ultra sensitive instruments. An observatory operating at a distance sufficient to be continuously beyond the shadow of the Earth provides an enormous increase in observatory efficiency as well as scientific data quality. Coherence time on Earth is short, and the exposure time is very limited. A few minutes of observatory operation well out into space would equate to a year's operation on the ground. The NRC Space Science Board also take into consideration the usefulness of a large space based telescopes in the *Astronomy and Astrophysics* report (Don88, p. 52) "A large space telescope is uniquely suited to stellar astrophysical programs in the ultraviolet. In fact, the ultraviolet resolution would be better than 0.3 pc in M101 and less than 0.8 pc in Virgo. Thus, the whole area of spectroscopic studies of stellar populations in fields much too crowded for HST become possible."

2.0 High Energy Mission Applications

> **Space Applications Missions**

This is a mission category having neither direct science nor manned exploration objectives but which are either supportive of those missions or are supportive of other aspects of NASA's goals.

2.2.10 LUNAR

<u>Mission Description</u> - Three lunar missions which are large consumers of electrical power that can be anticipated in the future include:

- materials processing,
- environmental maintenance for living quarters considering a manned presence, and
- electromagnetic transmittal of energy for propulsion and power applications.

As discussed in other sections, the settlement of planets may well depend upon the development of terraforming concepts and the utilization of local planetary resources. The moon serves as a convenient laboratory for the conduct of the enabling technology for habitation and for performing work functions at other planetary bodies. The moon serves as a source of helium-3 which can be used for Earth power as well as for space transportation and power. The idea to extract the helium-3 from mining of the lunar surface was originated by the University of Wisconsin. Approaches to make it feasible are under study at the Wisconsin Center for Space Automation and Robotics.

The moon provides a natural laboratory for the accomplishment of research and the development and qualification of the technology to settle planets. That work will require the acquisition and conversion of local ores, a large consumer of energy, to build the materials needed for construction. As one example, the use of lunar research for fusion equipment is discussed in Appendix B. The "Utilization of Local Planetary Resources" is the focus of the University of Arizona's space center. Along a corollary note, development of a suitable atmosphere, maintained at the proper temperature to provide a suitable environment for sustaining life on another planet, is another mission function requiring large energy demands for an airless, cold moon. It may be possible to develop isolated, encapsulated agricultural "fields" as a safeguard technique to test and perhaps even qualify new, genetically altered plant life without endangering the Earth's atmosphere – a critical single failure point for life here.

If one catastrophic event were averted by such a lunar laboratory, the expense would be worthwhile.

The moon provides a wealth of resources for construction, not only there, but for Earth satellite applications as well. The airless, low gravity moon makes feasible a safe, quick method to transport lunar manufactured material by rail guns or some by the other electromagnetic transportation means (beam power- Appendix B), as discussed in the Mars section. Propulsion benefits are discussed in greater depth in Section 2.2.9. The remote production of chemical propellants for space missions is another benefit of the moon. Having an adequate energy supply to meet resident lunar mission needs is essential, and fusion is an attractive energy source option, particularly with the presence of ^{3}He there. Calculations of the magnitude of power levels for the aforementioned missions were not undertaken here; it is suggested that some additional study be undertaken to better define the magnitude.

Mission performance analysis

The primary applications of fusion energy to the lunar mission are any which involve requirements for 10's of megawatts of electrical power and higher. Alternatively, if the means to produce a high thrust-to-weight propulsion system were developed, the advantages of fusion would be multiplied many fold. A propulsion system with a thrust-to-weight ratio high enough to lift substantial payloads from the moon or any other planetary moons would be a worthwhile endeavor for the reasons stated earlier concerning safety, environment, and performance.

2.2.11 STATIONARY (LUNAR) POWER FOR SPACECRAFT PROPULSION AND ELECTRICAL POWER

Mission Description

Although this title may seem inherently contradictory, it describes a new, different approach to propulsion. This idea pursues an answer to the question, "How can we transport objects more economically and safely in space?" It takes advantage of the local planetary resources or, as in the case of the airless moon, the lack thereof. The capability to transport the energy electromagnetically has been alluded to in this report's discussions on Mars and on the Earth's moon. Because of the concept's generic application to a wide variety of space developmental applications, a separate section is being devoted to this topic which contains Dr. Logan's in-depth examination of the concept in Appendix B. Essential to the implementation of the electromagnetic transmittal of energy through space is the availability of multimegawatt power supplies. Fusion energy can provide that power solution.

2.0 High Energy Mission Applications

Applications

There are several applications to consider: power for rail guns, electromagnetic transfer of power to a spacecraft for conversion to thrust via an ion engine, electrical power for a ground-based laser system which transfers energy directly to a spacecraft using an ablation driven propulsion system, and direct transmittal of electrical power via electromagnetic waves. A conceptual schematic of the laser transmitted energy - ion propulsion system is shown in Fig. 2.55 (Log88).

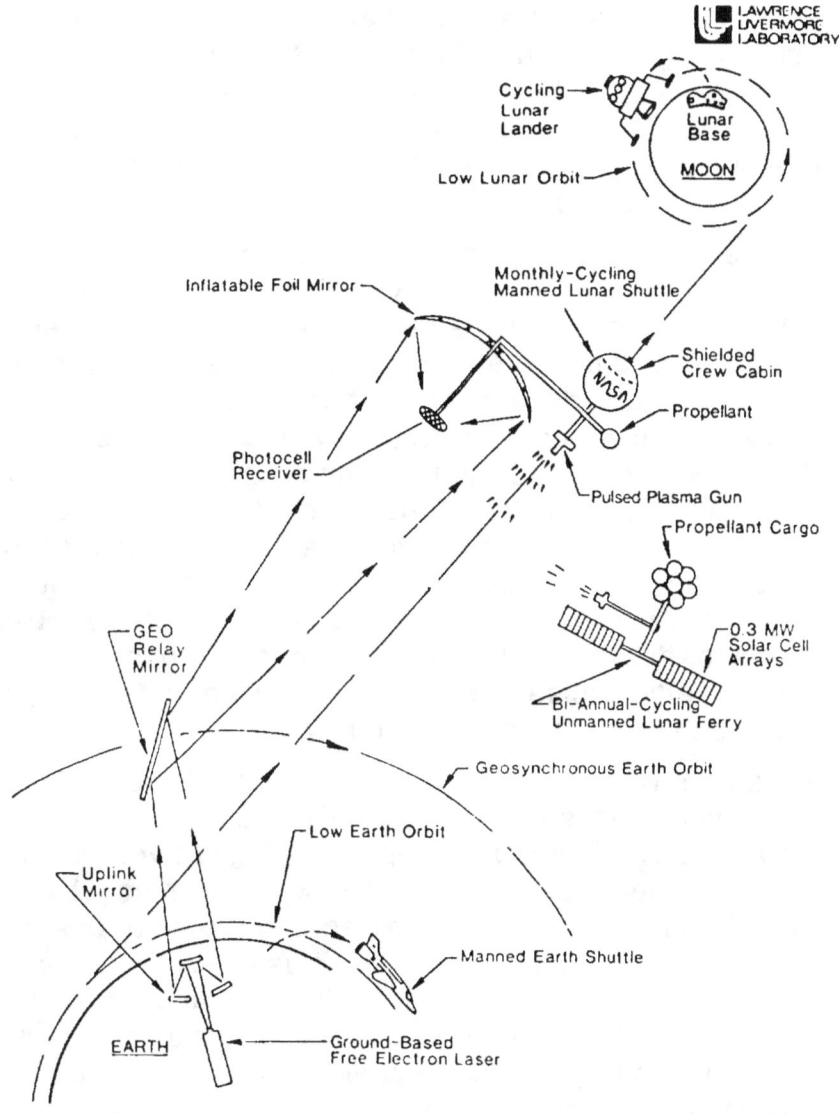

Fig. 2.55. An economical Earth-moon transportation system based on pulsed plasma guns remotely powered by free-electron-lasers. (Log88)

2.0 High Energy Mission Applications

There are advantages to a large source of lunar based power. The airless moon makes feasible a safe, quick method for material transport by rail guns or some of the other electromagnetic transportation means. If we seize the advantage of alternative sources of power, the environmental impacts on Earth are reduced. Total launch efficiencies are gained from the utilization of extraterrestrial materials rather than to transport heavy materials from the Earth.

Lower overall energy requirements to accomplish space construction projects can be anticipated because of the reduced lunar gravitational field and the absence of an atmosphere. This is a superior scheme in terms of energy efficiency. The reduced gravitational advantage can be used to even greater advantage if local resources are employed for the missions. This subject requires a systems engineering task for an overall evaluation. Lunar resources should be considered a part of any trade wherever space programs on a grand scale are concerned. If we were to advance the lunar materials processing technology to a sufficient degree of sophistication, construction of the reactor's bulk mass can be accomplished in-situ.

Consider the means by which supplies can be simply transported back and forth to Mars. The source of kinetic energy for the Martian shuttle vehicle can be a free electron laser for an ion engine drive or alternatively, a laser heat driven ablation rocket. The anticipated megawatt power requirement can be accomplished by either a surface or orbital based fusion reactor system. The reactor could also similarly be employed to power payloads back to Earth from Mars using the same propulsive system design concepts. These are concepts, the merits of which would have to be given further consideration after completion of system studies in order to determine the economic trade for comparisons with other technologies. The initial program costs would be high, but operational costs would be expected to be relatively low because the propulsive energy resides in a stationary location rather than on a flight vehicle permitting greater efficiencies, and obviously with safety improvements achieved compared to using manned transport vehicles.

The types of missions which can be supported using this concept are manned Mars missions using unmanned transport logistic spacecraft. Manned missions during extended stays on Mars require support by logistic vehicles, the concept being forwarded here being a flight vehicle powered remotely by a stationary energy source. For these unmanned vehicles, the trip time, energy, and cost trades can be made to optimize payload mass. With fusion energy, a broadened capability exists by which space missions can be optimized, that is, we can transport smaller objects to and from Mars by fusion powered lasers and chemical propulsion, and large, massive payloads powered by on-board fusion energy. The concept is equally applicable between the remote moons of Jupiter, for example.

The other key point is that there is a specific power trade which one can make to arrive at an optimal design. In the event that the flight fusion reactor systems are unable to initially achieve the high specific power required for flight performance, one alternative is the use of fusion energy from a stationary power source for the propulsion system energy source. In an operational mode, fusion as a power supply for a laser driven spacecraft, offers a competitive advantage, not so much in specific power as in offering a potentially high fraction of materials utilization from local lunar materials. This subject is discussed further in Appendix B.

Electrical energy beamed from the lunar surface to Earth orbiting satellites for the purpose of providing electrical power may also prove to offer significant alternative economies since attenuations and aberrations by an atmosphere through which ground based beams must pass are eliminated using lunar energy transmittals. The lunar power station can provide power to Earth orbiting satellites for meeting occasional peak power demands or to serve as a steady state supply of energy. An adequate energy supply to meet the power needs is essential for future mission options, and lunar fusion provides an attractive alternative, particularly with the presence of helium-3 there. The magnitude of the required power levels was not calculated in this study; it is suggested that additional analysis be undertaken.

2.2.12 MISSION LIFE CONSIDERATIONS

Some comment on mission success should be noted. With the greater duration flight times involved in outer planet explorations, the increased opportunity for hardware failure exists causing either reduced science data or loss of science objectives altogether. We have been very successful in accomplishing science missions, but the availability of more energy will permit the use of greater payload design margins and the use of more redundancy. Science instruments have been shown to result in the highest failure rate category (Blo78) of cataloged space equipment, thus demonstrating the need for this capability. Unpublished data taken from one of the NASA centers' spacecraft performance shows serious loss in achieving a useful life — i.e., one in which major mission objectives are no longer met — as the mission duration extends from 3 up to 10 years (Fig. 2.56).

2.0 High Energy Mission Applications

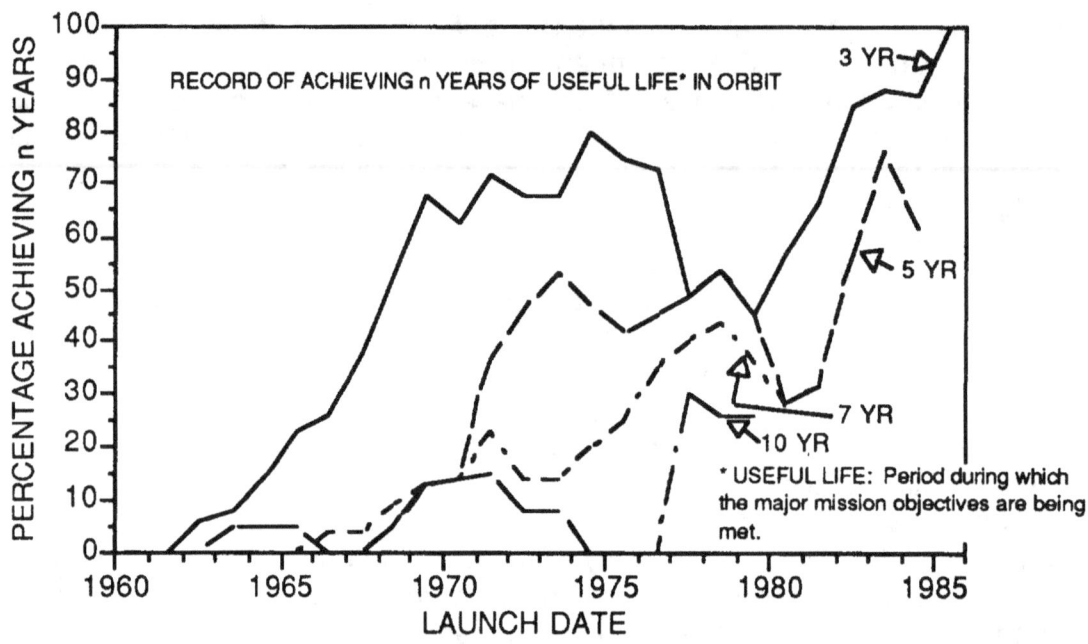

Fig. 2.56. Spacecraft performance life history.

These numbers must be carefully considered since the intent of the mission may have been to last for a significantly shorter life. There are two points to be derived from the curves: (1) there is a degradation of operational life as a function of mission duration, and (2) there are measures that can be taken to improve upon the operational function of the payload. That is exhibited by the initial positive slope, presumedly as a result of a design learning curve, and two, a relaxation of the standards learned as shown by the negative performance slope exhibited by the 1977-80 interval. Actions taken as a consequence of that negative trend show a positive slope which has endured up to the time limit of the current data base of performance characteristics. Science is the prime rationale for those missions. The conclusion is that the life performance of the science instruments is a concern although these are typically treated as lesser important functions of space missions. Further, there are measures that can be taken to lengthen life performance. One approach is increased margins. That requires an additional propulsion system capability – a subject where the science instrument payload performance can be increased by high specific power.

2.0 High Energy Mission Applications

2.2.13 SMALL SPACE FUSION REACTORS

Mission Description

One useful mission category which would find an immediate application in the space program is the group that could be accomplished by small fusion reactors, namely, those on the order of several megawatts or less. These reactor's may be very difficult, if even possible – see Appendix C. Power of that range would be compatible with the present planning of the advanced missions which are considering missions in the SP100 reactor (100 kW) and higher range. At the present, these more appropriately fall into the domain of fission reactors.

These missions include shuttling payloads between the Earth and moon, to geosynchronous missions between Earth and the moon, plus others within Earth orbit. If developed, such small reactors would serve as a Space Tug. This concept was not expanded upon in this study because at the present time the space application which fusion is most likely to fulfill more quickly are those falling into the high energy mission category. At some point in the future, however, the technology could evolve to the point where smaller reactor designs can be developed. This category of fusion reactors will be developed more likely for space rather than for terrestrial applications where the economies favor the centralized gigawatt power stations, although some specialized remote territorial location uses can be expected to occur. It is anticipated that LEO missions could be required if applications for large plane changes or large beam power in LEO were required.

2.2.14 AERONAUTICAL

Mission Description

To lower launch costs for commercialization of space and to improve safety in launch to orbit, a new technical approach is indicated. One of the unique thoughts forwarded for consideration is the use of fusion for aeronautical propulsion.

If space is ever to become a strong commercial enterprise, costs have to be decreased by orders of magnitude and safety substantially enhanced. NASA made a serious effort to significantly lower launch operational costs in the Shuttle Program but found that a large number of ground support personnel is essential to its launch and flight operations. Studies had been conducted to make it airline-like in an effort to lower flight operational costs, but even with those extensive activities, transportation costs to orbit remain far out of reach for the general population.

2.0 High Energy Mission Applications

Development of simpler, less ground personnel intensive spacecraft, ones having high mass fractions, are needed to reduce costs to orbit. Airlines succeed as a commercial endeavor because they offer a safe product at a reasonable cost. The cost is achieved in part by a quick turnaround and by a high payload mass fraction, using equipment which is serviced by a small number of flight and ground crew per flight. The ground servicing staff is orders of magnitude greater to launch the Shuttle as is the preflight readiness time. In the Shuttle program, the payload delivered to orbit comprises approximately 1% of the gross vehicle weight, whereas with the large wide body aircraft, it is 50 to 60%. The commercial aircraft is totally reusable; the Shuttle is not. The Shuttle maintenance requirements, as for example with the main engines, are more demanding and expensive. Space transport vehicles will require basic operational system specifications similar to airlines, or even more stringent, if space flight is ever to become economical.

Would fusion energy benefit the commercialization of space? Perhaps, but in a more distant time. The major requirements to achieve are high thrust and low mass (Section 8). As an option, fusion energy, in combination with the aid of aeronautical lift, may possibly be used to transport man and materials from Earth's surface to low Earth orbit without requiring vehicle thrust levels on the order of millions of newtons. This is a far reaching but potentially highly significant proposition. If technically feasible, the commercialization of space would be a great benefactor. The vehicle's lift-off thrust level requirements are lowered by a couple orders of magnitude where aerodynamic lift is used. The objective is to use a single stage to orbit or alternatively a high altitude, air-launched orbital stage, if that concept is determined to be more practical from a design and safety viewpoint. For lower altitude flight regimes, the reactor would serve as a heat exchanger of air to attain the required mass flow for high thrust. Thrust conversion to an internal propellant supply would occur at high altitudes. It should be pointed out that this may not be a far out concept when one realizes that in the 1960's, the United States had successfully tested nuclear fission ramjets. From a technical viewpoint the ramjet performed. However, the safety and environmental problems from fission product and radioactivity leakage to the air prevented its use. The employment of aneutronic or nearly aneutronic fuels may provide a practical solution for this application. It certainly deserves a review and a detailed study for consideration.

Thus, it is clearly far too early to make such promises for fusion, but one is driven by the appealing thought that if a sufficiently high thrust-to-weight reactor system can be designed for aeronautical lift purposes, safety in the launch phase would be substantially enhanced. The use of magnetic field hardware to provide energy conversion does not require the use of high speed equipment, such as turbopumps. Substitution of liquid helium for the highly explosive liquid hydrogen-liquid oxygen combination is another very positive step for safety. Perhaps some fusion research developments at some later point may negate the safety opinion, but at the present time it does offer an attractive approach worthy of consideration. Safety is discussed in greater depth in Section 9.

Application

This is a totally new concept, and a literature review did not reveal any prior study of this application.

Fusion reactors are not high specific power machines, at least not by the design concepts now being considered in comparison with the traditional specific power standards set for aircraft engines and spacecraft engines. The Space Shuttle's main engines operate at 1550 kW/kg for example. Hence, any such undertaking that provides large accelerations for aeronautical lift, or which provides even lesser amounts from the lunar surface or other small planetary bodies, must be considered to be extremely ambitious for fusion. The technical feasibility of high thrust fusion is unknown because this has never been considered as a fusion requirement.

The power plant for aeronautical lift must operate at a power level sufficient to produce on the order of 500 kN thrust. According to a fusion plasma model developed by Dr. Kernbichler (University of Illinois/Graz Technological University), the plasma would have to be in excess of 3 meters in diameter to produce approximately 10 kN, and that was considered to be an optimistic model. Use of multiengines and operation in the thermal transfer mode, discussed below, should be studied for feasibility. Otherwise the application will await the development of more compact, high specific power reactors.

The reactor design must provide not only a high thrust level, but it must do so at a high specific power. There are two ways of increasing the mass flow rate, direct injection of a diluent into the plasma or passing a fluid across a heat exchanger. The diluent is injected directly into the plasma at a sufficiently high flow rate to maintain the reaction. From a technical view point, the fundamental reactor design issues are: establishing the means for producing a uniform mixing of the plasma, avoiding extinguishment, and maintaining the required high specific impulse. A high β reactor design is essential (see Section 7.0). The means to accomplish high β is a key matter for a space fusion research program. Without the additional mass flow, the plasma alone is not sufficiently thrust producing, and fusion cannot conceptually be considered for aeronautical propulsion. The reactor must be of a relatively small size, too, in order to successfully accommodate airframe integration and to minimize drag.

One additional challenge that must be met is the neutron flux. As discussed in Section 4, purely aneutronic reactions are unlikely to be developed for the early reactor designs due to the inherent physics of the reactions involved. The advanced fuels under consideration are still neutron emitters, although at a very substantially reduced level. Neutron emissions from flight vehicles while operating in Earth's environment require a lot more care than in space. Neutrons are activating particles and surrounding aircraft structures will become radioactive. If tritium is used, the unburned products of combustion will proliferate radioactive materials into the atmosphere, probably at a level to its exclusion. Perhaps since hydrogen is light and tritium is a weak beta emitter, the magnitude of the hazard may be acceptable, the main concern being tritiated water. That is a subject requiring more research.

Flight operational limitations of the fusion powered aircraft may be a solution to some of the environmental issues raised. Takeoffs and landings may possibly be limited to remote sites. Neutrons decay rapidly, and the neutron fluxes from D-^{3}He may possibly be in a range that can be handled from an engineering viewpoint.

In summary, the attractiveness of a high payload capability, using a single stage to orbit is so attractive that this concept warrants further consideration. Without breakthroughs in specific power and thrust conversions, this application is not considered an early application of fusion energy. A study into this subject, including high payload mass fractions, is necessary to define concept feasibility. The goal of the program is to provide a safer, high payload mass fraction means of transportation to LEO.

2.2.15 FISSION REACTOR WASTE DISPOSAL

Mission Description

This particular suggestion is very speculative. But, in the event that progress can be made toward aeronautical fusion propulsion and if the opinion is verified by flight experience that fusion propulsion powered aerospacecraft provide a very safe, reliable mode for travel to space, perhaps one of the most significant contributions will be to provide space disposal of radioactive debris from the existing commercial fission power plants. Studies have been conducted to compare the relative cost and safety merits of Earth based and space based radioactive waste disposal. Space has been costly to access and the safety of such nuclear waste disposal missions in space will certainly be subject to heavy questioning, even more so now in view of the Challenger accident. On the other hand, there is always a concern of the risk affiliated with underground disposal – particularly earthquakes and leaching processes – and locations are becoming more difficult to find. No one finds a nuclear waste disposal facility in their back yard to be an attractive feature. Space offers an attractive alternative under circumstances where a safer, more economical means for transportation to orbit is achieved as discussed in Section 2.2.13.

2.0 High Energy Mission Applications

2.2.16 ELECTRICAL POWER AT SPACE BASED FACILITIES

Mission Description

Electrical power requirements for space based facilities is a subject about which there is a lack of analysis and data and which is therefore assumed to have received little attention and study. Space-based large power applications have been referred to throughout various sections, and it seemed appropriate to consolidate these concepts into a category of their own.

One significantly large power application is that for the manufacturing of propellants on planets. The generation of propellants for propulsion purposes is a topic of investigation by the University of Arizona's Utilization of Local Planetary Resources program. Manufactured propellants extracted from the lunar surface would significantly aid the performance capability of the chemical propulsion systems and provide power to run lunar factories.

As discussed earlier, the chemical propulsion systems are inefficient, that is, the low specific energy from those fuels requires copious quantities of mass just simply to move the vehicle's propellant from one location to another. On Mars, after extended stay times and settlement have been implemented by the space program, there will be a necessity to obtain in-situ propellant production capabilities. Sources of fuel and propellants at refueling stations along the way are exponentially more efficient than transporting outbound the vehicle's return supply of fuel and propellants. After the desirable ores are found, a large amount of power can be anticipated to be required to process the fuels whether chemical or fusion.

Helium-3, generated by the solar fusion furnace, can likely be found on any moon, asteroid or space body having a sufficiently high gravitational field to maintain a substantial dust layer and which lacks an atmosphere and magnetic field, permitting its deposition by solar winds. Fusion could, therefore, well serve as that power source for mining and for manufacturing helium-3. It could provide the power to extract chemical rocket propellants for low energy missions as well as to provide oxygen for a breathable atmosphere. The extraction of ores from elements and materials processing for fabrication purposes can be anticipated to constitute a large appetite for energy.

Other large power consuming functions include life support in terms of a breathable atmosphere and environmental temperature conditioning. An attractive technique for making energy available for local transportation is to beam microwave or laser energy to spacecraft or space-based outposts. That technique should be capable of providing a science enabling capability resulting in reduced flight operational costs and a simpler, quicker means for retrieving data. A more recent consideration to the subject of laser propulsion has been given to this subject by Dr. Logan (Log88) (Appendix B). This new look involves the latest thinking in laser technology which was not available when this idea was first explored in the 1960's. This new version involves a

2.0 High Energy Mission Applications

free electron beam to transmit energy to produce a translunar manned vehicle capable of delivering 12 MT to the moon. A multi megawatt power station on the lunar surface could provide power for cheaper payload transportation to Mars.

2.2.17 OTHER BENEFITS

New benefits to space will result from the availability of additional energy as discussed in Sections 1.0 and 2.0. The higher thrust levels will shorten flight times and provide the capability to conduct more extended periods of scientific exploration.

- Greater science return

    Because more massive payloads can be carried to and from the planets, more and larger samples which have been obtained from a wider exploration range on the surface can be returned to Earth for analysis. The additional payload capability allows one to consider large, remotely stationed laboratories. These could be unmanned with a highly advanced degree of artificial intelligence and equipment for real time in-situ analysis and decision making with decision execution by robotics. A large inexpensive energy source has the capability to provide the enabling capability to repair and to perform hardware changes and flight system upgrades quickly and with cost effectiveness.

- Helium-3 Mining

    In addition to the moon there are other sources for helium-3, but much more difficult to retrieve:

    - Jupiter: 10^{22} kg of ^{3}He are estimated to reside in the atmosphere of Jupiter. By a clever design of balloons and thermal cycles for liquefaction, the recovery of ^{3}He may be possible. This is a mission requiring a large power capability.

    - Uranus: The possible recovery of ^{3}He is one potential application of a high energy system mission to Uranus. Whether this yields a net economic benefit for expenditures is a subject for further study. Neptune is also a possibility.

2.0 High Energy Mission Applications

- Asteroids and comets

    There are, too, some far reaching, but practical, applications of high energy. If a large asteroid or comet were to approach Earth, revectoring the body via a high energy source would be mandatory to avoid a catastrophic loss of life, as theorized by the periodic life extinctions. Smaller ones are more frequent - in 1991 Earth was narrowly (~100K miles) missed by a small one having the destructive potential of an atomic bomb.

    Another high energy mission application is the use of the comets as a natural resource. They offer the potential for a natural supply of water for planets where the addition might be needed, as for example, Mars or the moon. Although it is too early to do little more than speculate regarding the vitality of the asteroids as a source of space based minerals for functional space use, that application could become an important aspect for space exploration. The first task is to characterize asteroid's composition. In the event that rare or greatly needed elements are discovered, it may be economically feasible to provide the propulsive means using a high energy vehicle to transfer the asteroid intact to the moon, Earth orbit, or to Mars for convenient extraction of ores. All of these missions require the expenditure of very large quantities of energy plus a significant amount of time.

2.2.18 SPACE COMMERCIALIZATION AND SAFETY APPLICATIONS

Mission Description

If the determination is made to mine helium-3 from the moon and if the terrestrial reactor concept, the tokamak, can be designed to burn it, then fusion could serve to power the lunar mining facility. Perhaps at some point, the economic trade would favor the manufacturing of materials there as well as the fabrication of structure and equipments of all categories, particularly where large masses are used in the space construction projects.

The moon, if developed, could serve as a valuable remote quarantine facility. Samples of other planetary materials, particularly any suspected of having extraterrestrial life threat potential, could be analyzed in great safety on the moon. Dangerous or potentially dangerous chemicals could be researched or manufactured there without upsetting the delicate environment of the Earth. The absence of convective wind currents reduce global hazards from such sources to zero. As mentioned earlier, it would be an ideal spot to conduct biological experiments to examine the effects of genetic engineering under controlled and isolated conditions. Thus, some of the controversial genetically engineered products now being considered for use here on Earth could have

2.0 High Energy Mission Applications

already been safely qualified prior to use on Earth. With fusion power available, those advanced mission concepts could at least receive further study for consideration. Avoidance of some biological catastrophe on Earth would in itself pay for fusion and the space program.

2.2.19 EARTH ORBITAL APPLICATIONS

Mission Description

As shown by the performance comparisons at the beginning of this section, high performance systems do not typically offer significant benefits to LEO missions since most of these applications are low energy missions. There are some possible LEO applications, however, where it could be useful to consider a high performance capability.

- Large out of plane LEO maneuvers require large velocity changes. It is possible, for example, to build a single large Earth observing spacecraft, the cost and mass of which is sufficiently great to desire that it be translatable for changes in orbital inclination in order to provide the desired ground coverage after placement into an initial low Earth orbit.

- Missions to GEO are also very energy demanding. Single large spacecraft can be placed there and serviced by a high energy logistics vehicle. There is concern regarding the eventual accumulation of expended spacecraft to cause a GEO debris problem. A high performance spacecraft could be used to remove derelict GEO spacecraft.

- Similarly there is concern that LEO debris can become a hazard to Space Station Freedom, EVA astronauts, and other spacecraft. The potential exists for use of a high energy spacecraft to economically remove the debris.

- A high energy electrical power space based station can transmit energy beams to meet peak demands to any of a large number of Earth orbiting spacecraft to reduce their power requirements and thereby reduce costs or to serve in an emergency back-up operational mode in the event of a loss in power.

- A high energy vehicle placed in low Earth orbit could also serve as a personnel rescue vehicle, assuming that the thrust can be at a sufficiently high level to provide a rapid response. This is not necessarily to be a fully dedicated vehicle, i.e., a space ambulance, but one serving a dual function. It could also serve as an orbital servicing and repair spacecraft for unmanned spacecraft.

- We could place a multimegawatt power supply at Space Station Freedom to perform materials manufacturing and to serve routine maintenance and sustenance functions. Large scale biological experiments such as plant growth and natural atmospheric revitalization for future long duration missions could be accomplished.

The feasibility and practicality of those suggestions have to be studied to determine their performance and economic viability. The development of smaller reactor technology will greatly benefit the Earth orbit application.

2.2.20 SPIN-OFFS

As with any research program many unanticipated applications fall out without regard to the original intent. The unanticipated "spin-offs" can be significant. Clearly there is potential for the space and the terrestrial high energy programs to complement each other. Medical benefits are frequently derived from space technology. Improved materials can result. Perhaps there will be technology to result in improvements in ground transportation systems, particularly if magnet technology is advanced to a point where it could be applied to high speed train applications.

2.3 SUMMARY

A number of potentially attractive high energy space missions have been identified where high energy, particularly fusion energy, could be of great benefit and which could be used now if available. One of the greatest benefactors is the Manned Mars Mission where the shortened flight time can be expected to contribute significantly to flight safety and to provide a substantial savings in launch and operational costs. Equally important are the science missions enabled by fusion.

Two high energy workshops should be held to explore the high energy mission class in greater depth. One would consider high energy missions. The other's function is to establish an optimal space fusion energy conversion program. A High Energy Mission Space Science Workshop of scientists in the space related disciplines, and any other appropriate areas, would be in a position to more fully forward and develop new high energy mission concepts. The Fusion

2.0 High Energy Mission Applications

Workshop would be instrumental in establishing a viable space fusion energy conversion program. These are discussed further in Section 14.

The key mission performance parameters and figures of merit for Manned Mars Missions (133/61 MT outbound/inbound payloads), outer planetary sample return (20/10 MT), and asteroid (20/10 MT, 3 to 6 visits) are summarized in Tables 2-7 and 2-8. The tables include mission data using fusion reactors designed to specific powers of 1 kW/kg and 10 kW/kg and variable, high specific impulse ranging from 5,000 seconds to 10^6 seconds. Table 2-7 presents key mission parameters for typical missions that comprise reasonable minimal flight times for reduced initial vehicle and propellant masses to achieve economical flights. Table 2-8 summarizes the times and penalties associated with those high velocity-high Δv missions. Manned exploration of the outer planets can be accomplished within reasonable flight times, values ranging from 3.5 to 5 years.

The key fusion parameter for space is specific power which should be above 1 kW/kg for the solar system exploration missions; for the stellar and Oort Cloud missions, above 10 kW/kg. The high specific power capability must be complemented by a variable, very high specific impulse, ranging to 10^6 seconds. Alpha Centauri could then be reached in a fly-by mode in 180 years, or in 290 years in a rendezvous mode, using a reasonable quantity of propellant. Those times could be decreased to 150 and 240 years respectively, but at an enormous increase in consumption of propellant. Manned exploration and settlement beyond the solar system is not feasible using fusion as currently envisioned. A deeper understanding of and application of the nucleus's strong force may perhaps be of value in that regard.

TABLE 2-7a. Performance summary for typical manned Mars, outer planetary sample return, and asteroid sample return missions. Performance data are referenced to a manned 131 MT outbound/61 MT return payload, unmanned payloads of 20 MT outbound and 10 MT return. The times shown are the total flight times exclusive of stay durations at the target.

Specific power = 1 kW/kg

Mission	t, years	Mo, MT	Mp, MT	γ, %	Pj, MW	<Isp>, seconds	Δv, km/s
Manned Mars	0.50	613	335	22	145	10,610	90
Europa	1.56	320	243	6.3	57	16,690	209
Titan	2.99	74	36	27	18	26,200	196
Miranda	5.34	60	26	33	14	35,680	233
Triton	5.85	108	62	19	25	35,130	314
Charon	7.42	81	41	25	19	40,530	317
Asteroids: 3 visits	1.72	163	107	12	36	18,550	185
Asteroids: 6 visits	3.39	162	105	12	36	26,120	254

2.0 High Energy Mission Applications

TABLE 2-7b. Performance summary for typical manned Mars, outer planetary sample return, and asteroid sample return missions. Performance data are referenced to a manned 131 MT outbound/61 MT return payload, unmanned payloads of 20 MT outbound and 10 MT return. The times shown are the total flight times exclusive of stay durations at the target.

Specific power = 10 kW/kg

Mission	t, years	Mo, MT	Mp, MT	γ, %	Pj, MW	<Isp>, seconds	Δv, km/s
Manned Mars	0.50	185	30	72	227	35,770	90
Europa	1.56	32	6.8	63	50	64,070	209
Titan	2.99	29	5.3	68	40	81,180	223
Miranda	5.34	26	3.4	77	27	117,509	233
Triton	5.85	27	3.8	74	30	129,620	283
Charon	7.42	27	4.1	73	32	137,069	317
Asteroids: 3 visits	1.72	44	15	45	96	57,020	257
Asteroids: 6 visits	3.39	141	12	49	83	86,735	329

TABLE 2-8a. Performance summary for a fast manned Mars, outer planetary sample return, and asteroid sample return missions. Performance data are referenced to a manned 131 MT outbound/61 MT return payload, unmanned payloads of 20 MT outbound and 10 MT return. The times shown are the total flight times exclusive of stay durations at the target.

Specific power = 1 kW/kg

Mission	t, years	Mo, MT	Mp, MT	γ, %	Pj, MW	<Isp>, seconds	Δv, km/s
Manned Mars	0.44	1,041	681	12.8	227	9,440	98
Europa	1.43	976	843	2.0	113	14,720	223
Titan	2.11	858	733	2.3	105	17,750	264
Miranda	3.48	809	687	2.5	101	22,860	337
Triton	4.59	1,031	895	1.9	117	25,780	393
Charon	5.49	797	677	2.5	100	28,570	420
Asteroids: 3 visits	1.44	955	819	2.1	116	14,600	217
Asteroids: 6 visits	2.82	992	852	2.0	120	20,300	301

2.0 High Energy Mission Applications

TABLE 2.8b. Performance summary for a fast manned Mars, outer planetary sample return, and asteroid sample return missions. Performance data are referenced to a manned 131 MT outbound/61 MT return payload, unmanned payloads of 20 MT outbound and 10 MT return. The times shown are the total flight times exclusive of stay durations at the target.

Specific power = 10 kW/kg

Mission	t, years	M_o, MT	M_p, MT	γ, %	P_j, MW	$<Isp>$, seconds	Δv, km/s
Manned Mars	0.18	1,034	676	12.9	2,225	18,870	196
Europa	0.81	92	50	22	219	42,210	352
Titan	1.20	99	56	20	234	50,650	437
Miranda	1.93	112	66	18	261	63,300	576
Triton	2.87	77	39	26	185	79,820	609
Charon	2.76	237	171	8.4	464	70,134	816
Asteroids: 3 visits	0.65	898	766	2.2	1,119	30,920	455
Asteroids: 6 visits	1.30	796	670	2.5	1,053	43,920	631

3.0 HIGH ENERGY SOURCES FOR SPACE

3.1 CANDIDATE HIGH ENERGY SOURCES

One study objective, in the end, is to investigate the means to provide a new approach for total program risk reduction in the accomplishment of high performance missions. In Section 2.0 candidate advanced missions requiring high energy levels were presented. The significance of high energy space missions for future programs and the performance requirements for their achievement were analyzed. In this section we examine the candidate energy sources for performing those missions and discuss the relative advantages and disadvantages to evaluate risk reduction approaches.

Major considerations for the selection of any high energy source include its mission enabling capability, its mission enhancing performance, the safety implications, the cost effectiveness of the system on flight operations, and overall acceptability including such factors as environmental effects and reliability. It is not sufficient that high energy missions be accomplished with only slight improvements for the expenditures committed. A cost effective, high performance system will inherently compel its development as a mission enabling and enhancing instrument; a system providing a slight performance increase will not. Hence, energy sources (Section 3), mission performance (Section 2), and cost (Section 10) became an integral, interwoven part of this study. Two other critical aspects of high energy source acceptability are the degree of safety or magnitude of the hazard which is offered and its kindness, or hostility, to its operational environment, topics of Section 9.0. The means and feasibility for converting the energy sources into propulsion and power systems are discussed later and, for fusion, in Sections 7 and 8.

Consider the demands on energy conversion systems placed into space to accomplish the kinds of missions discussed in Section 2.0. In those applications the flight system must:

3.0 High Energy Sources for Space

Flight system properties:

- minimize propulsion system mass,
- meet long system life time requirements of years,
- provide greater inherent system safety,
- provide a remote, reliable, and efficient space restart capability,
- use only radiation for cooling,
- be designed for the presence of a "free" continuous vacuum,
- provide power for variable propulsive thrust and specific impulse requirements,
- provide power for the generation of electricity,
- be designed to operate in a low acceleration environment (low thrust-zero gravity) or in the absence of gravitational loads (zero thrust-zero gravity),
- produce a very wide range of output power levels (throttable),
- be designed for long operations despite a lack of ready access for maintenance.

With consideration to the above requirements, an examination was made of the potential energy sources that could best fulfill those requirements. The energy source options discussed in this report include:

- fusion
- fission
- chemical
- matter-antimatter
- strange matter
- others.

Each energy source is discussed below.

Relative energy yields from the above energy sources are compared in Table 3-1 (Bor87), excluding strange matter. The specific energy release for fusion (deuterium and helium-3) is shown to be an improvement over the best chemical (hydrogen and oxygen) source by nearly 7 orders of magnitude. In comparison with fission (^{235}U), it is only slightly better while matter-antimatter is over two orders of magnitude greater than fusion.

TABLE 3-1. Specific energy release comparisons.

ENERGY SOURCE	SPECIFIC ENERGY, J/KG
Matter-antimatter	9×10^{16}
Fusion	3.52×10^{14}
Fission	8.20×10^{13}
Chemical	1.35×10^7

For space missions it is useful to consider the energy requirements, expressed in terms of Δv, for the more typical advanced missions. These are shown in Table 3-2 (Gar 88).

TABLE 3-2 Typical mission velocity requirements (impulsive burns).

Mission	Δv, km/s
Earth surface to LEO	7.6
LEO to GEO	4.2
LEO to Earth escape	3.2
LEO to lunar orbit (7 days)	3.9
LEO to Mars orbit (0.7 year)	5.7
LEO to Mars orbit (40 days)	85.0
LEO to Neptune orbit (29.9 years)	13.4
LEO to Neptune orbit (5.0 years)	70
LEO to solar escape	8.7
LEO to 1000 AU (50 years)	142
LEO to Alpha Centauri (50 years)	30,000

The authors considered the capabilities of chemical, nuclear thermal propulsion, and nuclear electric propulsion (NEP). The Inertial Upper Stage (IUS- a solid propellant powered stage) is capable of imparting a velocity of 5.2 km/s to a 1000 kg mass; the liquid propellant system is sufficient for 7.6 km/s. With staging, the upper limit for chemical propulsion is 20-30 km/s. The solid core fission, to be safely operated, is limited to a specific impulse of 850-900 seconds due to the inherent limitations of the thermal properties of materials. The maximum Δv is ~20 km/s from this technology. Ion engines do not operate under the limitations of thermodynamics and are considered capable of yielding 65 km/s in the near term and 115 km/s in the far term. This study examined missions with Δv's from 90 to over 15,000 km/s.

3.0 High Energy Sources for Space

Clearly a new approach to energy sources for propulsion will have to be acquired if more ambitious space missions are to be conducted and, to even implement in a practical sense, the plans now being generated for a permanent presence of man on Mars. While chemical and nuclear thermal propulsion can provide the transportation energy for the mission, the logistics supply system will become cost prohibitive unless high specific power and high specific impulse systems are developed. Fusion offers a very promising approach. Also, another interesting concept using fission has been suggested by Dr. G. Chapline (Lawrence Livermore Laboratory) termed "Fission Fragment Rocket Concept."

We discuss below each of those energy sources, how they can be of benefit in a practical sense to space flight fusion operational systems, the technology status, and limitations.

3.2 FUSION

Fusion is nature's choice of large energy release, it being the process by which the stellar furnaces have continued to operate for billions of years. It is efficient. The sun has used it effectively for 5 billion years, and current estimates show it proceeding for another five. Without fusion, a sun energy release based on chemical/gravitational processes has been estimated to endure for only a few tens of millions of years.

There is a class of high energy missions that fusion can uniquely serve for which either the chemical or fission energy solutions are too impractical in terms of cost, time, or technical feasibility. The energy produced by fusion is predicted to be at high multimegawatt power levels, and the space program benefits gain strongly by economy of scale. It can perform missions requiring high power levels which no other energy source is capable of performing, excluding the possible total energy conversion of matter-antimatter reactions.

A discussion of the background and status of the fusion energy program follows.

3.2.1 BACKGROUND

The national fusion energy program was commenced in 1952. The program has one purpose – the production of abundant, cheap, and safe commercial electrical power. There is no space fusion energy conversion program, although a low funded activity existed at the Lewis Research Center, the content of which is described in Appendix A. The DOE technology program wherein some concept might have space application was cancelled in 1991.

One key conclusion which evolved in the conduct of this study, and which is its primary product, is to show clearly the importance of fusion energy to NASA's space flight programs that are anticipated in the 21st century. It presents the rational for a new research program initiative and offers a strategy for its fulfillment, one that is independent of the terrestrial power program. Although many rationale are provided throughout the report, the top level considerations are:

(1) there are significant mission performance differences between that of NASA and DOE, and

(2) there are different mission objectives between the two organizations.

The first determines the applicability of the current research to the space application; the second determines the agency's program priority and, therefore, the program commitment given to research on the technology. Both factors combine to determine its availability.

The space fusion applications differ uniquely from the commercial terrestrial power applications. Mission differences result in significant differences in system requirements which the reactor design must reflect. Therefore, a successful terrestrial commercial fusion program does not necessarily equate to the successful application of fusion to space. The terrestrial program's mainline experiment, the tokamak, can be expected, when breakeven is demonstrated, to provide a great psychological encouragement and incentive to press forward; but the technology will not have space flight applicability. Unfortunately, the situation which may occur concerning funding for fusion research is a reduced emphasis on potentially useful space applicable research and experiments. That undesirable situation can be anticipated to be the casualty of a very funding-constrained fusion research program as DOE focuses its critical resources toward the production of net power from the tokamak at the expense of alternate confinement approaches. This undesirable situation has indeed occurred in 1991.

The NASA hardware for space missions demands quite different trades from those made for the ground based utility reactors including: mission and program objectives, system performance requirements, safety, reliability, and costs. Even the fuels and reaction physics may differ as a result of the two different applications.

3.2.2 STATUS

The technical status of the fusion program is described in detail in Section 8.0.

The demonstration task for terrestrial controlled fusion has turned out to be more difficult than originally anticipated, and the results from the terrestrial energy program are being obtained later than the program participants would have desired. Although the degree of difficulty has been greater than envisioned, those more difficult problems are being resolved. In part that is the rationale for the conclusions reached in this report concerning the timeliness and ultimate demonstration of net fusion energy. It would, therefore, be difficult to accurately project a schedule of success for a space fusion program, although the report offers a suggested schedule in Section 14. However, the large body of experimental, analytical, and computational tools developed for the terrestrial fusion program provide a good degree of understanding. Those points are developed later in this report, Sections 7 and 8.

Since fusion for space applications has not been under serious consideration by NASA, no major space fusion flight vehicle system studies have been conducted. As with any venture, costs are a major consideration before proceeding with implementation of an activity. Without a demonstrated fusion system, NASA lacks any valid means to establish dependable cost figures of merit. An inertial confinement approach was recently examined (Section 2.2.1), that being the most recent study conducted in the fusion technology for space (Ort87). Detailed system studies are eventually expected following positive developments in the terrestrial fusion program, but not in the near term unless some earlier decisions are made regarding the implementation of a space fusion program. NASA is currently operating on the principle that once the terrestrial energy program has completed the basic technology, NASA will commence developments in fusion for space using the terrestrial program results. By this strategy, however, much valuable time will have been forfeited; and the critical space issues still remain to be addressed. There will still be the need to develop a space reactor since the terrestrial design is not expected to apply to space.

Thus, in the final analysis the use of fusion energy in space may not be determined as much by the mission user requirements, i.e., terrestrial versus space, but by the hard results of a successful fusion reactor design and test program to demonstrate that man can provide a machine capable of controlled fusion such that it will have a meaningful application to space. It will also be determined by the economics of the operating system, the subject of Section 10.0.

The common energy denominator in advanced mission planning is currently to rely on chemical systems, or, subsequent to the commencement of this study, fission energy sources. Based upon our current state of knowledge of physics, however, fusion is theoretically the optimal source for engineering high specific energy fuels and will remain so for quite some time. Chemical systems, although they have serious performance limitations and safety concerns (Section 9), prevail since these systems are developed and are readily available. The attraction of fission is its higher performance compared to chemical systems and the fact that it has been demonstrated to operate in the propulsion mode. There is a major penalty for continuing to rely upon these systems however. The development of space fusion is, therefore, absolutely crucial if we are to accomplish man's dream of the exploration of the universe beyond our current visionary limitations. It will not be quick, nor easy, and the goals of these more advanced missions even assume international implications, as discussed later, Section 11.

Two types of confinement techniques are being researched: magnetic and inertial confinement. This report places emphasis on the magnetic confinement approach. ICF is too classified to explore openly although we can state several key issues. The performance characteristics of fusion are elaborated upon in Section 5.0.

3.3 FISSION

Fission has the advantage that propulsion has been demonstrated and small reactors have been flown in space. Fission will at least provide an intermediate power level up to approximately several megawatts. Perhaps fission can exceed those levels in space flight systems, but that has not been demonstrated. All fission machines carry the concern for the overall public safety due to the potentially severe radioactivity hazards and the effects on Earth's environment. Extensive measures are currently employed to ensure that power plant accidents are avoided to prevent the inadvertent release of radioactive materials into the Earth's environment. For flight systems, system safety measures impose performance restrictions. The public will be reminded of the Chernoble and Three Mile Island accidents, and the loss of Challenger. These accidents beat the "odds." In addition, the public will be reminded that Russian and U.S. reactors have reentered the Earth's atmosphere. A public perception will have to be considered and properly addressed prior to use.

Several fission energy conversion approaches for propulsion have been initiated or proposed as presented below.

3.3.1 NUCLEAR ELECTRIC PROPULSION (NEP)

In summary, there has been extensive testing of reactors and ion engines, including an ion engine flown in space as an experiment. A reactor can be safely placed into LEO although there is an extensive launch approval process involved. The system's consumables, the propellants, can be retanked in orbit for reuse for short duration missions although the fission fuel is not replenishable.

From flight operational perspectives, the NEP performance level is too low for the category of missions considered in this study to effect the savings offered by fusion or even to perform the missions which fusion is capable, as discussed in Section 2.0. The size of reactors becomes large – 100's of megawatts which is much larger than the designs currently researched. Current technology development is being performed on the SP-100, a 100 kW reactor. Accomplishment of the same flight time for the same payloads having the same M_o as fusion is not possible using NEP. Significantly greater mass in LEO is required, and the result is longer flight times. It uses large quantities of the noble gases, the supply of which is not established and may not be available. A large number of the sample return missions, those beyond Jupiter, were not practical for NEP. Even for Jupiter, reactors and thrust level requirements were 2 orders of magnitude above the current anticipated performance capabilities. Even then NEP can only perform a mission that requires twice as long as those accomplished by fusion. It does offer mission improvements for smaller payloads in comparison with chemical systems.

There is a safety hazard from the presence of reactors in LEO, although parking it in a high, nuclear safe orbit should be a satisfactory solution. Launch of an unactivated reactor, unlike RTG's, should be safe. With respect to reuse, the reactor is life time limited, thereby restricting its use from the long duration missions. "The operational life of these reactors is expected to be about 10 yr., 6-8 times longer than the life of the thrusters shown in Table 2 [12,000-15,000 hours]. Multiple sets of thrusters will, therefore, be required to fully utilize the energy stored in the nuclear power supply." (Gar88) Because refueling of the reactor cannot be simply accomplished in LEO, replacement of the core remotely and/or robotic disposal are required for safety purposes. The alternative is to totally dispose of the entire vehicle which is less cost effective than the reusable systems.

For missions of the energy level considered in this report, NEP appears to be limited to the moons of Jupiter and even there, a significant technical challenge remains. The performance level required for the sample return missions from Titan and beyond exceeds fission's power, at 7.6 MW. It exceeds life requirements and the performance and life expectations of ion engines. Those are not $\alpha_{p0.067}$ missions.

3.3.2 Nuclear Thermal Propulsion

Above and beyond the performance level of chemical propulsion systems, this approach is the best developed, the major engine parameters having been demonstrated. This technology was terminated in the 1970's due to the lack of a mission (Section 2.0). So although it is "developed" 20 years have elapsed since the project was worked on. Nuclear fission thermal propulsion systems were demonstrated to yield a specific impulse of approximately 850 seconds at high thrust levels. The NERVA engine (shown in Fig. 3.1a and 3.1b) program had demonstrated technical feasibility at the time of termination.

3.0 High Energy Sources for Space

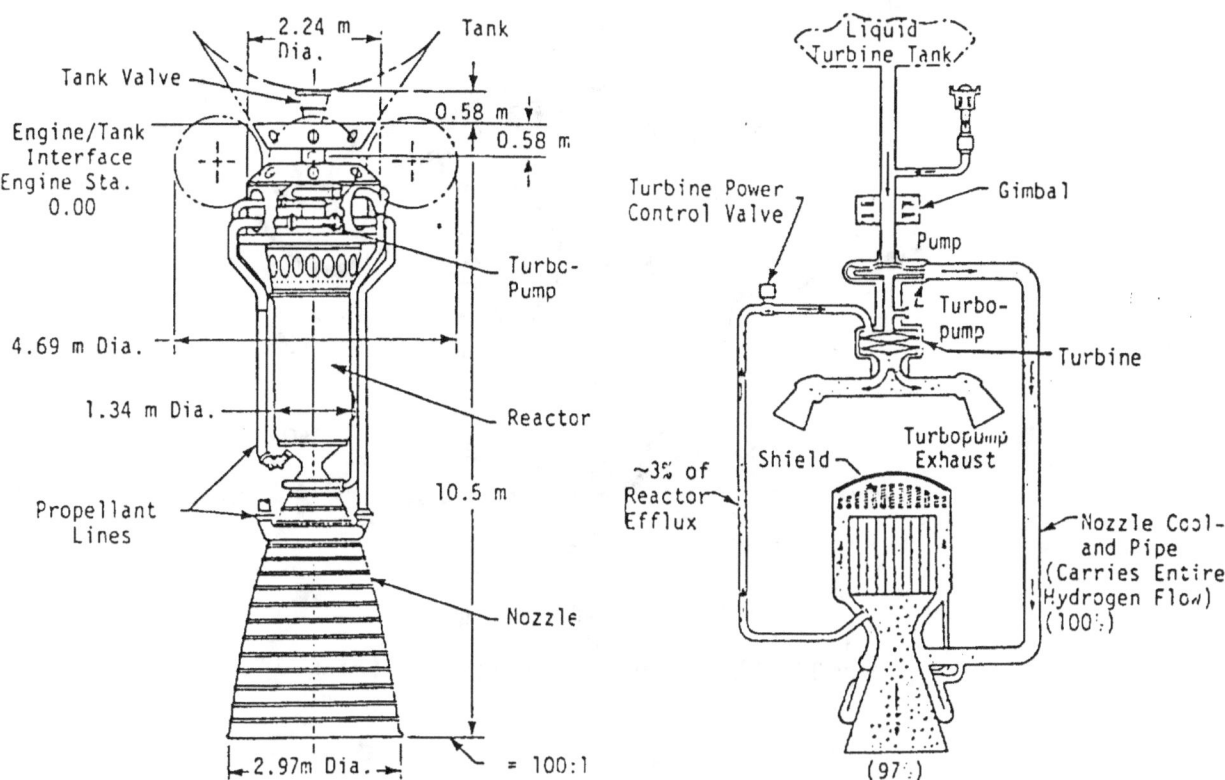

Fig. 3.1a. NERVA flight engine.   Fig. 3.1b. NERVA hot bleed cycle schematic.

Although these systems exhibit performance improvements over chemical systems, they inherently have all of the operational disadvantages that result from the emission of radioactive exhaust particles, a radioactive core, and a complex cool down system required for maintaining the structural integrity of the reactor core. Flight operations are complicated by the need for complex core cooling schemes which require a large number of cooling "burns," or thrusting pulses. For probes to the outer locations in the solar system where there is no threat to Earth's environment from an errant returning reactor, this engine could be particularly beneficial.

3.3.3 GASEOUS CORE REACTOR

This is a suggested concept having an open fission fuel cycle system. The concept is illustrated in Fig. 3.2.

3.0 High Energy Sources for Space

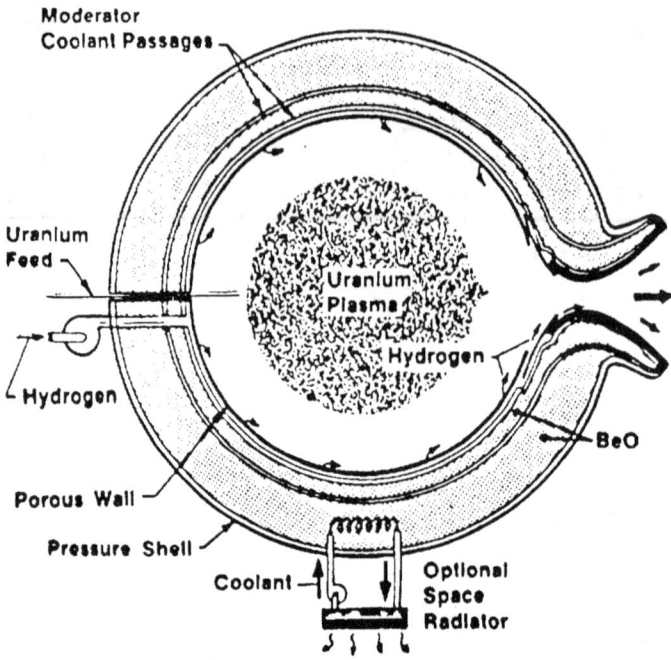

Fig. 3.2. Fission gaseous core reactor concept.

The material limitations inherent with the nuclear thermal systems is reduced by employing a gas core approach, permitting temperatures from above 4,000C to over 20,000 C. The temperature equates to specific impulses of approximately 1,500 seconds to 6,000 seconds. The advantage is a high thrust, high performance engine which permits short trip times and reduces M_o. Aerobraking can be eliminated which makes for a simpler, safer system transportation vehicle. Performance-wise, this is a very attractive concept. The exhaust products will be radioactive, thereby restricting use. For example, it could deposit radioactive wastes on the Martian and lunar surfaces or on the Space Station Freedom. Safety is a significant concern and this engine would be expensive to test on Earth in a manner that would avoid impacting the Earth's environment.

This concept has not been researched nor tested. It is anticipated to be a very difficult new technology to safely develop. Stability of the uranium plasma is a concern.

3.3.4 FISSION FRAGMENT ROCKET

The Fission Fragment Rocket concept mentioned earlier has significant potential if it can be verified to be practical.

> Fission fragments are directly utilized as the propellant by guiding them out of a very low density core using magnetic fields. The

> very high fission fragment exhaust velocities yield specific impulses of approximately a million seconds while maintaining respectable thrust levels. Specific impulses of this magnitude allow acceleration of significant payload masses to several percent of the velocity of light and enable a variety of interesting missions, e.g., payloads to the nearest star, Alpha Centauri, in about a hundred years or very rapid solar system transport. The parameters reported in this paper are based on a very preliminary analysis. Considerable trade-off studies will be required to find the optimum system. We hope the optimum system proves to be as attractive as our preliminary analysis indicates, although we must admit that our limited effort is insufficient to guarantee any specific level of performance. (Cha88)

The work is at a very preliminary analysis conceptual stage. A critical mass of fissile material in the form of 5 mm diameter fibers serves as the energy supply and propellant. The proposed fuel is ^{245}Cm or ^{242}Am. Success is predicated upon keeping ejected mass at a minimum particle size for optimal propulsion performance while still maintaining a critical mass for the fission reaction. The supply of available fuel for use on this mission is currently at 10 tons with more expected from the operation of commercial reactors. For propulsion the rocket uses fission fragments directly as propellant. Magnetic fields of 10^3 G direct the fragments out of the reactor.

POTENTIAL PERFORMANCE

To illustrate the extraordinary possibilities opened up by the fission fragment rocket let us consider a mission to the nearest star, Alpha Centauri, 4.1 light-years from the solar system. The device would start in a sufficiently high orbit so that the fission fragment exhaust will not return to earth. We assume that we have an americium fueled rocket and include a mechanical concept such that the fission fragments that are trapped on the carbon wires, along with the spent wires, are discarded periodically. We also assume a 10-GW reactor operating for about 40 years, the system coasting thereafter. For a fission fragment escape probability of 50% we can deliver a mass of payload plus structure of six metric tons in 100 years, fifteen metric tons in 121 years, or thirty metric tons in 148 years. If we could increase the escape fraction to 70%, we can deliver ten metric tons in 87 years, twenty metric tons in 101 years, or thirty metric tons in 113 years. It is thus easy to see that it is worth an increase in structural mass to increase the escape fraction if the goal is to minimize the transit time of the useful payload. Thus, an americium-powered fission fragment rocket holds the potential of a less than 100-year mission to the nearest star if the payload and structure mass can be kept sufficiently small—somewhat longer mission times are required if the mass is large.

Fortunately, the mission duration is not overly sensitive to a small reduction in reactor power. For a 30-metric ton payload and structure system, a 20% reduction in power would increase the trip time by only about 5%. As the power is reduced below 5 GW, the trip time begins to lengthen rapidly, but obviously, a power level somewhat below 10 GW is acceptable if it is found that the cooling requirements at 10 GW impose too much mass. Of course, the reactor operating time increases proportionately with a decrease in reactor power.

It should be noted that none of the components of the fission fragment rocket requires a new technology, except for the organic moderator if that is used. In addition, a significant infrastructure development would be required to produce large amounts of ^{242*}Am. Less stressing missions, such as deep but rapid interplanetary travel, would be much easier, of course. Such missions, could be done with a plutonium-fueled, or maybe even a uranium-fueled, rocket. Indeed, we believe that with sufficient funding a prototype fission fragment rocket using ^{239}Pu as the fuel could be flown by the end of the century. In Figure 3 {Fig. 3.3} we show an artist's conception of a prototype fission fragment rocket. (Cha88)

3.0 High Energy Sources for Space

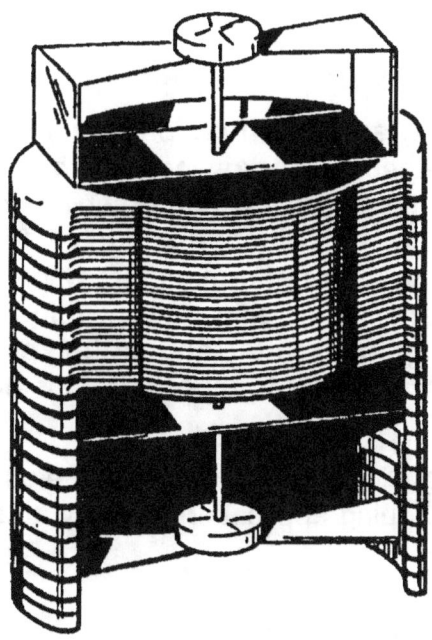

Fig. 3.3. Fission fragment concept.

The specific impulse is quoted as exceeding 10^6 seconds.

> A propulsion scheme utilizing this fact would allow acceleration of interesting payload masses to velocities approaching 10% the speed of light. For example, 100 t of fissile fuel might be used to accelerate a 500 kg payload to 1/20 the speed of light and reach Alpha Centauri approximately 100 years after launch. (Cha88)

For a trip to Alpha Centauri, the 15 GW reactor would operate for 25 years and coast for 75 years. This concept exhibits the level of performance which is needed for future missions.

Safety of the vehicle during developmental testing and operation is not discussed except to state that the reactor would not be operated until it is in a high Earth orbit where the fission fragments would not return to Earth. That design approach would obviously create concerns over the possibility of radioactive particles contaminating the Earth. For remote space flight missions, if the assumptions and preliminary analyses are valid, this would offer an attractive concept for high energy missions, i.e., where the exhaust contaminants would not be a concern. The fuel availability is another matter which has to be pursued as well as the environmental effects of fuel production since it will be necessary to produce additional quantities beyond the capacity made available from commercial reactors. The concept is discussed in further depth in Cha88-UCRL-99474 (Cha88).

3.4 MATTER-ANTIMATTER

Although matter-antimatter (proton-antiproton) mass reactions theoretically provide the greatest specific energy conversion (Table 3-1), the research requirements and the test demonstrations which remain to be performed in order to advance to the experimental state of fusion are of monumental magnitudes. The total system performance may prove matter-antimatter to be less advantageous than the specific energy release would indicate. Major difficulty in harnessing the matter-antimatter produced energy can be anticipated. It is a reaction that is inherently unsafe, reacting by contact as hypergolic propellants.

While there are very significant impediments to the use of $P\bar{P}$ as a space energy source, there is one intriguing application, namely, to serve as a fusion ignitor.

3.4.1 BACKGROUND

The concept of using antimatter reactions dates back to 1953 when Sänger proposed the use of antielectrons and electrons as reacting fuels (San53). The nuclear matter-antimatter annihilations are observed as a product of high energy particle physics where all of the matter is converted to energy, yielding $E=mc^2$, making possible a theoretically very light weight system at high energy levels.

The reaction produces energy in the following manner. Each proton-antiproton reaction produces 1,876 MeV energy in the form of relativistic neutral and charged pions. The neutral pion decays into high energy gamma rays. A significant percentage (approximately one third) of the $P\bar{P}$ energy appears in the form of hard gamma rays which are difficult to convert to thrust without using substantial mass for heat conversion. The charged pions decay into neutrinos and muons, each respectively carrying 275 MeV and 973 MeV of the pion's 1,248 MeV energy. The muon, as discussed in Section 4.0, decays into a positron, an electron, and 2 neutrinos. The positrons and electrons react, producing 0.5 MeV gammas.

Antimatter is extremely difficult to obtain. It has only been observed in nuclear reactions produced from very high energy accelerators. That is the only method by which researchers are able to obtain antiprotons for experiments. Antiparticles have not been observed naturally. Thus, antiprotons, as a fuel for this source of energy, would have to be manufactured, particle by particle which would not be cost effective to consider any time soon, even if suitable mass annihilation reactors had been designed and proven. The ratio of energy required to manufacture a unit mass of antimatter to the energy extractable from the reaction is about 10^4 to 1. Current estimates by some are projecting this to decrease to 100:1.

3.0 High Energy Sources for Space

From a safety viewpoint, it should be noted that if any of the antimatter ever came into contact with matter, the results would be devastating. Working with it should prove to be quite interesting!

Very limited study has been accomplished to perform thrust conversion directly from the reaction and also as a heat exchanger for a working fluid. Magnetic nozzles are employed for thrust. Magnetic containers and system components will be required to store, control, and work the antimatter. The engine's operation is described below.

> This rocket design concept uses a static magnetic field configuration in the shape of a conical rocket nozzle. The magnetic field is produced by the turns of a coil that increase in radius and separation so that the magnetic field lines form straight lines, all of which emanate from a common center on the axis. Within the field is space vacuum except for the antiproton beam, the hydrogen beam, and the annihilation products.
>
> The beam of antiprotons enters from the left and collides at a right angle with a beam of hydrogen coming from below. If the two beams are 2×10^{20} ions/s each, then 95% of the antiproton ions are annihilated.[7.23] The ion current in each beam is approximately 30 A.
>
> The charged pions produced by the $p\bar{p}$ annihilation follow paths that are along a cone whose vertex is the common center point of the magnetic field lines and whose surface is defined by the initial velocity vector of the pion. The vertex angle of the cone depends upon the velocity, charge, and mass of the pion and the strength of the magnetic field at the point of tangency.
>
> The dynamics of the motion of the pion in the magnetic field confines the pion to the surface of the cone. If the pion velocity vector is to the right, the pion will spiral out of the engine to the right and produce thrust. If the pion velocity vector is to the left, it will spiral toward the vertex of the cone, circle around just below the tip, then reverse direction and spiral back out to the right and exit the nozzle. Only the small fraction of pions with a velocity vector nearly parallel to magnetic field line at its point of origin will be able to travel up the throat and out of the engine the wrong way.
>
> The specific impulse of this engine is the velocity of the pions at their time of formation. For the mean kinetic energy of 250 MeV, this is a velocity of 94% the speed of light or a specific impulse of 28,800,000 s! The energy from the 30 A of antiproton ions will run this engine at the same power level as the three Space Shuttle Main Engines, 24 GW. With the high specific impulse, however,

3.0 High Energy Sources for Space

this 24 GW of power only produces 70 N of thrust. Such a design is probably best suited as the last stage in an interstellar probe design. (For83)

Refer to Fig. 3.4 for a description of the engine.

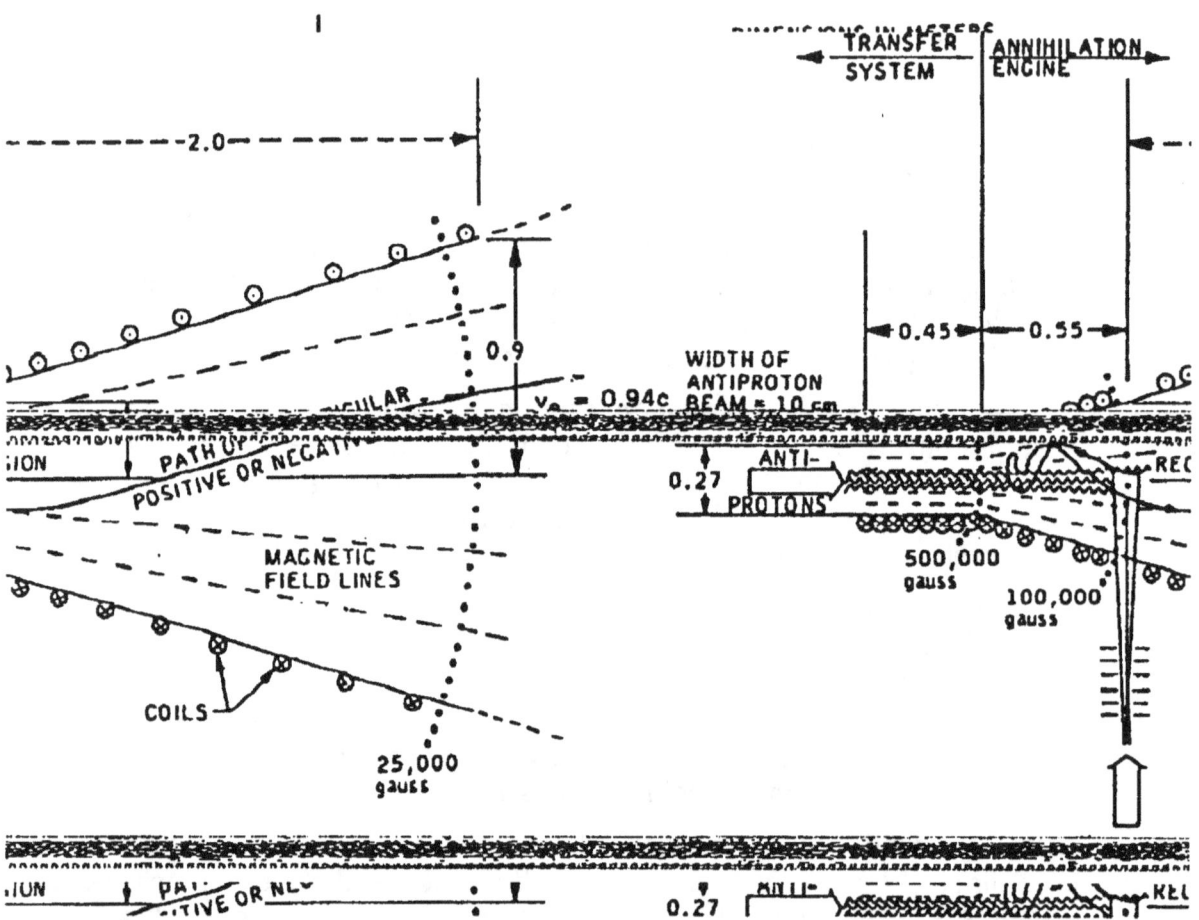

Fig. 3.4. High exhaust velocity matter-antimatter propulsion system concept. (For83, Fig. 7-9)

In any consideration of comparative evaluations of this nature, it is essential that all system aspects be taken into account. The greatest drawback from a space flight systems design utilization perspective is that 40% of the energy is in the form of gamma rays. That will decrease the design realization of the effective specific power potential.

Clearly, research should be performed to examine how real this fuel is as the "ultimate" energy source, but it should be considered only as conceptual at this time, nowhere nearly as developed as fusion.

The design solutions for a flight system comprise one reason why this energy source is not as attractive as indicated by the two order of magnitude increase in specific energy. Heat balance and shielding can be expected to impose very difficult conditions for a flight system using this form of energy. There has been some study accomplished to determine a technique to convert thrust directly from the reaction and also to serve in a dual mode as a heat exchanger for a working fluid. Magnetic nozzles could be employed for producing thrust. To store, control, and work the antimatter, magnetic containers and system components will be required. Flight performance, mission design analysis has not been conducted.

There is a particularly attractive feature of matter-antimatter which could benefit space, namely to serve as an ignition source for fusion. The high energy content, ease of ignition, and the high energy release would potentially favor a very light weight design that should make this concept at least worthy of study.

3.4.2 STATUS

Limited basic studies on that energy source are underway, primarily by the Air Force. Dr. Forward in 1983 completed a survey of advanced propulsion concepts which could show promise for application in the next century (For83). Among the concepts considered was matter-antimatter. In another activity, Dr. Borowski of the Aerojet Propulsion Research Institute presented a paper at the AIAA/SAE/ASME/ASEE 23rd Joint Propulsion Conference comparing fusion with matter-antimatter systems for interplanetary space travel (Bor87).

The concept is considered by Dr. Forward as "feasible," but "expensive." (For83,1-17) The first concern is the production of fuel. Methods for trapping and cooling of the atoms are also critical technology issues that remain. The antimatter must be retained in a stable magnetically levitated antihydrogen ice configuration. Inherent system design efficiency is a big question since there will be significant shielding needed to protect man and material from the gamma radiation. The studies reviewed did not reveal any specific power studies conducted for space flight. Two different operational modes have been examined: high thrust ~10^6 lbf coupled with low specific impulse, ~2,000 seconds and low plasma thrust with high specific impulse. The latter will be more difficult to achieve. The annihilation energy requires quick conversion of the plasma into propulsion energy.

One conclusion reached by Dr. Borowski was that "On the basis of preliminary fuel cost and mission analyses, fusion systems appear to outperform the antimatter engines for difficult interplanetary missions." (Bor87)

3.0 High Energy Sources for Space

The analytical understanding of $\bar{PP}$ energy conversion for a fight system is ostensibly nowhere as nearly developed as fusion. There will be significant cost and safety concerns with the development and use of antimatter. It is not clear that when system considerations are taken into account, whether the spacecraft system design will realize the higher specific energy yield, so this concept was not pursued further.

3.5 STRANGE MATTER

Strange matter, like antiparticles, have not been observed in a natural state. Their detection has only been through high energy particle physics. Strange matter consists of equal numbers of up, down, and strange quarks. There is a theoretical possibility that strange matter might become stable if it can be grown to a sufficient size. If so, it would provide a basis for a very compact source of energy. At the present time, only experiments are being suggested to validate the theory.

Strange matter is included to illustrate the current thinking and to indicate where advanced energy sources might be headed. The energy level would be high, quoted as <938 MeV (energy/baryon). A large percentage of the energy release will be in the form of gamma radiation which, as discussed earlier, is not as desirable as purely charged particles for space applications (Sha89).

3.6 OTHER

Dr. Forward's report, Section 3.4, had evaluated 64 concepts for advanced propulsion energy for space flight. The study concluded that 28 were sufficiently well defined for preliminary technical analysis. Four were recommended for more detailed study in the second phase of the contract: solid metastable helium, solar heated plasmas, perforated solar sails, and antiproton annihilation.

Solar heated plasmas:

Work has been performed on solar heated plasmas at the University of Washington. This concept uses solar energy to heat an alkali metal to a plasma which captures the solar energy and transfers it to a hydrogen working fluid. This concept is limited to a specific impulse of ~1,500 seconds. Further, it is limited to near sun solar system exploration applications and will not meet the high energy mission requirements of Section 2.0.

Perforated solar sails:

This is a concept proposed for a reduced mass solar sail, perforated with holes smaller than a wave length of light.

Solid metastable helium:

This concept uses lasers to make excited helium stable and magnetic fields for forming it into a room-temperature ferromagnetic. The theoretical impulse is calculated to be ~3,500 seconds, probably ~2,000 seconds in an operational engine.

Whereas the solar energy systems have many advantages from environmental, and perhaps from some safety perspectives, that source of energy is inadequate for the far distant missions where the solar energy is low. It could be a very useful concept nearer the sun. Those concepts were suggested as new ideas without any developmental research activities to back the performance of these systems. The payload mass delivery capability of those propulsion systems has not been addressed, which was not the study intent. Additional details and bibliography are available from the reference.

3.7 SUMMARY

Based upon the performance requirements of missions in this study, the conclusion drawn is that fusion is the most viable source of high energy for space programs. As a consequence, considerable effort was placed on further defining the unique potential of fusion energy in accomplishing future space missions.

Space fusion research, however, has received a negligible effort worldwide. The other basic energy sources, namely, fission, chemical, electrochemical, and solar, will continue to have space applications, and space research programs are appropriately being pursued for those energy sources. Matter-antimatter could theoretically exceed fusion in performance, but suffers from a less established data base and will be more difficult to produce a safe, economical, mass efficient flight system.

Opportunity exists for a United States to take a leadership role in space fusion research. The program would implicitly provide, too, an alternate confinement approach to the mainline DOE program.

4.0 GENERAL DISCUSSION OF FUSION REACTIONS

A brief discussion of the main principles of fusion energy is provided to present fundamental principles and key parameters used in the report.

4.1 PRIMARY NUCLEAR FUSION REACTIONS

In fusion reactions, under the right set of conditions, light weight nucleons join to form other nucleons, referred to as "ash." Some of the ash is burned in secondary reactions although this is usually a small contributor to the total fusion power. The conversion of mass to a specific quantity of energy is determined by the mass loss between the initial reacting mass and the residual rest mass of the reaction products in accordance with the equation, $E = mc^2$. The energy appears as kinetic energy of charged particles and/or neutrons depending upon the fuels selected for the reaction. The challenge in achieving controlled fusion has been in designing a satisfactory confinement scheme to contain the high temperature plasma (10^8-10^9 °K) in such a fashion that a net positive yield of energy results. The status now is that we have currently reached 4×10^8 K. Other conditions must simultaneously be met, however, before net power controlled fusion occurs, as discussed below.

The number of nature's elements which will fuse are indeed quite numerous. To fuse nucleons, several conditions must be met. Sufficient kinetic energy must be imparted to the ions to overcome the mutually repulsive Coulomb forces and to penetrate their respective nuclei. Hence, the quantity of energy to initiate fusion reactions is large, requiring greater than 10 keV in comparison with a few eV to initiate chemical reactions. Whether or not two nuclei fuse is a statistical matter of nucleons colliding at the proper point of impact and with a sufficiently high energy (velocity) to result in nucleon penetration. The reaction rate parameter of interest here is $<\sigma v>$ which is the average product of the fusion reaction's nuclear cross section (σ), cm^2, and the relative ion velocity (v), cm/sec. The reaction rate parameter is plotted as a function of plasma temperature in Fig. 4.1.

4.0 General Discussion of Fusion Reactions

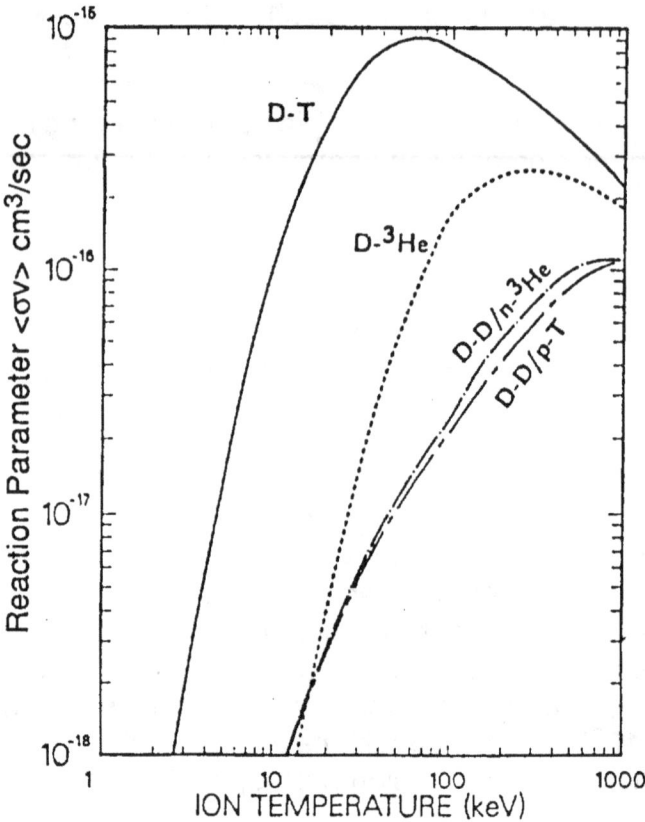

Fig. 4.1. Reaction parameters and cross sections for various fusion reactions (Mil76).

The product of the reaction rate coefficient with the energy per reaction determines the energy density. The plasma, in addition to being driven to high ion temperatures, must be confined for an adequate time (τ) at a sufficiently high ion density (n) to sustain burning. The confinement figure of merit of a plasma is usually measured by the parameter $n\tau$ (Lawson criteria), where n is the plasma density (ions/cm^3), and τ is the energy confinement time (s).

Neutrons, as typical reaction products, are immediately lost from the plasma without a transfer of energy to the plasma. The charged fusion products, i.e., ions, are slowed by the background plasma, and their energy then serves to heat the plasma and any cold fuel input. When the product of fuel confinement time and fuel density ($n\tau$ product) is sufficiently large ($\geq 5 \times 10^{14}$ cm^{-3}-s for D-T and $\geq 2 \times 10^{15}$ cm^{-3}-s for D-^{3}He, for example), the charged fusion product heating can balance plasma energy losses from conduction, convection, and radiation as bremsstrahlung and synchrotron radiation. When this condition occurs, the plasma is said to be ignited and the burn can proceed without further input of energy from external auxiliary heating systems. The approximate distribution of energy lost from the plasma in the form of neutrons and radiation for the three main fusion fuel cycles of interest for space is shown in Fig. 4.2 (Men89, San88) and compared with the charged particle energy available.

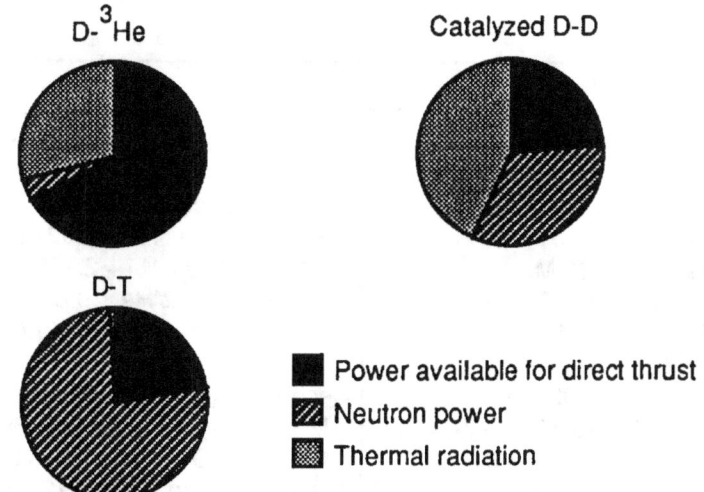

Fig. 4.2. Approximate distribution of energy among charged particles available for direct thrust, neutrons, and thermal radiation which appears as surface heat.

Let us now pursue how these energy sources can be made available to do useful work. Only the charged particle power is available for efficient direct thrust. To be useful, the energy from radiation and neutrons must be converted to electricity by a less efficient thermal cycle and used to drive an ion or plasma thruster, as in fission electric propulsion systems. Waste neutron heat must be radiated to space by the spacecraft's radiators. Hence, for space applications, neutrons are an impediment to the required high performance properties since disposal of energy wastes requires a radiator mass. Some of the heat may be extracted for a small on-board electrical power mainframe capability. The goal would be to maintain the highest temperature to keep the radiator mass as low as possible.

Analyses of losses are essential since they can determine whether the reaction can ignite and sustain burning. From an efficiency viewpoint, one would like to achieve the maximum fuel burn-up in the smallest space possible. The actual amount of fuel burned is strongly dependent upon the reactor design, and, to give an indication, approximately 5% to 30% of the fuel is consumed in the reaction.

For magnetically confined fusion devices, the fusion power density scales roughly as $(\beta^2 \times B^4)$, where β is the ratio of plasma pressure to magnetic field pressure, and B is the magnitude of the magnetic field. The particular value of ion temperature, T_i, and $n\tau$ for ignition depends primarily on the chosen fuel cycle and, to a lesser extent, on the reactor configuration as discussed in Section 8.0. The parameter, β, however, is a plasma energy density parameter

4.0 General Discussion of Fusion Reactions

more closely tied to engineering design considerations; and the reactor's β capability varies widely among plasma confinement concepts.

As mentioned earlier, there are numerous light element fusion reactions which can occur (Mil76 & McN82), Table 4-1a.

TABLE 4-1a. Fusion Reactions (McN82)	
Deuteron-Based Fusion Fuels	Proton-Based Fusion Fuels
<u>Primary Reactions</u>[a] $D + T \rightarrow n + {}^4He + 17.586$ MeV (3.517 MeV) $D + D \rightarrow p + T + 4.032$ MeV (4.032 MeV) $D + D \rightarrow n + {}^3He + 3.267$ MeV (0.817 MeV) $D + {}^3He \rightarrow p + {}^4He + 18.341$ MeV (18.341 MeV) $D + {}^6Li \rightarrow 2{}^4He + 22.374$ (22.374 MeV) $D + {}^6Li \rightarrow p + {}^7Li + 5.026$ MeV (5.026 MeV) $D + {}^6Li \rightarrow n + {}^7Be + 3.380$ MeV (0.473 MeV) $D + {}^6Li \rightarrow p + T + {}^4He + 2.561$ MeV (2.561 MeV) $D + {}^6Li \rightarrow n + {}^3He + {}^4He + 1.796$ MeV (~1.134 MeV) <u>Secondary Reactions</u> $p + T \rightarrow n + {}^3He - 0.765$ MeV (–) $T + T \rightarrow 2n + {}^4He + 11.327$ MeV (~1.259 MeV) $T + {}^3He \rightarrow n + p + {}^4He + 12.092$ MeV (~6.718 MeV) $T + {}^3He \rightarrow D + {}^4He + 14.319$ MeV (14.319 MeV) ${}^3He + {}^3He \rightarrow 2p + {}^4He + 12.861$ MeV (12.861 MeV)	<u>Primary Reactions</u> $p + {}^6Li \rightarrow {}^3He + {}^4He + 4.022$ MeV (4.022 MeV) $p + {}^9Be \rightarrow {}^4He + {}^6Li + 2.125$ MeV (2.125 MeV) $p + {}^9Be \rightarrow D + 2{}^4He + 0.652$ MeV (0.652 MeV) $p + {}^{11}B \rightarrow 3{}^4He + 8.664$ MeV (8.664 MeV) <u>Secondary Reactions</u> ${}^3He + {}^6Li \rightarrow p + 2{}^4He + 16.880$ MeV (16.880 MeV) ${}^3He + {}^6Li \rightarrow D + {}^7Be + 0.113$ MeV (0.113 MeV) ${}^3He + {}^3He \rightarrow 2p + {}^4He + 12.861$ MeV (12.861 MeV) ${}^4He + {}^9Be \rightarrow n + {}^{12}C + 5.702$ MeV (0.439 MeV) ${}^4He + {}^9Be \rightarrow n + 3{}^4He - 1.573$ MeV (–) ${}^4He + {}^{11}B \rightarrow p + {}^{14}C + 0.784$ MeV (0.784 MeV) ${}^4He + {}^{11}B \rightarrow n + {}^{14}N + 0.158$ MeV (0.011 MeV) $p + {}^{10}B \rightarrow {}^4He + {}^7Be + 1.147$ MeV (1.147 MeV)

Fig. 4.1a. [a]Energy release is $Q(Q_+)$, where Q is total energy release including the energy of the neutron and Q_+ is the charged-particle energy only (p = protium, D = deuterium, T = tritium, n = neutron).

We shall be concerned primarily with just three reactions, i.e., those listed in Table 4-1b, during the discussions on space energy fusion fuel applications.

TABLE 4-1b. Fusion Reactions for Consideration for Space Applications

A. The most important fusion reactions for space applications

1. $D + {}^3He$ = p (14.68 MeV) + ${}^4He$ (3.67 MeV)
2. $D + D$ = n (2.45 MeV) + ${}^3He$ (0.82 MeV) (~50%)
       = p (3.0 MeV) + T (3.67 MeV) (~50%)
3. $D + T$ = n (14.07 MeV) + ${}^4He$ (3.52 MeV)

B. Aneutronic Reactions

4. $p + {}^{11}B$ = 3 ${}^4He$ (8.7 MeV total)
5. ${}^3He + {}^3He$ = 2p (5.7 MeV each) + ${}^4He$ (1.4 MeV)

C. Potential side reactions (for reference)

6. $p + p$ = e^+ + D + 1.42 MeV
7. $T + {}^3He$ = D (9.5 MeV) + ${}^4He$ (4.8 MeV) (41%)
         = p (5.4 MeV) + ${}^4He$ (1.3 MeV) + n (5.4 MeV) (55%)
         = p (10.1 MeV) + ${}^4He$ (0.4 MeV) + n (1.6 MeV) (4%)
8. $T + T$ = 2n (10.03 MeV) + ${}^4He$ (1.3 MeV)

For the remainder of the elements, the reactivity is low, i.e., there is a low reaction cross section associated with reactions that were not selected. Additional reactions are shown in the table for reference since they comprise unintended side reactions of the primary fuels or of their reaction products. The p-^{11}B and ^{3}He-^{3}He reactions are noteworthy because they are purely aneutronic, that is, occurring without the production of neutrons, although there is a trace of radioactive ^{14}C present in the p-^{11}B reaction. The ignition temperature of p-^{11}B is 300 MeV, a level where bremsstrahlung radiation losses become important, and also increased by higher atomic number. Use of these fuels would be predicated upon extensive study and research. However, the reaction of these ions is only marginally sustained against radiation losses due to bremsstrahlung and as a result are not likely to achieve energy breakeven, i.e., they would need to be driven with a perfectly efficient external energy source and are therefore not suitable for space energy generation.

Each fuel combination has its advantages and disadvantages. The D-^{3}He reaction has the major advantage of producing a large percentage of its reaction products as either of the two charged particles, i.e., a proton and alpha particle. As shown by equation (1) it is nearly an aneutronic reaction while still having sufficient reactivity to ignite. The source of neutrons results from the statistical side reaction of D-D. The D-^{3}He reaction is calculated to be at least 95% neutron free under typical reaction conditions. It can be over 99% neutron free if some advanced reactor configurations prove to be sound.

The D-^{3}He reaction, however, is more difficult to ignite than D-T. Another consideration is that ^{3}He is not readily available on Earth and will require lunar mining or other expensive alternative sources (Wit86). Lunar mining operations were discussed in a dedicated ^{3}He workshop held at Cleveland, "The NASA Lunar Helium-3 Fusion Power Workshop" (Anom88). Lunar ^{3}He mining has been extensively studied at the University of Wisconsin Center for Space Automation and Robotics (WCSAR). Studies indicate feasibility and practicality.

The mainline terrestrial Department of Energy (DOE) reactor, the tokamak, is designed to burn D-T. That fuel cycle is the easiest to ignite, and the fuels are relatively available on Earth. Drawbacks of using tritium are its radioactivity (12.3 year half-life) and the fact that it produces 80% of the released energy in energetic neutrons. Deuterium is readily extracted from sea water, and tritium is made from neutron reactions with ^{6}Li in a blanket surrounding the D-T plasma in the blanket. The Aries III study early in 1991 showed that the advanced tokamak will not burn D-^{3}He.

The basic D-D reaction (equation 2) is not attractive due to a low fusion power density and a significant neutron production. To provide additional energy for D-D, a "catalyzed" D-D (cat-D) fuel cycle has been suggested in which the D-D reaction products are assumed to be recovered and reinjected into the plasma until complete consumption is achieved. That reaction can reach power densities comparable to that of D-^{3}He, but it still produces a large fraction of its energy as neutrons. This is a theorized cycle which lacks a plan for demonstration at this time.

The Air Force Studies Board for the National Research Council reached the conclusion that

> ... some fusion systems potentially offer specific powers (kilowatts per kilogram) and total power characteristics that make them candidates for both space power and space propulsion applications" and that "D-^{3}He clearly offers the best combination of low neutron yields and relatively modest confinement conditions. (Mil87)

Ideally for space, one would like to achieve a purely ionic product, i.e., one totally devoid of neutrons. That is desired since efficient direct thrust and power conversions are made possible from charged particles, and since the freedom from neutrons eliminates the energetic, damaging particles which have potential for energy conversions in space only by the generation of heat, an inefficient design solution for space flight. Neutron bombardment of the surrounding first wall structure stimulates radioactive product generation. This bombardment results in deep first wall penetration by neutrons. That deep penetration will displace the first wall's atoms from their normal position causing the wall material to ultimately lose its strength. This system aspect will be accounted for on the terrestrial reactors by a maintenance and logistics program which for many space applications will not be available due to lack of access;

and, for the remainder which are accessible, replacement will always be difficult, hazardous, and expensive. Secondary reactions of the neutron exposed material also creates a new hazard.

The life time for a reactor operating in space without maintenance will, therefore, be a function of the neutron flux and the mission duty cycle. Any D-T system will require either the installation of massive shielding and/or separation of neutron and radioactive sensitive equipment, including man, from the radioactive sources. D-^{3}He ameliorates, but does not totally eliminate this problem. This is a topic oriented more towards design solutions rather than reactions and is discussed in Section 7.0. In spite of the problems with neutrons, the first three fuel cycles are likely to remain the reactions of choice due to their greatest reactivity.

4.2 OTHER NUCLEAR FUSION REACTIONS

In the interest of being comprehensive in this discussion of nuclear fusion reactions, two other fusion reaction categories should be noted, but they are not recommended for space at this time.

Muon catalyzed reactions provide a very attractive option for cold temperature ignition of fusion fuel (Raf87). In principle, during these reactions a muon attaches to the hydrogen atom in lieu of an electron, the objective being to increase $<\sigma v>$ via a reduction in the Coulomb forces. Because the muon carries a charge equivalent to the electron, but with a mass that is 207 times greater, it in effect causes the hydrogen atom to assume the electrostatic appearance more of a neutron due to Bohr radius contraction, reducing the requirement for high ion temperatures. The reaction has actually been demonstrated at liquid hydrogen temperatures in a laboratory environment as early as 1958. Muon catalyzed fusion appears valuable only with D-T as fuel at the present time. Muon catalyzed D-^{3}He has been examined, but the resulting atom is too large to sufficiently enhance the reaction rate, since the atomic number of helium is 2.

One limitation is the short half-life of the muon which decays to an electron and positron plus two neutrinos in 2×10^{-6} second. This short time period is sufficiently long for about 150 fusion reactions to occur (Jon86). The second limitation pertains to relative energy output, or power gain, from the system. Muons are generated from π-mesons produced commonly by high energy protons (300 MeV) from high energy particle accelerators. The formation of muon hydrogen atoms is statistical, that is, the capture of the muon would have to occur in a cloud of deuterium and tritium. In principle, a high muon flux is generated, sufficient to allow capture and to allow the fusion reaction to occur within the muon's half-life. Subsequent to the fusion reaction, the muon is released and allowed to attach with another nucleon, an event which occurs about 150 times during the life of the muon. The concept, although attractive, is not practical for space consideration at this time since the energy to make

4.0 General Discussion of Fusion Reactions

muons is substantially greater than the amount released by all the muon catalyzed reactions which are essential in achieving a self sustaining power balance. The ultimate success of this form of cold fusion may be determined by the development of an efficient light weight accelerator.

Another approach involves **spin polarized reactions**. Nuclear reaction cross sections are a function of the nucleon spin just prior to the reaction. The plasma reactivity for D-T and D-^{3}He is calculated to be increased by 50% using polarized nucleons, i.e., the D-^{3}He reaction seems to have the same spin dependence as D-T. Of great interest, too, is the potential to depress the D-D reaction, and hence, neutrons. This, of course, has a very positive benefit on reducing system mass. Studies are being performed to efficiently produce polarized fuels that would provide an overall net benefit to fusion.

> The atomics part is aimed at establishing efficient methods for producing coherently polarized fuel nuclei and for bringing them inside the plasma, where collisions and wall effects must be sufficiently weak so as to insure that the depolarization time is sufficiently longer than the particle fusion time (see e.g. [2]). The conclusion was reached that, if appropriate materials can be used, sufficiently long depolarization times may be hoped for. At the same time the necessity was stressed of a direct experimental test under fusion conditions. (Peg87)

The initial thought was to effect a reduction of D-D reactions by 100; now it appears that only a factor of 2 reduction is achievable, although some controversy remains.

This fusion concept has received significant consideration and investigation by Dr. Kulsrud of the Princeton Plasma Physics Laboratory. It was proposed in 1982. During the meeting in Italy in which muon catalysis and polarization were discussed, the conclusion reached in the Kulsrud advanced fuel paper (Kul87) was that

> Of all the possible benefits from spin-polarizing the fusion plasma the most attractive would be the suppression of the neutrons to make a nearly neutron free reactor. It is unfortunate that nuclear physics is not able at this time to tell us definitely whether this is possible. It is hoped that new experiments will lead to a resolution of this question.

The conclusion reached here is quite clear that polarization cannot be counted on at this time. If it does become a reality for the terrestrial application, system studies would be required to examine the net benefit to space applications. Recirculation power requirements will be a topic of great interest to the space program if spin polarization is proven beneficial.

4.3 SUMMARY

For space flight applications where mass is always at a premium, the most efficient systems are those permitting the minimal system mass. For fusion the theoretical physics has been well established, the difficulty being in the physical attainment of extremely challenging physical parameters. Of the various nuclear fuels possible, the space application benefits greater than the ground program from reactions that maximize the production of charged particles and minimize neutrons. The optimal fuel is, therefore, considered to be deuterium and helium-3.

5.0 THEORETICAL PERFORMANCE CAPABILITY OF FUSION ENERGY CONVERSION FOR SPACE

Fusion propulsion and power systems are theoretically capable of continuous operation at high energy levels. In fact, they are large energy producing devices of the megawatt and greater category only, as envisioned presently, and generally are not currently considered suitable for small energy applications where the more conventional systems will yield better, more economical performance characteristics. Fusion energy has five key desirable performance characteristics with respect to space energy applications:

(1) very high specific energy and high power density, exceeded only by matter-antimatter annihilation,

(2) moderately high thrust levels,

(3) variable high specific impulses which can range up to the limit as determined by the fusion product energy,

(4) high levels of efficient electrical power production, and

(5) creation of much more energy than is required to produce the fuel.

(1) <u>High specific power</u>:

The specific energy for the D-^{3}He fusion reaction is 3.52×10^{14} J/kg. As shown in Section 2.0, high specific power and variable high specific impulse are mandatory characteristics for propulsion systems to be used for practical interplanetary and interstellar space travel. The potentially high specific power capability of fusion (≥ 1 kW/kg) leads to attractive parameters for space. One estimate shows that a Spherical Torus Tokamak design could yield a specific power of 5.75 kW/kg, while a Spheromak Compact Toroid could provide 10.5 kW/kg; both designs assumed that neutrons could be suppressed by using polarized D-^{3}He fuel. A conceptual Inertial Confinement Fusion (ICF) system of producing 100 kW/kg (Bor87) has been designed at a 20 GW power level; it burns D-T. More detailed analysis to determine potential specific power capabilities for space would be beneficial.

(2) <u>Moderately high thrust</u>:

While operating in the thrusting mode, the maximum thrust values of MCF reactors are ultimately limited on the high thrust end by the low plasma mass density of approximately 10^{15} ions per cc that currently can be maintained in a magnetically confined fusion (MCF) reactor. By the addition of propellants to the plasma, the fusion reactor can yield higher thrust from the higher mass flow rate but at a decrease in specific impulse. That characteristic is beneficial for space applications. The capability to vary the specific impulse offers advantages for mission mass optimization. For aeronautical flight higher thrust and specific power are needed as discussed in Section 2.0.

The plasma power density, determined by the reactor's design characteristics for a given confinement scheme, is approximately proportional to the square of the quantity, β. β defined as the ratio of the plasma pressure to the magnetic field pressure, and thus high beta designs are inherently favored for space applications where high power densities offer mission performance advantages.

Conversion of fusion plasma energy to thrust using the D-^{3}He fuel cycle has received attention from many researchers. In a recent report on the results of an analysis of a linear tandem mirror reactor, Dr. Santarius at the University of Wisconsin shows the thrust to mass ratio of 3×10^{-4} to 3×10^{-2} being attainable (San88). The reactor size is 2000 MW, and the jet power per unit power system mass is 1.2 kW/kg. That power level would produce thrust levels on the order of 0.1 to 40 kN. The propulsion system is long (113 m) and massive (1250 MT). As mentioned earlier, the physics of fusion energy utilization scales towards large sizes. Refer to Fig. 7.7 for a scaled schematic drawing of this propulsion engine.

Higher mass fraction (payload mass to the initial vehicle mass) designs for fusion vehicles are possible since the amount of fuel required is smaller than other concepts. A large part of the reactor's total mass in MCF configurations results from the large magnets required for plasma containment and for the magnet's load carrying structure. An economical space reactor may be designed using stronger light-weight load carrying materials not suitable for mass insensitive commercial power. Upon the development of efficient driver technology, inertial confinement (ICF) concepts are expected to provide an alternative confinement concept. That achievement, however, requires the advent of small, lightweight, and higher efficient drivers.

The confinement time, $n\tau$, during which the plasma is maintained is an important parameter. As the thrust levels increase, the fuel burn-up becomes less complete and, therefore, less efficient with decreasing confinement times. Confinement time under high thrust level operation is a research subject which the space program will have to undertake to obtain for its use.

As in all propulsion systems, for a given power level there is a trade-off between specific impulse and thrust (Eng62). This performance trade is shown for a typical fusion propulsion system design in Fig. 5.1 (San88) in terms of specific impulse versus the ratio of thrust-to-weight.

5.0 Theoretical Performance Capability

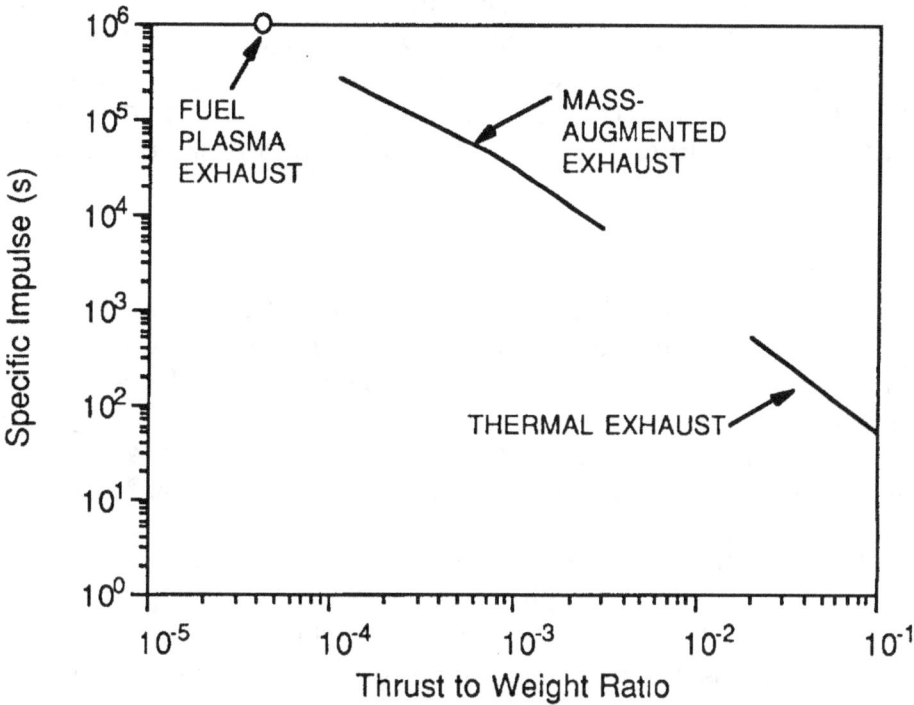

Fig. 5.1. Specific impulse versus thrust to weight (San88).

This range of specific impulses and thrust-to-weight values optimizes M_o and allows either fast, manned missions or cargo missions with high payload fractions. Greater mission operational flexibility and mission efficiency can be attained.

An important aspect of fusion propulsion systems is, thus, the inherent design feature of permitting planners to tailor a thrust and specific impulse program to any given mission. The missions discussed in Section 2.0 are considered as specific impulse unlimited missions in which the reactor can produce the optimal specific impulse to minimize the total propellant and propulsion system mass (Moe72). They are moderately low thrust powered vehicles that utilize low specific impulse at higher thrust levels early in any given mission while increasing the specific impulse and decreasing the thrust as the flight progresses in order to maintain a constant acceleration, the approach used in Section 2.0 to fly those missions. The values of specific impulse shown are the averaged values over the mission, unless specifically stated to the contrary.

Fusion energy meets those requirements. Fusion propulsion systems for different missions are likely to be similar in design but will be tailored to operate in somewhat different modes so as to accommodate the specific impulse variations. The means to accomplish thrust augmentation and specific impulse variations by a uniform mixing of the effluent into the

5.0 Theoretical Performance Capability

plasma exhaust is, therefore, a key ingredient in the development of fusion propulsion to implement this capability. Variable thrust modes, along with specific impulse, are discussed under the specific impulse subsection which follows. The generation of thrust and throttling capabilities, at highly efficient specific impulse values, are very key, specific space research activities which a terrestrial fusion energy program will not pursue.

(3) <u>Extremely High specific impulse</u>:

The high specific impulses available from plasmas and the possibility of continuous thrust operation were among the main reasons that, over thirty years ago, a small research effort on fusion energy for space application was begun by NASA (Mas59). Refer to Appendix A (Sch91). This space research activity, primarily at the NASA Lewis Research Center but including independently funded work at Aerojet-General Nucleonics and at the Air Force Office of Scientific Research, made a number of important contributions. Very significantly, the D-^{3}He fuel cycle was identified as one of the most attractive options (Eng62). Much earlier, the benefit of continuous thrust, even at low acceleration levels, was recognized in 1929 by Dr. Oberth as allowing substantial gains in payload mass fraction and mission duration. A comparison of fusion propulsion performance characteristics with chemical propulsion systems in terms of selected mission applications is given in Section 2.0. In addition, it was considered important for comparative purposes to perform a similar set of calculations for a vehicle having a propulsion system designed to a specific power expected to be ultimately produced by a nuclear electric propulsion system (NEP).

Three separate reactor operational modes will be designed to vary thrust and specific impulse: direct plasma exhaust, mass augmented (diluent) plasma exhaust, and thermal exhaust, (Fig. 5.2, San88).

5.0 Theoretical Performance Capability

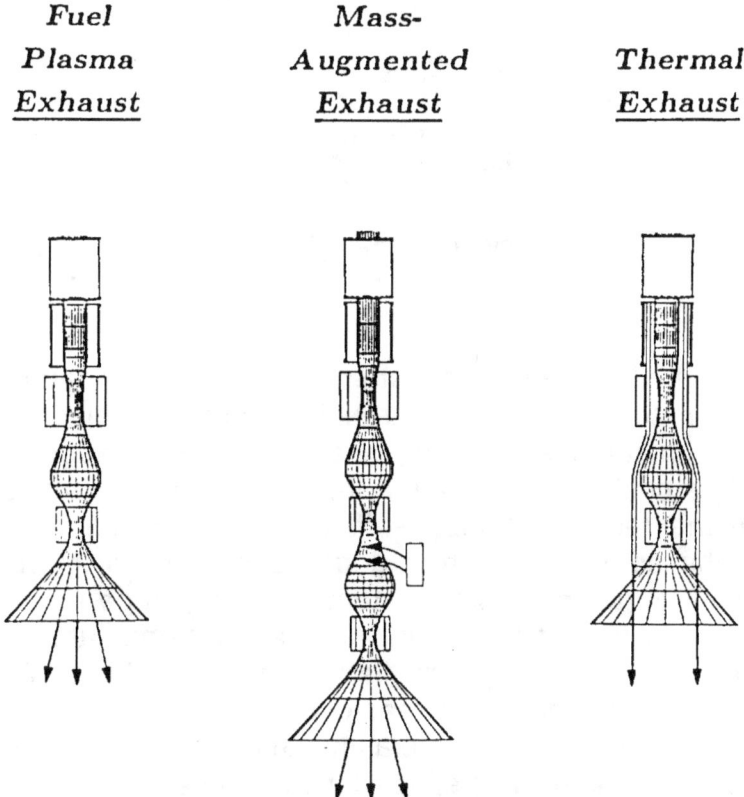

Fig. 5.2. Performance operational modes for a linear fusion propulsion system (San88).

The plasma temperature in the power-producing region of a fusion reactor will range from about 10 keV to 100 keV (1 keV≈10^7K), while fusion products will have energies up to 14.7 MeV. In the Tandem Mirror reactor shown in Fig. 5.2, a reduced electrostatic potential at the nozzle end allows some high energy plasma to escape directly, yielding very high specific impulses as a result of the high temperature plasma. Specific impulse varies as the square root of the temperature.

$$\text{Isp} = C_F c^* = \frac{C_F}{g}\left[\frac{\sqrt{2gRT}}{k\sqrt{[2/(k+1)]^{(k+1)/(k-1)}}}\right]$$

The value for the direct mode specific impulse is approximately 10^6 seconds. Because of the relatively low ion density of the plasma (10^{15} ions/cc), the reactor's mass flow rate, and therefore the reactor thrust level, is low.

For ready reference the key equations of rocket performance are provided below:

5.0 Theoretical Performance Capability

$$F = \frac{d}{dt}(mv) = \dot{m}v$$

$$Isp = \frac{F}{\dot{m}g} = \frac{v_e}{g}$$

$$P(W) = 1/2\dot{m}v_e^2 = \frac{g}{2}FIsp$$

For producing thrust a low field strength magnet is added to the exhausting end of the reactor and gasses are injected into the plasma to produce the mid-range thrust and specific impulse, i.e., the mass augmented operational mode. This cools the plasma but increases thrust by the larger mass flow rate. In the reactor's "scrape-off layer," the plasma's boundary which protects the core plasma from degradation due to interactions with low energy neutral particles, plasma temperatures range from about 1 eV to 1 keV. In magnetic confinement fusion, these plasmas are insulated from the first wall material by magnetic fields which can be designed to provide direct thrust. By varying the reactor's operating mode through the direct injection of a diluent into the plasma and exhausting the cooler plasma through a magnetic nozzle, specific impulses can be selected within a wide range of values from about 5,000 seconds to 1,000,000 seconds (Englert). "Diluent" refers to the inert propellant mass in this report to distinguish between the two nuclear fuels and a non nuclear (inert) reacting fluid.

In the high thrust, low specific impulse mode, the low end range of specific impulse can be extended down into the 100's of seconds regime by using the fusion energy thermally. Gases flowing over a heat-exchanger are heated and expelled as in the NERVA type propulsion systems or as heat exchangers in a chemical propulsion system. As with those systems, the thermal fusion propulsion mode's temperatures are limited by material properties to about 1200K.

Refer to Fig. 5.1 (San88) for a description of the variation of specific impulse. If part of the scrape-off layer (low energy plasma outside of the core) is exhausted directly, there will be only a small impact on the plasma power balance; however, if the core plasma is partially exhausted, the energy confinement requirements for the remaining core plasma will become more stringent. The understanding of mixing of propellants (thermalization) with the fusion plasma is an undeveloped technology, but a very critical one for the use of fusion propulsion. Research and development are needed to assure good mixing of the diluent and high temperature plasma.

To compare with fission, the energy produced by fusion is predicted to be at high power levels (>100 MW) and to gain strongly by economy of scale. Multigigwatt reactors will be well within the realm of fusion reactors. Some estimates indicate that it may be possible to have reactors as small as 10

5.0 Theoretical Performance Capability

MW (Rot89), although this is a topic which requires considerable study and experimentation. Others in the field would consider this too low. Fission energy being developed for space now is on the order of 100 kW with an upper limit of several megawatts considered achievable for space in the SP 100 approach. The NASA fission reactor is designated as the SP-100 reactor to indicate a 100 kW power output for a space power reactor application. A system using nuclear fission generated thermal power (800-900 seconds) and electric propulsion offer an attractive option from an Isp standpoint (10^3 to 10^4 seconds), but the conversion inefficiencies and large inert mass requirements may render it less effective from a systems perspective. Its specific power is on the order of 0.03 kW/kg with projections to the future research work increasing the value to 0.067 kW/kg. The Isp limit is ~10,000 seconds whereas fusion is considered capable of delivering an Isp up to 10^6 seconds and a specific power of 0.5 kW/kg to 10.0 kW/kg at power levels above 100 MW.

(4) <u>High electric power</u>:

Increased electrical power requirements can be anticipated for future space missions. Typically, the amount of energy for electrical power will be much less than that needed for thrust. Power levels and uses are the subject of Section 2.0. A system trade study needs to be performed to analyze whether a relatively low mass system can be added to the flight vehicle for conversion of a part of the "waste" heat into use for powering the electrical system aboard the spacecraft as an option to provide on-board maintenance power during powered flight.

Various methods of converting fusion energy directly into electricity have been investigated. For example, similar to the ability to use fusion energy for directly producing thrust, a portion of the reactor's plasma can be diverted to produce electrical power directly by having ions intercept conducting plates of selected voltage (Pos69, Bar83). The plate electrostatically slows plasma ions directly converting their kinetic energy into electricity (Fig. 7.17). Electrons will have been reflected by a negatively biased grid. Another option is to directly convert synchrotron radiation energy, in the form of microwaves, from the plasma by the use of rectennas (Log86). These direct electrostatic conversion fusion electrical power systems are high megavolt voltage, high current (tens to hundreds of amps) systems.

The interest in charged particles from fusion reactions is based upon the capability to convert the particles directly to electricity, thereby achieving high efficiencies resulting in improved specific power vehicle designs. Compare the relative efficiencies (rejected power to output power) in Fig. 5.3.

5.0 Theoretical Performance Capability

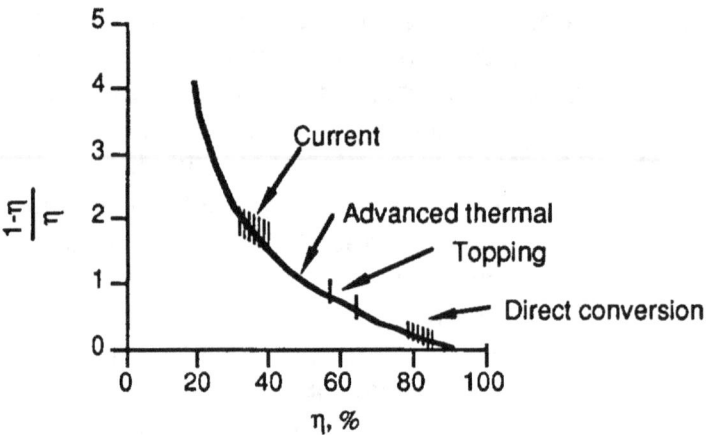

Fig. 5.3. Reduction of waste heat with increasing efficiency. Here $1-\eta/\eta$ represents the rejected-to-output power ratio. (Fig. 1.4, Mil76).

The mass required to convert neutron energy to thermal energy, and the thermal conversion to electricity results in a design having approximately a 40% efficiency level. Direct conversion has been researched at Livermore. The single stage design reported in Bar83 produced a net efficiency of 48%. Efficiency is increased by the addition of stages. Efficiencies of 60% to 70% are predicted (Mor73). More experiments and testing are essential to realize the high efficiency of direct conversion. A recent paper on this subject states that

> ...The path to high efficiencies is not at all simple or straightforward. Every percent increase above the current 49% power plant value will be difficult to obtain. ... Thus ultra high efficiencies (above 60%) will require almost perfect mating of the components. (Per88)

One area for increasing the efficiency is to devise methods to collect electrons also for direct conversion.

An important consideration for scientific outposts will be the choice of a dual function reactor. In this proposed concept, the fusion system power level can be reduced for the propulsion mode, thereby operating the reactor at a lower plasma density and negating the need for an alternate energy system at low power operation, once the mission destination has been reached. Some fusion system configuration options may allow switching between propulsion and power production at a modest penalty in total mass and complexity, although no detailed design with this objective in mind has yet been accomplished. This is a new consideration.

(5) <u>Positive energy production</u>:

Fusion reactions produce net power gains, in contrast with matter-antimatter annihilation reactions, where large amounts of energy (4 orders of magnitude) must be expended to create a small amount of energy as a space fuel. Thus, with fusion the burden on terrestrial power plants or other space power sources is greatly reduced. In addition, fusion fuel sources in the outer solar system exist, in particular, the gas giant planets and their satellites. Fuels which are to be applied to the advanced space missions should not be a planet Earth monopolized fuel, that is, we must be capable of using those fuel resources and making them available from other local planetary resources.

5.1 SUMMARY

Fusion, with vastly reduced radioactivity problems in comparison with fission, plus its characteristics of higher specific power, higher thrust, and more efficient electrical conversion system, warrants significant attention as an energy source for space missions. It is the most attractive form of high energy release systems presently known since significant cost and safety implications with matter-antimatter render it a less desirable alternative for the two order of magnitude increase in specific energy gained. That advantage may not be usable when one takes into account the system mass penalties which will ensue once the entire set of system requirements have been determined.

6.0 FLIGHT SYSTEM CONSIDERATIONS AND REQUIREMENTS

Operational simplicity, economics, and practicality are paramount for use of any space flight system. An overly complex system from an operational perspective is not likely to be accepted unless the performance is absolutely mandatory. Operational complexity was a major reason for rejection of the NERVA during the 1960's. Another was inadequate performance gains for the applications considered. The missions could be accomplished more simply and less costly using chemical propulsion. Hence, the importance of the functional and flight operational attributes expected of a fusion engine is sufficiently great that flight vehicle system design and flight operational factors must be addressed at concept initiation.

Fusion technology has advanced, as discused later in this report, such that confidence in its ultimate capabilities was developed in the course of this study. This confidence must be tempered to the degree that one can make judgments on a preliminary evaluation of the status, from projections of the performance that is ultimately anticipated and from the capability of that performance level to meet requirements for the integration of a fusion powered propulsion/electrical power system into the flight vehicle.

We have seen in Section 2.0 a range of missions requiring high energy levels. We examined the nature of fusion reactions in Section 4 and the extraction of power from fusion in Section 5. Before examining confinement concepts, it is essential that we obtain a grasp of the important system aspects of a flight fusion system, namely, those requirements which are expected of the fusion system and to make judgments on the degree of difficulty with meeting those requirements.

The earliest conceptual research or developmental phase constitutes the optimal time to pursue the system design challenges and thereby assure minimal program impact. For predevelopmental technology, the system considerations indicate research endeavors which, if accomplished in a timely fashion, will provide the most effective program management approach towards achieving a successful, safe, reliable high energy mission capability. A well planned, predevelopmental research program can reap enormous dividends.

Thus, along with the requirements derived from mission considerations in Section 2.0, "Mission Applications," (such as reactor power level and burn times), it is important to define those additional requirements which ultimately will be placed upon a fusion reactor from the space systems perspective for its successful use. Therefore, this section identifies the key system parameters and the desired value trends. For the purpose of this discussion two flight elements are separately considered, the space reactor and the interfaces of the space reactor with the flight vehicle system.

6.0 Flight System Considerations and Requirements

6.1 SPACE REACTORS

This section presents those anticipated unique requirements imposed on fusion reactors for the application of fusion energy to space missions.

6.1.1 SPECIFIC POWER

As shown in Section 2.0 this parameter, combined with variable high specific impulse, is key for the successful operational use of space fusion energy. The goal is clearly to develop the maximum attainable for propulsion and electrical power systems. A specific power value of 10 kW/kg is preferred for the planetary missions. Values much below 1 kW/kg would provide little functional flight value but could serve as large space based stationary electrical power plants. For those beyond planetary distances, that is the minimum value needed to permit reasonable mass and flight time characteristics, considering the distances involved.

6.1.2 THRUST

Three somewhat arbitrarily established thrust levels are appropriate for consideration:

(a) low: 1 N to 10,000 N

(b) medium: 10,000 N to 50,000 N

(c) high: 50,000 N to 500,000 N

The low thrust range applies to the low thrust, low vehicle mass planetary missions considered in this study.

The mid-range is expected to be of value for high thrust use for the massive, multiple stage vehicles operating beyond the solar system to the Oort Cloud and to Alpha Centauri. This thrust level is analogous to the OMV (Orbital Maneuvering Vehicle), Space Tug, or orbit-to-orbit vehicle category.

If a reactor having light-weight, high power level, high specific power, and low volume characteristics can be achieved, the high thrust level will power aircraft to provide a single stage-to-orbit capability and lift from the surface of planetary bodies.

6.1.3 SPECIFIC IMPULSE

To achieve effective performance and to realize an economical return from the development of fusion, the system's capabilities must exceed that of existing propulsion systems or provide the same capability at a reduced total overall system cost sufficient to warrant the developmental costs. To exceed ion engine performance, that would indicate a requirement for fusion systems to produce Isp>10,000 sec at 2 N thrust on a scale of sufficient magnitude to warrant the research and developmental expenditures. From a fission rocket engine point of reference (NERVA) that would imply fusion system performance in excess of an Isp of 860 sec at a 40,000 N level of thrust.

From the calculations in this study, the specific impulse must be capable of being highly variable to provide a mass optimized mission. The specific impulse requirements in the study were strongly mission dependent, ranging from less than 10^4 seconds to over 10^6 seconds.

Consider the Manned Mars Mission as an example illustrative of the range in variations in the optimal specific impulse. Although one would never fly the ten year mission, the average Isp for it optimized at 55,000 seconds; but for the 0.44 year mission, it decreased to about 10,000 seconds. The optimized "value" of Isp is actually a range to reflect changes in the spacecraft's velocity. In this report the averaged valued is stated except for some examples where the values are presented to provide a better understanding of the variation ranges. Those values apply to a reactor propulsion system specific power of 1 kW/kg. If we were to fly a specific power propulsion system of 10 kW/kg, then an Isp of greater than 30,000 seconds will be required to optimize the vehicle mass for a flight time of 0.44 years. For the 10 year mission, by comparison, it is more than 170,000 seconds. The greatest Isp demand noted was in excess of 3×10^6 seconds for the stellar mission.

6.1.4 FUEL CYCLE

To achieve a mass optimized reactor design, the purely aneutronic fuel cycles would be preferred if they could produce energy breakeven and sufficiently high power density to meet mission requirements. Lacking that option the next preferred selection is those producing the minimum neutron flux. That is deuterium - helium-3. Refer to Section 7.1.

6.1.5 BETA

This value should be as high as possible, to minimize total system weight within the bounds of safe, reliable, and stable plasma performance. Values on the order of 90% are considered to be possible, but no study has been performed to determine a minimum acceptable value.

6.1.6 IGNITION

The minimum power level and minimum plasma temperature are desired to simplify in-space starting system requirements and to reduce the inert mass. This will be determined by the choice of fuel selection and details of the plasma confinement scheme.

6.1.7 THROTTLE CAPABILITY

The reactor power output will be expected to vary to match changes in the thrust and specific impulse. Similarly, a variable electrical power output will be called upon to meet variations in the mission's electrical power system demands to effect efficient reactor operation for overall fuel economy. The values and ranges are considerations that will be defined from flight system, mission, and operational trade studies. Mission and preliminary design analyses need to be performed to establish definitive mission requirements. Whether pulsed or steady state operation is used, which is yet to be determined, will be a function of the reactor design, ease of ignition, and shutdown characteristics.

6.1.8 PLASMA STABILITY

Plasma stability is a reactor design dependent variable. At the present time, from the system user's perspective, the need is to design the reactor to minimize the reactor's sensitivity to deviations which could disrupt the plasma's stability characteristics. The design must also provide margins commensurate with the reactor's sensitivity to instabilities and the effect on plasma stability, if any, that the operational range of the vehicle may have.

6.1.9 POWER LEVEL

The power levels are defined by the individual mission designs and the specific power characteristics of the propulsion system being flown for these missions. Many of the planetary missions analyzed were accomplished using a 50 MW jet power capability. The manned Mars mission required 225 MW for $\alpha_{p1}/0.44$ and 275 MW using a $\alpha_{p10}/0.44$ propulsion system. The stellar (Alpha Centauri) rendezvous mission is the most demanding. In addition to the specific power and time dependence functions, there is also a stage factor. The power requirements for the stellar missions take on new dimensions. A specific power of 10 kW/kg places a 2.8 TW jet power requirement on the 4-stage system while allowing completion of the mission in 244 years. If a α_{p100} fusion system were possible, the magnitude of the first stage power output is 28 TW. A reduction of the number of stages to 2 diminishes the first stage power to 7 GW while allowing the mission to be accomplished in 308 years. Power levels will be decided by system constraints, such as mass, hardware design features, etc., traded against mission objectives, such as flight time, Δv, etc. The high reactor power level required for the propulsive phase of a mission would not be optimal for running such a machine continuously to produce electrical power unless megawatt power levels are continuously required for electrical spacecraft power generation as well. No such need could be anticipated. Hence, the system should have some means for storage of electrical energy for the reactor system restart energy and the vehicle's electrical power system.

To produce electrical power for data transmissions, and any other vehicle electrical power equipment, an output of up to 25-30 MW was identified for stellar missions. Restart is considered to be the power design driver.

6.1.10 ELECTRICAL POWER VARIABILITY

The desire is to conserve fuel and to extend reactor life through the means to vary the reactor power output to a level consistent with the mission's power load profile.

6.1.11 DUAL MODE OPERATION

A simple reactor control mode should be provided such that, either the flight crew or the vehicle, while operating autonomously, should possess the capability to operate the fusion system in either a propulsion mode or a power mode or in both modes simultaneously.

6.1.12 MASS

The minimum that is consistent with the overall system design, inherent high safety, and high reliability is the goal. This is an extremely critical performance parameter and is therefore one worthy of significant research attention.

6.1.13 EFFICIENCY

Unburned fuel in the exhaust is a waste of a valuable fuel, and the losses create additional burdens on the vehicle mass. Therefore, a highly efficient fuel burn utilization design is desired or, alternatively, the means provided to retrieve unburned fuel with a minimum use of recirculating power.

6.1.14 RECIRCULATION POWER

Designs which use a minimum power to operate the reactor are preferred from an efficiency viewpoint.

6.0 Flight System Considerations and Requirements

6.1.15 LIFE

The minimum time required for a round trip Manned Mars Mission is 4 months. A progression in requirements for reactor life capabilities up to 50 years is needed for space science exploration out to the nearest star.

6.1.16 MODES OF OPERATION

A continuous burning operational mode is preferred for reliability gains.

6.1.17 FAILURE TOLERANCE

The capability to accept any single failure without disruption in performance or safety to man or on-board equipment is mandatory.

6.1.18 SPACE ENVIRONMENT

The reactor must be capable of operation in a space environment.

6.1.19 SUMMARY

The status of the above space reactor requirements is provided in Table 6-1.

6.0 Flight System Considerations and Requirements

TABLE 6-1. Research status of key space-relevant fusion powered reactor parameters necessary to meet space flight system requirements.

SPACE REACTOR PARAMETER	RESEARCH STATUS			
	None performed	Little performed	Prior or active research[1]	Comments
6.1.1 Specific power		X		Limited study. Requires burning of fuels.
6.1.2 Thrust:				
(a) low: 1 N to 10,000 N	X			Limited conceptual work. Experiments required.
(b) medium: 10,000 N to 50,000 N	X			Has not been addressed.
(c) high: 50,000 N to 500,000 N	X			Has not been addressed.
6.1.3 Specific impulse		X		Limited conceptual work. Experiments required.
6.1.4 Fuel cycle			X	Considerable analysis performed. Burn experiments required.
6.1.5 Beta			X	Very limited analysis performed. Experiments required.
6.1.6 Ignition	X			Has not been addressed.
6.1.7 Throttle capability	X			Has not been addressed.
6.1.8 Plasma stability			X	Limited analysis done. Burn experiments required.
6.1.9 Power level	X			Burn experiments required. Lacks analysis of the space band of interest.
6.1.10 Electrical power variability	X			Has not been addressed. Burn experiments required.
6.1.11 Dual mode operation	X			Has not been addressed.
6.1.12 Mass		X		Very limited study. Requires burning of fuels demonstration.
6.1.13 Efficiency		X		Very limited study. Requires burning of fuels demonstration.
6.1.14 Recirculation power	X			A function of the reactor design. Very limited study. Requires burning of fuels demonstration.
6.1.15 Modes of operation	X			Work will follow net power demonstration
6.1.16 Low/no neutrons produced	X			Varies with fuel selection and reactor design.
6.1.17 Failure tolerance	X			Has not been addressed. To follow net power demonstration
6.1.18 Space environment	X			Has not been addressed.

[1] No space fusion program exists. "Prior or active research" refers to either work which the DOE terrestrial power program is or has pursued or, alternatively, work which the NASA fusion research program pursued. The NASA work performed is presented in depth in Appendix A.

6.0 Flight System Considerations and Requirements

6.2 INTERFACE OF THE FUSION FLIGHT POWER REACTOR SYSTEM WITH THE FLIGHT VEHICLE

This section examines the system aspects of the fusion power system and its interfacing functions with the flight vehicle.

6.2.1 SPACE RESTART CAPABILITY

The goal is no inherent limit on the number of reactor restarts. Start-up should be quick and simple. Recirculating power and inert mass should be minimized. This requirement addresses a key capability to the successful use of a Space Fusion Reactor (SFR). It will probably be the major parameter in establishing the vehicle's electrical energy storage requirement. It will be a key subsystem in determining the overall mass of the vehicle as well as establishing the vehicle's operational characteristics and sufficiency with regard to fulfilling mission objectives.

6.2.2 FUEL STORAGE CAPABILITY

A vehicle, operating during the extent of its intended mission duration, will be required to produce specification performance after being subjected to the anticipated environmental exposure, both natural and that which is internally generated by the system's operation. Operational environmental exposures have typically created the greatest technical challenge and threat in the space program. One of the key challenges for the fusion vehicle is to provide the capability to store liquid helium fuel for a period of years (centuries for stellar missions) while being exposed to the heat generated by the spacecraft and while maintaining a low mass cooling system.

6.2.3 RADIOACTIVITY

No direct ionizing or neutron induced radiation is the goal. While this does not appear possible at the present time, the design solution is aided by the selection of the fuel minimizing the neutron flux.

6.2.4 REUSE

A large number of reuse cycles is clearly essential from an economics perspective. But at this phase of space fusion technology development, it is too early to establish a realistic number with any high degree of confidence that would represent the optimal economic value or even a value that is readily technically achievable. The fuel selection of D-^{3}He is anticipated to significantly enhance the cycle life without refurbishment compared to that for D-T. This advantage of helium-3 is due to reduced neutron interactions which otherwise result in first wall material degradation from deep neutron penetration induced damage and the ensuing radioactivity.

Economics dictate that the reuse capability for manned mission reuses should be a substantial number, approximately 40-50 being suggested here, depending upon the duration of missions. It is likely that several classes of vehicles will be developed, each specialized to accomplish a specific set of mission objectives for which it has been designed. For example, the short trip times and flight frequency desired for Manned Mars Missions would indicate that a low specific impulse performance, but high power level reactor, would be required to be used 40 times. This assumes a 20-year life, flying the vehicle back and forth to Mars twice annually. For solar system science missions, where flight times are longer, varying from 1.5 to 8 years with lower jet power requirements and higher specific impulse requirements, 20 reuses would equate to a 100-year life for the reactor, assuming an average mission time of 5 years. Obviously a singular mission use reactor applies to the Oort Cloud and stellar missions but with up to 50 GW jet power, specific impulses of 4×10^5 seconds, and 600 restarts of a steady-state reactor for data transmissions if a biannual duty cycle is selected.

6.2.5 SERVICING

The system must be designed to be serviceable for replenishment of all consumables. This includes refueling either in space or on an extraterrestrial body.

6.2.6 ENERGY STORAGE

A capability must be provided to make available the requisite energy for reactor restarts and to act as an energy reservoir for mission phases where the demand exceeds the power supply. Fission systems are to be avoided to enhance safety. One estimate is that 1% to 3% of the reactor's power output will be required to start the reactor.

6.0 Flight System Considerations and Requirements

6.2.7 SIZE

For space flight, size is not a particular concern. For aircraft, the reactor system envelope is particularly critical in order to minimize drag and impacts on vehicle design, airframe integration, and operational costs as well as to maximize vehicle performance.

6.2.8 LIFE

A minimum of 20 years for those space vehicles serving as workhorse devices in the vicinity of the Earth, including manned Mars, should be a reasonable vehicle life goal where the use rate is high. Additional life for the planetary science class should be the goal, or 50 to 100 years using modularized updated technology for ease of orbital replacement where warranted. For stellar missions, the flight time can range to over 300 years, plus some allowance for the collection and transmittal of science data. Use of staging reduces the burning duration to approximately 50 years per stage.

6.2.9 MASS

Vehicle dynamic performance optimization, with the goal of minimization of the total system mass, is of utmost importance. Even with the enormous energy supply available from fusion, inert vehicle mass is still a consideration. Trades are required to design the optimal fusion propulsion system configuration to match the system's capabilities with the Manned Mars, planetary, and stellar mission categories as discussed in Sections 6.2.4 and 6.2.8.

6.2.10 MAINTENANCE

On orbit maintenance and repair is a design requirement. The goal is to avoid disassembly of a large reactor in space for return to Earth. This, too, would indicate a preference for D-^{3}He since the lower neutron flux will present a substantially reduced wear out rate and minimized operations which enhances safety and provides a more favorable economic trade.

6.2.11 PULSED VERSUS STEADY-STATE OPERATION

For overall simplicity in terms of vehicle system design, ease of ignition, reliability, and vehicle control dynamics, steady-state operation is clearly preferred. Analysis is required to verify the capability of any pulsed system to meet the reliability and performance requirements for such a flight system. Whether a device operates pulsed or steady state is not important provided other key parameters like specific power and reliability are met. For example, a technique of overlapping thrust rise and decay pulses from multiple engines is of course an option which could be accomplished by either shaping the thrust tail-off or by increasing the pulse repetition rate. This also implies a short duration for the engine start-up transient. A steady state system is considered to have inherent reliability advantages over one which has transient dynamics that result from frequent restarts in the pulsed systems.

6.2.12 POWER CONVERSION AND TRANSPORTABILITY

This capability refers to the reactor serving initially as a propulsion system with the capability for conversion into an electrical power generation system after reaching orbit around a planet or moon. By making use of a single purpose reactor, programs are provided with an enormous cost advantage in those missions where large power requirements follow the propulsion function. The system can either be used in orbit or transported to a planet's surface for the subsequent use of large electrical power production. The ease of operational conversion to electrical power production, the capability to withstand the dynamic loads, the inertial performance (Δv, thrust to weight, etc.), the on-board controls to permit operation under both modes, refueling, maintenance, etc. are some of the parameters to be explored.

6.2.13 EMERGENCY SHUTDOWN

The means to quickly and safely terminate reactor system's burning should be provided as a contingency capability.

6.2.14 HEAT BALANCE AND COOLING

The goal is to avoid designs requiring the expenditure of consumables for coolant purposes and to minimize the radiator mass. This includes the post-burn heat soak-back. Analysis of the spacecraft's thermal balance is very important in achieving a system capable of yielding the necessary thermal characteristics. For example, one of the critical fluids is cryogenic helium which must be cooled to 4K while other hardware must be maintained sufficiently warm to function without imposing large power drains. It will be necessary to remove a large quantity of heat from the reactor and from the direct power conversion system. (Refer to Fig. 4.2.) For example, the Manned Mars Mission requires ~270 MW jet power output. A 440 MW reactor would produce that level of charged particle energy (Fig. 4.2). The D-^{3}He thermal power dissipation requirement is thus 132 MW.

6.0 Flight System Considerations and Requirements

6.2.15 SOLID STATE PROPULSION

To achieve the required ultra high reliability requirement a "Solid State Propulsion" System (SSP) is the design goal. The SSP concept features no moving parts or components that are subject to erosive wear such as electrodes.

6.2.16 SELF DIAGNOSTICS AND CORRECTIONS

The space vehicle's long operational life and great distance requirements demand an autonomous system design approach. The system should contain the diagnostics to recognize an adverse trend and must be capable of initiating the means to control the errant parameter prior to the onset of a failure.

6.2.17 OPERATIONS

The reactor design should allow for simple space operations. Complex cool down schemes like those required for NERVA will dampen support for fusion and increase the vehicle mass. Operational restrictions because of radiation or neutron flux from any reactor will have a similar effect. Where design options exist, a careful selection must be made to provide for a _simple operation_, keeping in mind that operational simplicity may be a cause for acceptance or rejection of a technical concept. Maintenance requirements should be set at a minimal level, but accessibility in the space environment is a must, even where maintenance is not a planned operation.

6.2.18 SAFETY

The operation of the reactor must fall within acceptable degrees of risk to those on-board as well as the population as a whole. Similarly, on-board equipment and non-fusion powered spacecraft, like the Space Station Freedom which will be exposed to its operation, should not be adversely affected by fusion powered vehicles. Keeping the radiation emissions within the background levels would be the preferred goal.

6.2.19 REDUNDANCY

The system, because of its criticality, should be designed to tolerate a minimum of two failures, either hardware or operational, with the objective of avoiding either the loss of a system or causing a hazardous situation to occur under these conditions.

6.2.20 SPACE STATION COMPATIBILITY

The fusion spacecraft should be designed to be compatible with Space Station Freedom (SSF) in order to fully use its resources without generating new space logistics requirements. Orbital assembly, if required, should be accomplished by the SSF facility to avoid the requirement for a new orbital installation.

6.2.21 ENVIRONMENT

Its operation should not provide an impact to the Earth's environment nor contribute to space contamination in an unacceptable fashion. One particular focus is the effect of neutrons in general and of ions on the upper atmospheric chemistry due to the operation of large fusion vehicles. For example, would charged particles have an adverse effect, either of a temporary or lingering nature, upon Earth communications? The operation of any power plant on any planet can be anticipated to eventually become a matter of environmental concern. There should be no radioactivity added to Earth's environment above the background level.

6.2.22 RELIABILITY

System reliability is typically achieved by the design of redundant hardware, and the reactor design would have to lend itself to that feature. Alternatively, the use of significantly large design margins such that a high degree of reliability could be achieved without redundancy is likewise acceptable. Demonstration of the mean time between failure (MTBF) of years is a challenge that will probably only be met by carefully controlled flight operational experience.

6.2.23 TESTING AND QUALIFICATION

To minimize costs, an Earth based method of testing and qualification of the reactor and component designs is preferred to space based techniques. System qualification testing will be required in a space operational environment and probably not ultimately attained until years of flight test and flight operational experience have been obtained. The means for stress testing and accelerated life testing are needed.

6.2.24 SPACE BASED VEHICLE DESIGN

The mass, size, and costs that can be anticipated to be associated with returning the reactor to Earth plus the reflight back to orbit costs suggests that it remain on orbit throughout the life time of the reactor and vehicle. The design should be compatible with permanent exposure to the space environment. If time critical replacement of components is essential, space removal and replacement are requirements.

6.2.25 DISPOSAL

At the completion of its useful life, a means to satisfactorily dispose of the reactor is necessary. Disposal should be an initial design requirement in order to avoid undue problems at its life cycle completion. Perhaps it entails only a simple burn maneuver to send the vehicle outside of its hazard range. Disassembly and return to Earth or storage on the moon are other, but less desirable, options. Perhaps the vehicle design will be continuously updated such that this parameter has no significance. It, however, should be addressed at the start of the program.

6.2.26 ECONOMICS

All of the aforementioned system criteria, and others, will establish the final economics of the fusion propulsion and power system, as well as the overall ease of implementation of a space fusion system. The fusion spacecraft will comprise a very large initial investment, and therefore it will be incumbent upon NASA to adopt a philosophy to thoroughly understand the components and system design for nominal and contingent operations. To be cost effective, the system's flight operational costs must be minimized which can be accomplished by a high degree of autonomy to reduce operational costs and accidents due to human error. Also, for economy the development of highly reliable hardware is mandatory.

6.2.27 STATUS

The status of research for each parameter is presented in Table 6-2.

6.0 Flight System Considerations and Requirements

TABLE 6-2. Research status of key fusion system requirements.			
Parameter	Research status		
	None performed	Very limited	Prior or active research[1]
6.2.1 Space restart capability	X		
6.2.2 Fuel storage capability	X		
6.2.3 Radioactivity		X	
6.2.4 Reuse	X		
6.2.5 Servicing	X		
6.2.6 Energy storage	X		
6.2.7 Size	X		
6.2.8 Life	X		
6.2.9 Mass	X		
6.2.10 Maintenance	X		
6.2.11 Pulsed versus steady state operation	X		
6.2.12 Power conversion and transportability	X		
6.2.13 Emergency shutdown	X		
6.2.14 Heat balance and cooling	X		
6.2.15 Solid State propulsion	X		
6.2.16 Self diagnostics and corrections	X		
6.2.17 Operations	X		
6.2.18 Safety			X
6.2.19 Redundancy	X		
6.2.20 Environment			X
6.2.21 Reliability	X		
6.2.22 Testing and qualification	X		
6.2.23 Space based vehicle design	X		
6.2.24 Disposal	X		
6.2.25 Economics			X

6.3 SUMMARY

The system analysis shows that very difficult tasks need to be addressed, the solution of which will not be researched in the terrestrial program. The need for a space fusion R&D program is manifested by the absence of definitive data on critical parameters. One of the greatest is the requirement to provide for a space fusion reactor start/restart capability. The principal motivating force behind fusion other than for performance is the safety that it provides. If a fission reactor is required to start it, a significant advantage is lost in safety, in

6.0 Flight System Considerations and Requirements

operational flexibility, and in the wide range of space mission applications that fusion can serve. Two other very important system tasks which the terrestrial program will not address include designs offering a minimum maintenance for space environmental operations and firing durations over many years.

7.0 FUEL AND DESIGN OPTIONS FOR SPACE FUSION REACTORS

This section considers the options and optimal fuels for a space reactor application, the current inventory of fusion experiments, and the reactor designs which burn those fuels and which are most likely to have space applicability.

The national fusion energy program is dedicated to the terrestrial application of fusion energy for commercial electrical power production. The absence of space fusion reactor (SFR) designs and test data necessitated a review of the terrestrial designs with the objective of matching the space system requirements presented in Section 6.0 with the terrestrial designs. Consequently, lacking a space fusion activity, only a few conceptual fusion reactor design studies having potential for space applications emerged during this review. All of those studies which have been conducted for space have been very limited, and the level of detail has been low. One fusion reactor concept emerged as the preference, based upon the known features and a fairly limited experimental data base.

Each reactor design under consideration for possible terrestrial application has unique physics characteristics under which it operates and therefore unique physics issues to resolve although commonalities also exist between the two applications. SFR designs will differ from the terrestrial reactors as a consequence of:

- differences in application,
- the operational environments, and
- the mode of operation for mission applications.

Therefore, one must realize that the resolution of physics problems for the terrestrial program will not necessarily contribute to or lead to a SFR design. For example, the leading contender for the terrestrial program is the tokamak. But due to the tokamak's large mass and low beta which results in unacceptably low specific power, its use for on-board flight propulsion and power is not expected to meet the criteria for space reactor utilization.

One possible tokamak application remains, namely, its use as a lunar, or any land, based power source. One such concept is presented in Appendix B. In that application the reactor serves as a multimegawatt power source for laser propulsion, that being a unique space application of a terrestrial design. Except for that limited use, NASA will, of necessity, employ design approaches for space which differ from those for terrestrial power. These SFR's will be different and can, therefore, be anticipated to have a new and different set of physics issues to deal with and to resolve. Their operational requirements are in an entirely different operational regime. The solutions for fusion system problems for the terrestrial application may, therefore, not necessarily apply to space.

7.0 Fuel and Design Options for Space Fusion Reactors

This section first deals with the fuel of preference for space and its availability. Undertaken next is a brief description of candidate terrestrial reactor designs having potential application for space. Their status is reviewed and their applicability as energy devices for space power and propulsion considered. Section 8 provides a summary of program status of the major funded DOE programs, the preferred design approach for space, and program options for consideration.

7.1 FUEL SELECTION

Fusion reactions were discussed in Section 4.0. The three easiest to ignite were presented in Table 4-1b. This section analyzes those preferred fuel options for space. The features necessary for space applications are presented in Table 7-1:

TABLE 7-1. Preferred fuel characteristics.

1. Performance
    a. High power density
    b. Charged particles
    c. Ease of ignition
    d. Can be readily engineered for space use
    e. Long space storage life
    f. Available to meet mission requirements
    g. High power output reactor designs- gigawatts
    h. Space storable
2. Safety
    a. Nonradioactive
    b. No neutrons in the flux
    c. Non toxic
    d. Non-flammable
    e. No ionizing radiation
3. Economics
    a. Reusable design permitted
    b. Minimal maintenance
    c. Minimal mass
    d. Minimal environmental impact
    e. Low mission fuel costs
    f. Flight reliability

7.1.1 SPACE FUEL OPTIONS

Three fuels and their availability are considered for space use: deuterium-tritium, deuterium-^{3}He, and deuterium-deuterium.

7.0 Fuel and Design Options for Space Fusion Reactors

7.1.1.1 DEUTERIUM-TRITIUM

This fuel pair has received the greatest attention for the terrestrial power generation program application. Ideally the high energy yield from the D-T reaction appears to be highly suited to space. Another consideration favoring its use is its relative availability, although breeding tritium in space would be very difficult. The physics of the D-T reaction is the easiest fusion reaction to demonstrate. Another attractive feature is its superior plasma power density properties. Refer to Fig. 7.1(Mil76) which shows its reactivity, $<\sigma v>$, as a function of temperature and which compares it with other fuels.

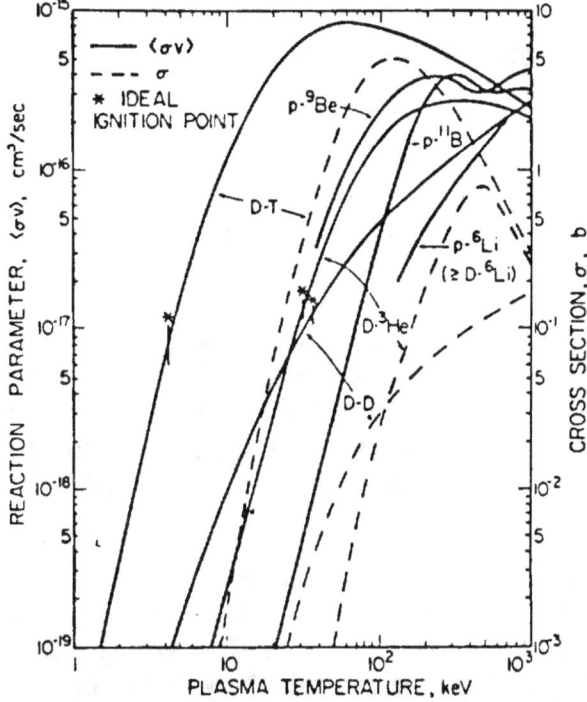

Fig. 7.1. Fusion fuel reactivity (Mil67).

The drawbacks to tritium are several. It is radioactive and will offer only limited relief from the expensive safety considerations to which space science payloads using fission power sources are now subjected. One motivating factor for the use of fusion energy is the circumvention of those problems. The weight penalty to protect the public from a large tritium payload release in the event of a launch accident is an undetermined quantity. Radioactive payloads are designed to worse case flight conditions, which could be a severe penalty if massive quantities of tritium are required as determined in the VISTA approach. The fuel radiation hazard during transport to orbit is nowhere nearly as significant as with an RTG. But public concern will be important, and massive quantities of tritium for Low Earth Orbit (LEO) space application may not be accepted in the final analysis. The neutron activated materials will be maintained at safe orbital altitudes. As a minimum, a very extensive public educational program would be essential.

7-3

7.0 Fuel and Design Options for Space Fusion Reactors

Tritium's 12.3 year half-life will ensure the need for a continual requirement for tritium production from lithium, a process requiring fission reactors or D-T fusion reactors and the attendant environmental impacts. Because it cannot be stored indefinitely due to radioactive decay, the production facilities must be capable of producing large quantities quickly. At the present time the United States production rate is estimated at 5 kg annually from the Savannah River weapons plant. The VISTA spacecraft used 40 MT of fuel for one manned VISTA Mars mission, half of which is assumed to be tritium. That would require the fabrication of ~4,000 Savannah River fission reactors to meet the fuel demand for that one mission on an annual basis.

An alternative is to breed tritium using breeding fusion reactor designs. These have been given consideration. A lithium blanket concept is the preferred approach:

$$^6Li + n \rightarrow {}^7Li \rightarrow T + {}^4He.$$

Obviously for a 200-plus year mission duration, tritium's short half life eliminates it as a fuel of consideration unless a mass efficient means of tritium production is provided aboard the spacecraft. Otherwise, the Δv for the spacecraft and its braking propulsive maneuvers or electrical power generation occurring 200-300 years after lift off would not be possible at the time of need, if tritium were used.

Mass in space flight programs is always placed there at a performance premium. As discussed earlier, 80% of the energy produced by this reaction resides in neutrons. Extraction of useful energy from neutrons is only achievable thermally using devices which typically operate at efficiencies not much greater than 40%. Cooling mechanisms are required to expel the residual heat. In space, this means an added mass penalty cost for radiators and consequently, a performance penalty. Probably the most significant concern is the high neutron flux of ~3 million watts per square meter (Hol88). This high neutron flux damages materials at a rate greater than an order of magnitude higher than in fission reactors. Neutron bombardment of the reactor's first inner wall leads to its demise in a time period that may be as brief as one year. There will be significant first wall radioactivity from the neutron activation products. These include a wide range of radioactive elements as discussed in the MIT safety and ESECOM studies (Section 9). The options are either total reactor replacement in orbit or reactor disassembly there for replacement of the exposed hardware and its disposal. Orbital disassembly of a large device, a difficult operation in the space environment without the radioactivity factor, will necessarily be accomplished remotely by sophisticated robotics having an advanced degree of artificial intelligence. That entire operation will also require some very well thought out contingency capability in order to be safely conducted – a very expensive operation.

Tritium is an option for solar system exploration, although not the preferred fuel.

7.1.1.2 DEUTERIUM–HELIUM-3

The D-^{3}He fuel cycle, while it does not entirely eliminate those problems, does very substantially reduce the risk or defer the onset. Chemically, helium-3 has the very desirable property of being an inert fuel without the flammability hazards of hydrogen or tritium. That fuel cycle reduces the magnitude of the flammability hazard but does not eliminate it since deuterium is also used. Unlike tritium, it is not radioactive. Therefore, the continual replacement problem of tritium due to decay is not a problem with helium-3. By comparison, on very extended space missions the half-life of tritium would make it useless for electrical power generation, particularly for the stellar missions without massive tritium breeding systems. The large penalties for neutron shielding and mass additions for resultant heat rejection systems are reduced substantially. The decay heat also requires the means for heat rejection.

Dr. Logan examined the question of relative merit by comparing specific mass comparisons between D-T and D-^{3}He. (Log88) A model was developed assuming that the maximum heat transfer is the design limiting factor and that other reactor design parameters were left unconstrained except to adhere to the inherent limitations of nature regarding material properties, etc. His results are presented in Fig. 7.2 which shows the better specific mass characteristics of D-^{3}He over D-T for a wide range of β.

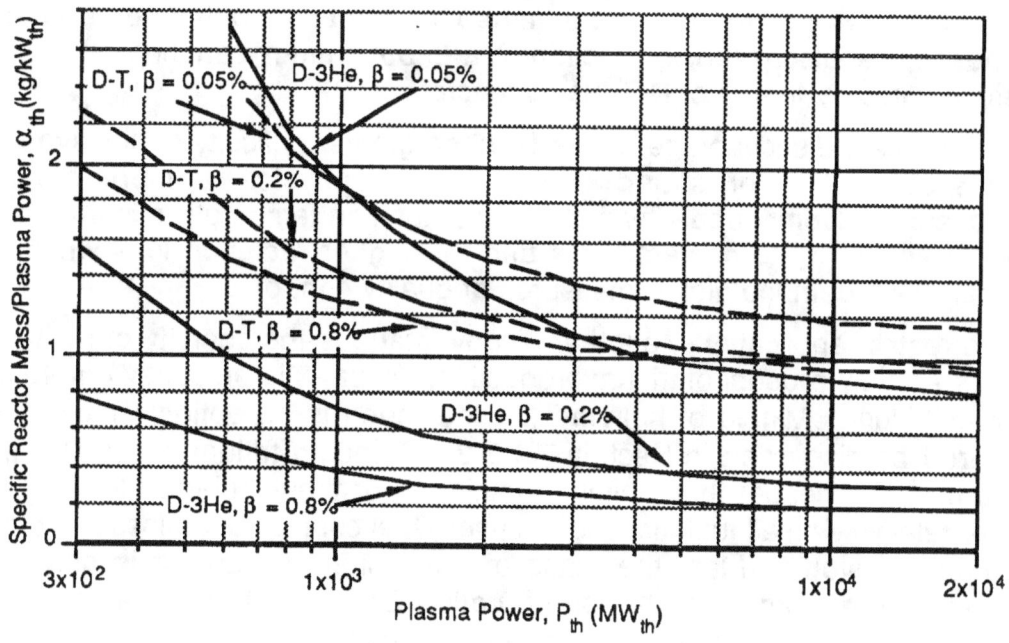

Fig. 7.2. Specific mass characteristics of D-^{3}He compared to D-T for a wide range of β.

7.0 Fuel and Design Options for Space Fusion Reactors

The substantially reduced neutron flux of 0.09 megawatts per square meter (Hol88) lowers the radiation hazard as well. Maintenance on orbit is significantly simplified from the reductions in the first wall flux. Consequently, the final parts of the solution – the engineering and performance aspects – are very significantly simplified by the selection of D-^{3}He. Simplifications to the system hardware and flight operations represent significant savings to the space program where the engineering and managerial costs to achieve flight readiness of programs are enormous, where the transportation costs to orbit are anticipated to decrease to ~$4,000 per pound of payload (assuming the anticipated Shuttle launch costs decrease from $320M to $250M to launch 27 MT into LEO), and where orbital operational costs are orders of magnitude higher than ground operational costs. The investment in developing the D-^{3}He physics appears to be an excellent one at this point in time. The space fuel preference topic was recently examined by McDonnell Douglas Space Systems Co. and General Atomics for the Astronautics Laboratory of the Air Force Systems Command. The same conclusion was reached (Hal89).

The D-^{3}He fuel cycle is particularly attractive since the reaction products are charged particles which readily produce thrust by propelling them as bleed off particles from the plasma through a magnetic nozzle. Fortuitously, more than 95% of the reaction's energy is present in the form of charged particles, namely, alpha particles and protons, the energy of which can be converted directly to propulsion and/or electrical power without the usual thermal and mass inefficiencies and losses. By the proper use of design parameters the neutron flux can be reduced to approximately 1%, based upon a sophisticated model recently developed (Ker89), although the concepts can be traced back to at least 1976 (Mil76) and to the NASA research work (Appendix A). With regard to its availability, helium-3 can be mined on airless bodies for space fuel purposes, as discussed in Section 7.1.2, but tritium can not. It has to be manufactured by fission or D-T fusion reactors.

The significant disadvantage of D-^{3}He is the greater difficulty to achieve ignition compared to D-T or stated differently, the physics is more difficult to demonstrate. Compounded with the higher ignition temperature is the plasma's greater sensitivity to contaminants and the higher radiation loss which could ensue. It will demand higher measures of quality control.

For terrestrial applications, another disadvantage is the scarcity of helium-3 on Earth. As a by-product of tritium production, helium-3 is presently one tenth the cost of tritium. Mining of helium-3 on the moon is an option which is being explored as discussed in Section 7.1.2.2. Breeding tritium in the D-T lithium cycle may be cheaper than helium-3 mining. For those solar systems missions where relatively small fuel masses are used, fuel cost is not an issue. To fly one manned mission to Mars the mass of the helium-3 required is only slightly greater than one kg. For the long duration Oort Cloud and stellar missions, there are no attractive fuel options except helium-3.

7.0 Fuel and Design Options for Space Fusion Reactors

Helium-3 has been reported to be available on the lunar surface in sufficient quantities that mining appears feasible (Wit86). A conference was held at Cleveland on April 25-26, 1988 to specifically address the possibility of lunar mining of Helium-3 and the feasibility of D-^{3}He fusion reactions. While it was considered an enormous mining undertaking simply in terms of the mass of material moved, no technical obstacle was reported. One of the conclusions drawn in the NASA Lunar Helium-3 Workshop (anom88), as reported in the Executive Summary, was "that lunar mining of ^{3}He is feasible." The lunar quantity present is calculated to be ~10^9 kg (Wit86) based upon the lunar samples analyzed from the Apollo and Luna missions. Perhaps the greatest obstacle is legal, i.e., who owns the mineral rights to the moon, rather than technical issues, but that is presumed to constitute the least obstacle to the United States, and it appears to be surmountable (Bil89):

> (8) Existing space law and other international arrangements suggest that an acceptable basis can be found for cooperative international production of He3. These precedents include various types of national mining laws; the Antarctic system experience; the Moon Agreement; and the INTELSAT, INMARSAT, and Space Station Agreements." (p 110).

A minimal cost, space logistics capability, can be attained by processing helium-3 on the Moon. NASA can take full advantage of the lunar presence of helium-3 by processing it for use on fusion powered vehicles directly from the moon without return of the element to Earth. Hence, it is expected that lunar activities will assume a major role in a fusion vehicle's operations, at least to the extent of extracting helium-3 there from the regolith and liquifying it.

It is important to understand that the helium-3 supply does not solely reside with the mining of the lunar surface. Dr. Miley has given this subject recent attention (Mil88). The first alternate choice to lunar mining is the use of ^{3}He produced by the D-D reaction and the extraction of all of the unburned ^{3}He from the primary D-^{3}He burning. One other alternative supply which can be considered is by the reaction:

$$p + {}^6Li = {}^3He(2.3\ MeV) + {}^4He(1.7\ MeV).$$

There is a question concerning whether the reaction is "energetically viable." There is a significant amount of helium-3 on Jupiter, too, estimated to be 10^{22} kg, the recovery of which has not been addressed. Dr. John Lewis at the University of Arizona has suggested its availability from the atmospheres of Uranus and Neptune. A sufficient quantity (600 kg) (Wit86) is available on

7.0 Fuel and Design Options for Space Fusion Reactors

Earth naturally from reactors in a sufficient quantity to provide a test program. Table 2 from Kul87 is reproduced here for reference as Table 7-2.

TABLE 7-2. Terrestrial Resources of ^{3}He (Kul87)

Source		He-3 content, kg	Equivalent MWe-(yr)[a]
a. Natural			
Natural gas wells			
Present storage		29	290
Known reserves		187	1870
b. Man-made			
a. U. S. Department of Energy			
MRC sales		1.3/yr	
MRC inventory		13.4	
b. CANDU reactors (the year 2000)			
Production		2/yr	
Inventory		10	
Weapons stockpile [b]		15/yr	
Total	Annual	18	180
	Inventory	239	2390

[a] 10 MWt-yr/kg - ^{3}He

[b] Estimate (Wittenburg et al., 1986)

MRC - Monsanto Research Corporation

CANDU - Canadian Deuterium Uranium

From the standpoint of fuel physical properties, liquid helium-3 is cooler than liquid tritium by 23K, requiring superior cooling techniques. At standard conditions, the gas-liquid transition, Table 7-3, for each are:

TABLE 7-3. Selected gas-temperature transition temperatures for fusion fuels.

Helium-3	3K
Helium-4	4K
Hydrogen	22K
Deuterium	24K
Tritium	26K

Its higher density offers some measure of compensation in terms of reduced containment weight and cooling requirements. Table 7-4 compares data to be taken into consideration in the selection of fuels.

TABLE 7-4. Comparisons of fusion fuel operating regimes.

Parameter	Fuel Options			
	D-He^3	CAT D-D	D-T	p-B^{11}
Physics Merits				
Plasma Temperature, 10^6, K*	640.0	170**	160.0	1740
Engineering Design Merits				
Neutron flux, 10^{12}, n/cm^2/s	0.25		100.0	0
Relative production, %	0.05	0.5	1	0
System Performance				
Relative Plasma Power Density, MW/kg***	0.013	0.005	1	-

* Peak plasma power density
** Will not ignite at this temperature
*** Useful power per unit mass

7.1.1.3 DEUTERIUM-DEUTERIUM

The space program can use the same source for deuterium as the terrestrial program, i.e., the abundant supply that resides in the oceans. One interesting possibility considered in this study was whether the moon would serve also as a source for deuterium, and if so, could its recovery be accomplished as a by-product of the mining of helium-3? As it turns out, this was a matter of considerable research during the examination of the Apollo lunar samples. Unfortunately, the results were not encouraging. Dr. Epstein, California Institute of Technology, reports that

> A maximum value for the deuterium concentration in lunar hydrogen gas (almost wholly of solar wind origin) has been estimated to be about 5 ppm. Taking into account the contribution of deuterium formed by cosmic-spallation processes, the D/H ratio of the solar wind therefore probably is no larger than 3×10^{-6}. (Eps71).

That compares to 157 ppm as representative of the mean value of D/H in ocean water. A large percentage had not been anticipated based upon contributions by solar winds since the mean reaction time for D + H → ^{3}He takes 4 seconds in the sun.

7.0 Fuel and Design Options for Space Fusion Reactors

There are some indications that a higher concentration of deuterium may be found elsewhere. The *Scientific American* featured a report on Halley's Comet. The authors note that "In the comets water deuterium is from five to 10 times more abundant than the interstellar average ..." (Bal88). According to a report by Robert and Epstein, "It has been observed in interstellar clouds that the D/H ratio of molecules like HCN is at least 100 times greater than the D/H ratio of terrestrial hydrogen (Wanner, 1980; Penzias, 1980)." (Rob81). If correct, that data would indicate comets could contain deuterium at up to 1000 times the concentration of terrestrial water.

7.1.2 SUMMARY OF THE FUEL DISCUSSION

The entire fusion flight prototype developmental time must be considered, that is, we are not just interested in the demonstration of breakeven, where the fusion power produced equates to the direct energy input to the plasma. The end objective is an engineered flight system with net power production. Whereas the D-T physics is easier to demonstrate "success" early, the engineering aspects are not. It is quite conceivable that the completion of the engineering portion of the program could be completed using D-^{3}He before D-T. Although this report indicates a preference for the D-^{3}He reaction, system trade studies based on realistic engineering conceptual designs would have to be conducted to numerically show the fuel selection's greater performance advantage.

7.2 REACTOR CONCEPTS

Two different plasma confinement approaches account for the DOE funded activity and worldwide too: magnetic confinement and inertial confinement. These two approaches are discussed below.

7.2.1 MAGNETIC CONFINEMENT FUSION (MCF)

The best developed DOE terrestrial experiments are presented in this section and discussed for space mission suitability. To provide a comprehensive overview, non-DOE concepts and experiments are also included.

A few general reactor characteristics and principles relative to space use can be stated, and we focus on those of greater interest. These include the following:

1. The space designs should provide a minimum specific power value on the order of 1 kW/kg to accomplish solar system missions, preferably greater. For stellar missions, a minimum of 10 kW/kg is needed, preferably higher.

2. It must be capable of burning the fuel cycle preferred, D-^{3}He.

3. A minimum reactor mass is essential. The major mass items are the magnet – particularly the structure for load carrying of the magnetic fields force as opposed to magnetic field producing plates – and the neutron protective shield protecting the magnets. Reactor self generated field designs have inherently improved specific power.

4. Recirculation power should be minimized.

5. The design should permit a simple conversion to direct thrust and electrical power.

A variety of reactor design configurations have been considered by the terrestrial program. This section will briefly discuss the more significant terrestrial reactors and comment on candidate configurations for space fusion power and propulsion. The DOE's program's test progress, plus inherent characteristics of alternate confinement experiments, provide confidence in the belief that fusion systems can be developed for space on a relevant time scale. The following section provides an overview and examines the status of experiments on configurations with respect to the postulated requirements for a viable reactor.

7.2.1.1 FIELD REVERSED CONFIGURATION

From this study, the most promising design for meeting specific power for space applications is the Field-Reversed Configuration (FRC). The FRC is a high β machine officially designated a "compact toroid," shown in Fig. 7.3. It combines some of the attractive features of both toroidal and linear systems. The closed inner field provides good confinement of the plasma while the linear topology of the external magnetic field lines would be conducive to direct thrust.

7.0 Fuel and Design Options for Space Fusion Reactors

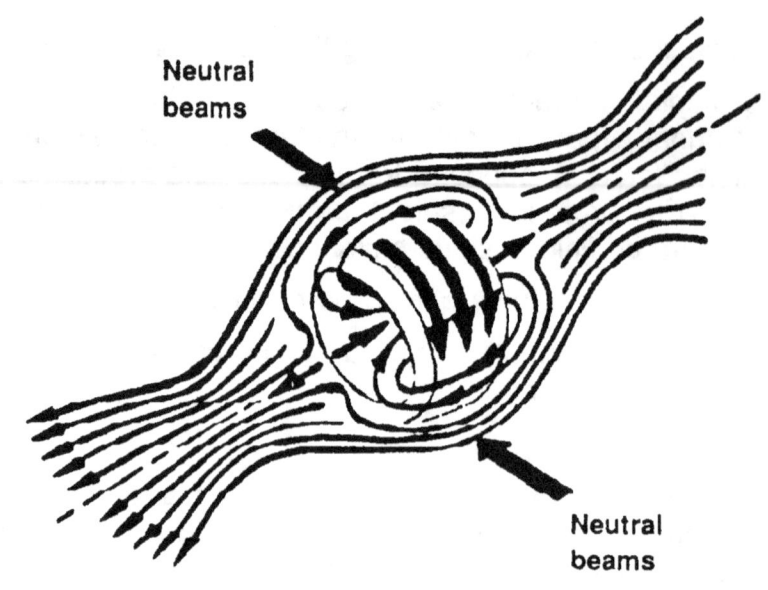

Fig. 7.3. Field-Reversed Configuration (FRC).

Because the FRC uses primarily solenoidal magnetic field coils and operates at high β, the configuration should lead to very attractive parameters for space applications – good plasma confinement, high power density, potential for steady state operation, and overall compact design. The steps to establishing the plasma are shown in Fig. 7.4.

7.0 Fuel and Design Options for Space Fusion Reactors

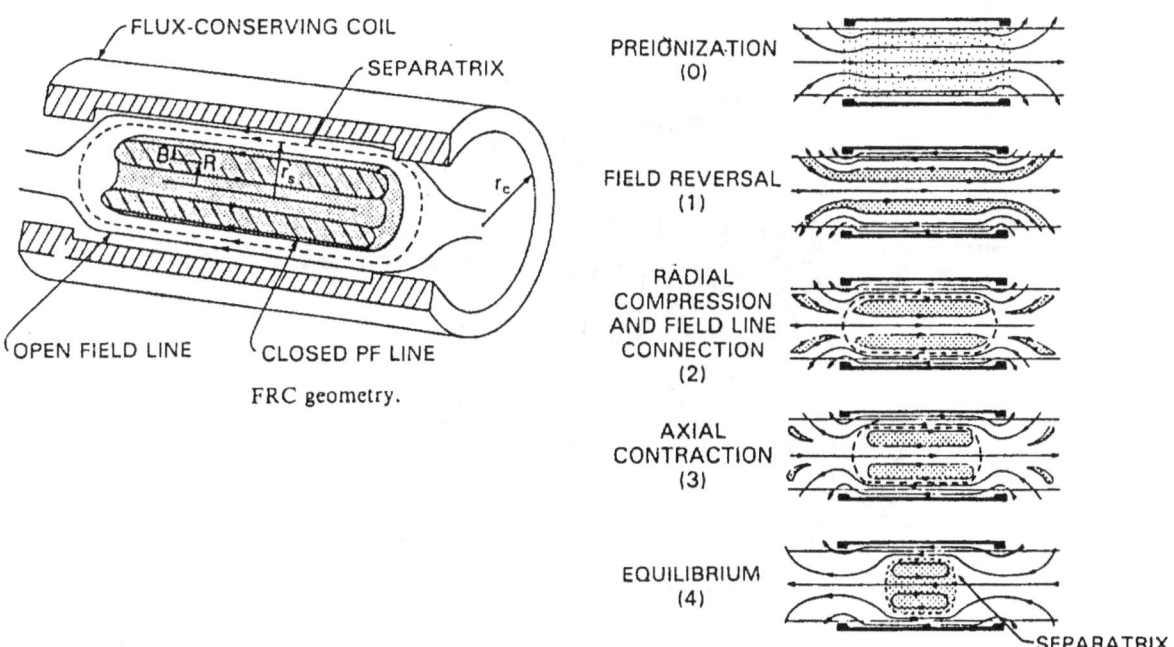

Fig. 7.4. Stages of FRC plasma formation.

The main difficulty in evaluating the concept is that the FRC is at an early stage of development, so that extensive experimentation and testing are needed. There has been only a modest worldwide research effort in progress but the Los Alamos experiment was terminated in 1990 and plans call for termination of the Spectra Technology experiment in 1991. (DOE eliminated the alternate confinement approaches in FY91 due to budget reductions). Because the FRC concept exhibits lower power operational characteristics than many of the other fusion reactor options, each of the developmental steps is likely to be less expensive. It should be noted that a FRC space reactor design has not been performed and that the reactor has not advanced toward the breakeven parameters as the tokamak. (Refer to Section 8.1.2.)

7.2.1.2 TOKAMAK

The tokamak, shown in Fig. 7.5, is the clear magnetic fusion research reactor leader worldwide for the terrestrial fusion program.

7.0 Fuel and Design Options for Space Fusion Reactors

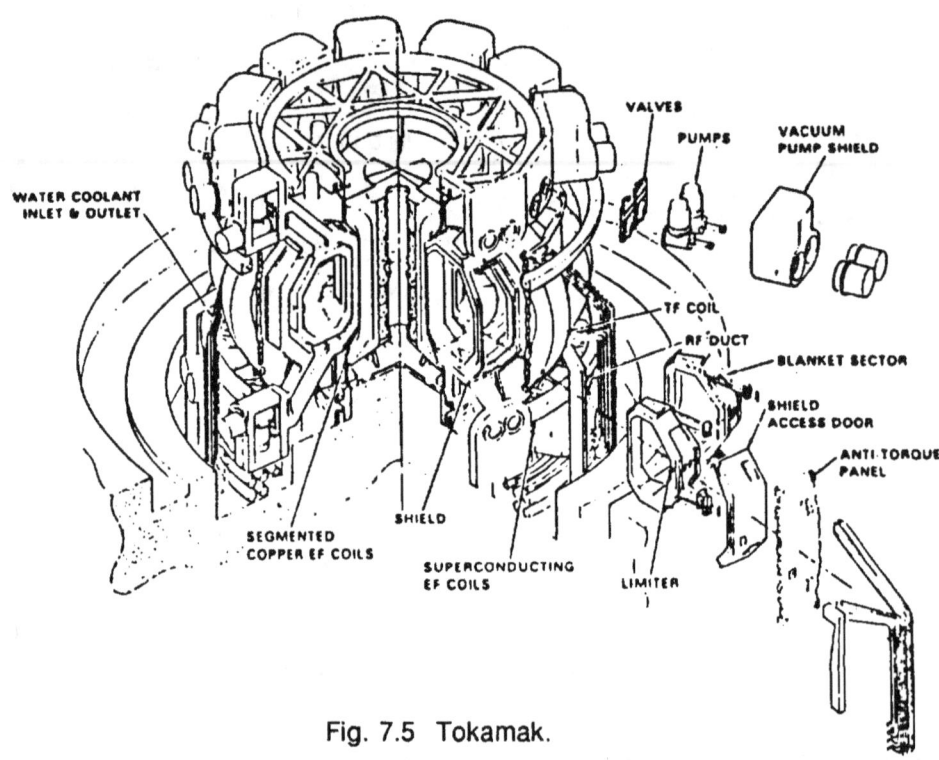

Fig. 7.5 Tokamak.

Historically, this design has had considerable success in achieving attractive values for the ion temperature and energy confinement time compared to the requirements for an ignited plasma. There remains little doubt in the fusion research community that a sufficiently large tokamak can reach the physics parameters of the reactor regime, but there is increasing doubt in the fusion research community that a sufficiently large tokamak can reach the physics parameters for economical commercial power production in this reactor's operational regime.

The key disadvantage of the tokamak with regard to space application is that the design and experiments presently achieve only low β values, leading to very massive magnets due to the insufficient utilization of the magnetic field; and, hence, an overall massive system results. A major effort in terrestrial tokamak reactor research is underway on "advanced" tokamaks, whose primary feature is higher β values. A difficulty for tokamaks, as in all other toroidal reactor designs, is that in order to produce thrust it will be necessary either to add a plasma extraction device – a "magnetic nozzle" to directly exhaust a plasma or to add a separate system for conversion of the fusion energy to electricity and then power some form of plasma or ion thruster. Both of these options are expected to require considerable extra mass and therefore are expected to result in reduced propulsion system efficiency. A detailed design study conducted by UCLA, ARIES III, concluded in March 1991 that an advanced Tokamak is not a suitable approach for burning D-^{3}He, the space fuel of preference.

7.2.1.3 SPHERICAL TORUS

Consideration has been given by Dr. Borowski (Bor87) to space applications of a tokamak variant called the spherical torus (Pen85), shown in Fig. 7.6.

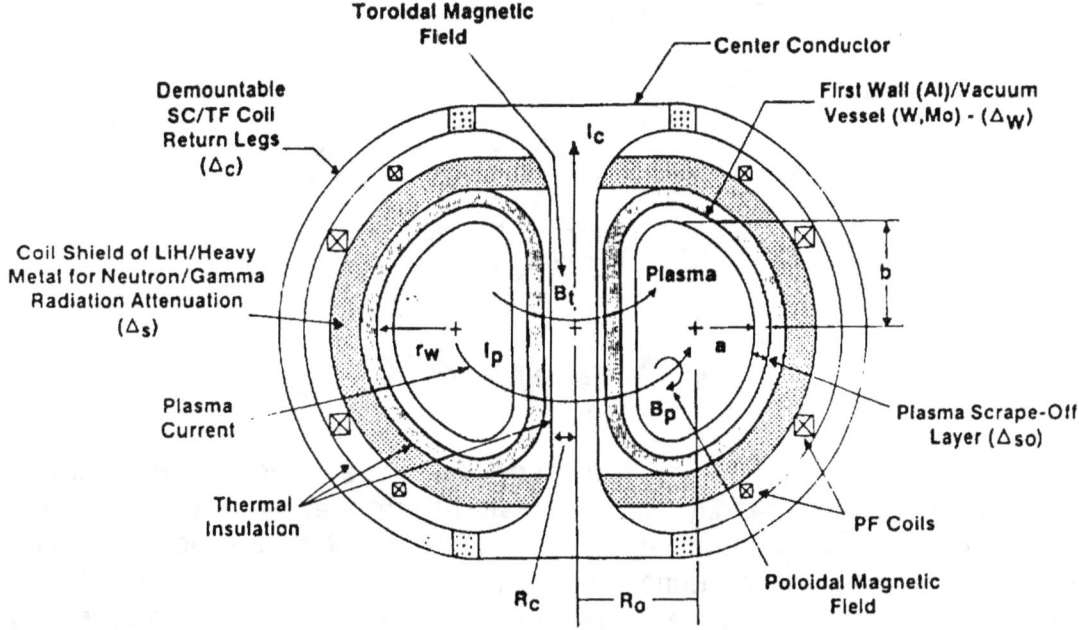

Fig. 7.6. Spherical Torus.

Attractive parameters (including $\alpha_p \rightarrow 5.75$ kW/kg) were postulated based on the possibility of polarizing the D-^{3}He fuel to suppress neutron generation. This design allowed the use of superconducting magnets rather than the copper magnets of most terrestrial spherical torus designs, which would eliminate the requirement for a large recirculating power fraction and the mass associated with power supplies to run resistive coils. There is no spherical torus program currently in operation.

7.2.1.4 TANDEM MIRROR

The tandem mirror, shown in Fig. 7.7, is the leading linear fusion reactor configuration and appears to be attractive for space applications based on preliminary conceptual designs (San88).

7.0 Fuel and Design Options for Space Fusion Reactors

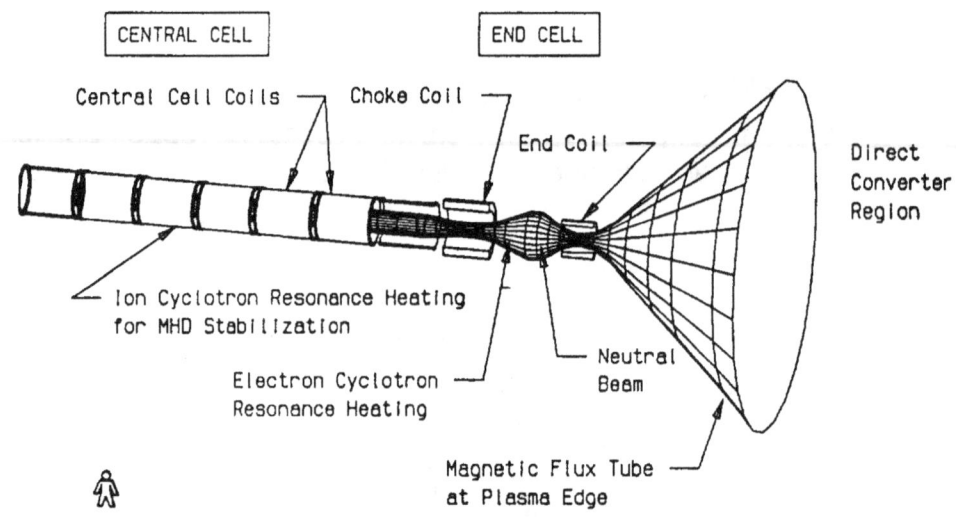

Fig. 7.7. Tandem Mirror.

Design values for the tandem mirror's specific power were greater than 1 kW/kg, and the designs readily accommodated the generation of direct thrust over a wide range of specific impulses and for direct conversion of fusion energy to electricity. A key difficulty with relying on developing the tandem mirror for space is that only a very small terrestrial research program presently exists for the concept, although there is a moderate experimental and theoretical data base available. This situation has arisen because budgetary constraints caused the Department of Energy to eliminate most of the mirror research program and to concentrate essentially all of its funding into research that supports the mainline tokamak program. This decision occurred in 1986 and, although some important "moth balled" experiments remain, a substantial effort would be required to develop a tandem mirror for space.

7.2.1.5 SPHEROMAK

The Spheromak, shown in Fig. 7.8, is also a compact toroid and shares with the FRC the desirable feature that lends itself to lower power designs, but it is at an early developmental stage.

7.0 Fuel and Design Options for Space Fusion Reactors

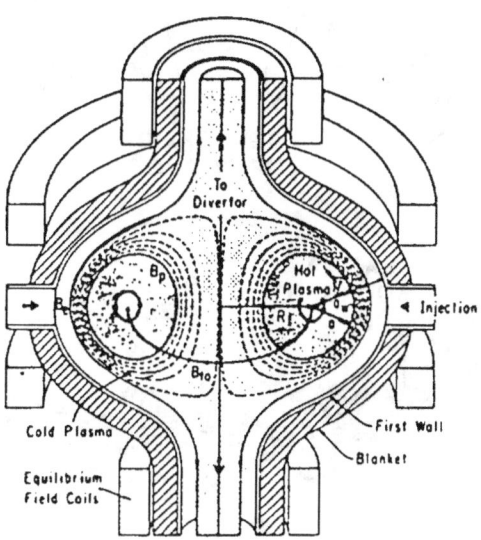

Fig. 7.8. Spheromak.

A space version of the Spheromak has been examined and found to warrant consideration for space (Bor87). Although the spheromak would operate at a β value two to five times lower than that of the FRC, its similar geometry would also allow the generation of direct thrust. The present worldwide effort on the Spheromak is relatively small, and most of the United States spheromak research program was recently terminated by the Department of Energy in favor of the FRC.

7.2.1.6 ELECTRIC FIELD BUMPY TORUS

The Electric Field Bumpy Torus (EFBT) (Fig. 7.9) was the major configuration pursued during the NASA Lewis Research Center's space fusion program (Rot72).

7.0 Fuel and Design Options for Space Fusion Reactors

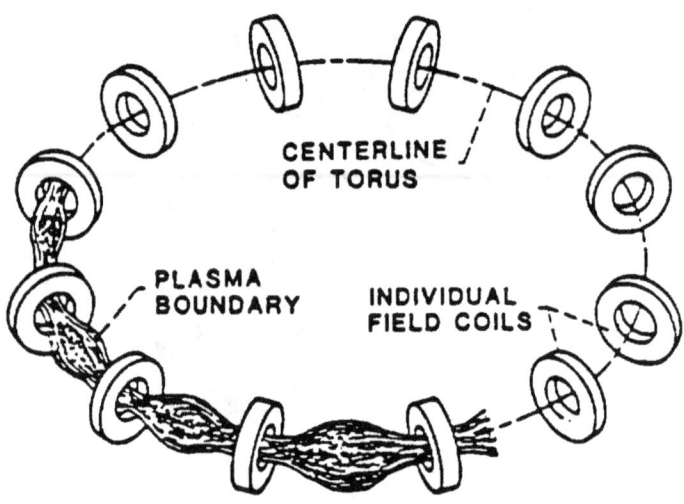

Fig. 7.9. Electric Field Bumpy Torus.

The key benefits of the EFBT concept are steady-state operation and solenoidal magnet geometry. This reactor combined electric and magnetic fields to confine the plasma and to provide for plasma heating and stability. The program was concluded in 1978 when NASA terminated its fusion research endeavors as a result of a reduced Agency budget. The NASA program is discussed in depth in Appendix A (Sch91).

7.2.1.7 ELMO BUMPY TORUS

This reactor (Fig. 7.10) was funded by DOE and was operated at the Oak Ridge National Laboratory.

7.0 Fuel and Design Options for Space Fusion Reactors

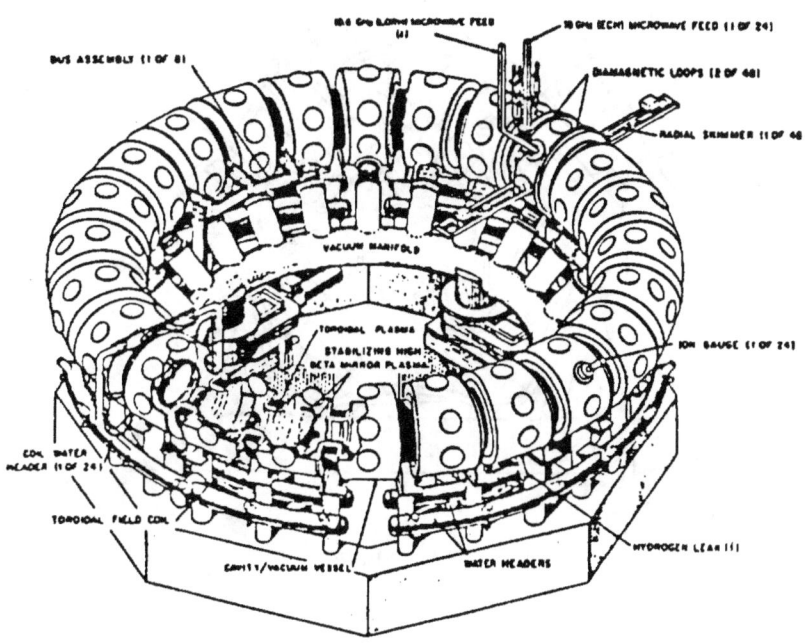

Fig. 7.10. ELMO Bumpy Torus (EBT).

This, too, is a steady state machine with β values up to 0.5. The principle is to use rf heating of electrons to a high relativistic energy level, which in turn heats the plasma ions. The significance of this reactor is that it is one of the few steady state, high β reactors. It is no longer part of the DOE fusion research effort, primarily due to problems in overcoming some inherent transport loss limitations, although a small worldwide effort exists. A postulated change to improve the transport properties of the initial EBT concept, the Bumpy Square (BS), was proposed but not funded.

7.2.1.8 OTHER MCF

Some other configurations presently pursued in the terrestrial fusion research program include the stellarator (Fig. 7.11) and the reversed-field pinch (RFP).

7.0 Fuel and Design Options for Space Fusion Reactors

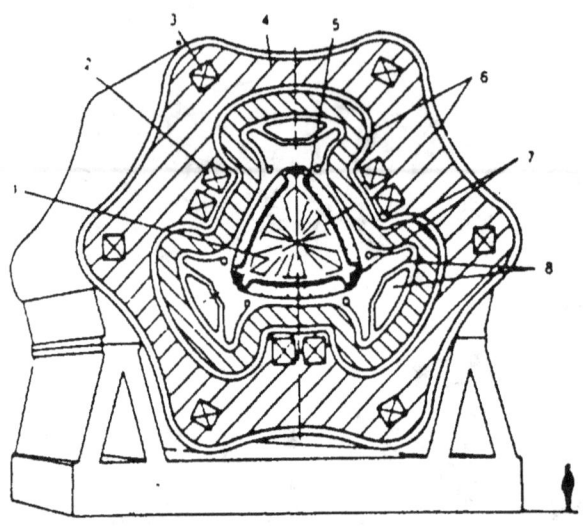

Fig. 7.11. Stellarator.

The stellarator would operate steady-state but is very large (massive). The RFP substantially reduces the magnetic field needed from that of the tokamak, but present designs require large recirculating power fractions and associated massive equipment. Although these concepts both have a moderately developed data base and significant experimental facilities exist, present terrestrial designs indicate that these devices would be intrinsically massive. However, breakthroughs may occur which would significantly enhance the feasibility of the concepts for space applications. Conceptual designs specific to space applications have not been performed.

There are many other confinement concepts funded out of the national program, but the most recent and most extensively developed ones have been covered.

7.2.2 INERTIALLY CONFINED FUSION (ICF)

Inertially confined plasma operates on the principle of concentration of large energies, usually ion beams or lasers, onto a small fusion fuel pellet target. The outer surface of a fuel mass is ablated away under the rapid high energy release, sending a shock wave toward the center of the fuel pellet, thereby compressing the fuel to meet fusion parameters (Fig. 7.12).

7.0 Fuel and Design Options for Space Fusion Reactors

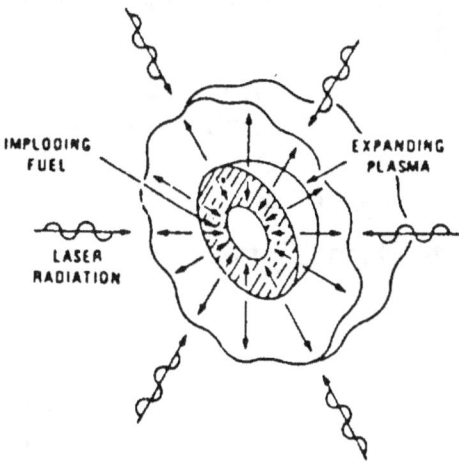

Fig. 7.12. Inertially Confined Fusion (ICF).

These reactors operate under the principle of satisfying the Lawson criteria by achieving very high densities for very short periods of time. Great success has been met with matching theory and predicted results with data, so in that sense it is considered to be better understood than MCF. It is not a new concept, one having originated approximately 20 years ago.

Obtaining insight on the status of ICF for space is difficult because of its strict classification nature. This work is funded by the Department of Energy. Much foreign information, paradoxically, is unclassified and is being pursued publicly by the Japanese, as are all fusion concepts. Whatever conclusions one draws from the United States ICF program has to be taken on the faith and enthusiasm of those who have been involved. Without that assurance, one would have to question the practicality of ICF as a viable space energy concept, at least for the near term.

Unlike MCF, in more recent times a study was conducted to examine the potential for ICF to perform interplanetary missions (Ort87). The vehicle was named "VISTA." It delivers a 100 MT manned payload to Mars and returns 100 MT in 100 days, including a two week stay. A vehicle having an initial mass of 6,000 MT was considered. A conceptual drawing is shown in Fig. 2.13.

Because the fuel pellet gain (ratio of energy out to energy in) is much higher with D-T than with D-^{3}He, the VISTA concept maximizes performance using the D-T fuel cycle, allowing neutrons to escape freely into space. The thrust level is high, providing minimal gravity losses, a factor taken into consideration in the analysis. The current status of gain is classified, and this is one of the key parameters essential to understanding its vitality as a fusion energy source. The achievement of a gain greater than 1,500 is required by this concept. The next laser facility, the Laboratory Microfusion Facility (LMF), is now on the

7.0 Fuel and Design Options for Space Fusion Reactors

drawing boards for operation in the next decade. It is expected to achieve a gain near 100 by the late 1990's.

The authors are to be commended for a well conducted study and thoughtful analysis. More experimental results would be necessary to support and substantiate such system analyses. Data are essential to better define the characteristics of fusion ICF spacecraft.

These, then, are some issues which ICF must address:

Fuel

The concept as now defined burns D-T fuel which, as stated earlier, is clearly not the preferred space fuel for MCF because of safety and space system performance and operational reasons. D-T was proposed in the VISTA study as the preferred ICF fuel because it down-sizes the laser ignition requirements. Tritium production will require a large number of reactors with the attendant environmental impacts.

Gain

To obtain the high specific power, a high gain reactor design is essential, assumed to be 1500. That is a considerably higher value than demonstrated, believed to be on the order of 4 orders of magnitude beyond today's capability.

Laser technology

ICF will require a significant increase in laser output technology from the current level of 0.1 MJ to the requirement of 5 MJ. The current lasers are almost 300 meters in length already, and clearly a lot of work is essential to scale down the size. There is no known technical reason prohibiting reduction of the laser size according to the proponents of the ICF technology. Advanced laser designs are on the drawing board to cause size reductions and to increase the output. This would require the use of the excimer gas laser to accommodate the increased output to 5 MJ. Because this work is weapons research related, however, little incentive exists to down-size the laser design for a flight application.

Start-up

A tremendous power supply is essential to power the bank of lasers, currently planned to be between 100 MJ to 200 MJ.

Ignition

The readiness of this concept to proceed further cannot be established without understanding the status of gain, the key indicator of viability. It is apparently advancing at a rate sufficient for those who have access to test results to suggest its consideration.

7.0 Fuel and Design Options for Space Fusion Reactors

Safety

The nature of nuclear reactions in the ICF mode is different from the low pressure MCF, thereby providing some reduction in the fraction of fusion energy emitted as neutrons. The neutron level, nevertheless, is still quite high, causing it to be a significant consideration in vehicle design and operations, particularly in the vicinity of SSF. (Refer to Rot90.)

Reliability

Another concern is the capability of the reactor to reliably deliver an accurately targeted, uniformly distributed, driver energy load to a moving, very small (1 mm) target over a period of days to months at a repetition rate of 30 hertz. The target pellets must be manufactured as the vehicle proceeds during powered flight.

Electrical Power

The high energy missions under consideration are also large consumers of electrical power. The inductor concept is new and requires a considerable amount of analysis to determine its ability to perform the intended function at a high efficiency. While that would not detract from the propulsion capability, there are overall system implications for the long duration missions and consequently implications on the strength of ICF as a good candidate energy source for those missions.

Operations

Operations in flight must be simplified or the concept will not be used unless this is a totally enabling technology. Simplified fusion flight operations, to name several, must include: minimum perturbations to the Space Station Freedom or to other flight operational spacecraft, minimum effects on the Earth's environment, minimal interface with Earth based control centers for space flight control, avoidance of complex cool down procedures as required with NERVA. The availability of the tritium mass required is another concern.

Testing

Many or nearly all of the system components should be verifiable on the ground, and this is expected to be possible. However, the vehicle as a system will require in-space qualification testing due to its size as well as a consequence of the neutron and x-ray emissions.

Radiators

The system requires a substantial improvement in the specific heat rejection capability of radiator design technology. Otherwise the specific power performance capability will not be reached.

7.0 Fuel and Design Options for Space Fusion Reactors

Plasma

There is a need to understand the plasma characteristics in an unconfined space such as that which VISTA uses and the plasma's interaction with the magnetic thrust chamber. The plasma kinematics, particularly with regard to cooling and recombinations, will need to be better understood if the quoted efficiencies are to be attained.

Further analysis and design studies on the aforementioned topics are warranted but will not be pursued for space unless NASA elects to undertake a fusion program. The exception is the demonstration of ICF gain which DOE will continue to develop for their applications.

This is an important concept, managerially speaking, from two points. One, ICF offers not only an entirely different alternative concept but an entirely different technical approach to fusion. It, therefore, provides a technology back-up in the event that MCF does not yield the desired results. The other reason is leverage. One of the most expensive issues to resolve is demonstration of gain, i.e., research which NASA can acquire without funding. We only need to simply monitor progress. The other major expense is down sizing the mass driver, a goal not high on the DOE priority list, one requiring space funding.

The ICF progress has been stated to be good; and, even more importantly, the experimental results are stated to match well with theoretical predictions, or at least better than with the MCF machines. For space the preferred fuel is D-^{3}He. Hence, the technology where NASA may have to contribute to ICF funding is in the demonstration of the higher gain values required for D-^{3}He and light weight drivers. Whether or not it will serve as a space power and propulsion system depends on the viability of the issues raised here, and that is something that NASA alone will have to explore.

The performance is certainly attractive for solar system mission applications since it inherently simplifies the first wall material engineering problems of MCF and the thermal constraints of mechanical chambers, but driver energies (50 MJ) for fusion of D-^{3}He must be developed into a package of sufficiently low mass that high specific powers can be designed. Other ICF space related research required consists of improved radiators and plasma/thrust development. An order of magnitude in specific mass radiator performance is necessary. The pellet formation, targeting, inductive recirculation power generation, and plasma thrust conversion are others. This concept will be unacceptable for long duration missions because of tritium's short 12.3 year half life. Breakthroughs in research permitting the generation of tritium during flight could become an option.

7.2.3 OTHER CONFINEMENT CONCEPTS

The following concepts have been examined in varying depths, but they are not pursued under the Department of Energy's fusion research program. All of these options have only a very small or nonexistent data base at present and require substantially more extrapolation of parameters to reach the reactor regime. Therefore, they can only be considered as speculative at present and will remain so until testing has been conducted to validate or invalidate the concept. There are also others, like the Dense Z Pinch, a small LANL program of ~$300K, which have not even been listed.

The intent of this paper is to indicate where their "maturity" stands relative to the mainline concepts. One valuable management tool which a SFP could provide is to make available the program funding by which experiment confinement options such as these could be explored. The funding is too great for individuals to sponsor and the risk too high for capital venture. Some other examples of non DOE funded concepts follow.

7.2.3.1 RING ACCELERATOR EXPERIMENT (RACE) (LLNL INDEPENDENTLY FUNDED ACTIVITY)

In this design, the plasma is confined without the use of heavy magnets but is physically compressed by a magnetic coaxial gun. The plasma rings are produced at the breach end of the accelerator by a magnetized coaxial plasma gun. These plasma rings are initially accelerated by a breach end magnet and then accelerated axially down the coaxial barrel to velocities on the order of 1000 to 3000 km/sec by a discharge of current from capacitors (260 kJ, 120 kV) producing **J X B** forces with the torus's magnetic field. At the end of the barrel, physical compression of the ring plasma results from a converging section at the gun's nozzle end. Refer to Fig. 7.13 for a description of the reactor.

7.0 Fuel and Design Options for Space Fusion Reactors

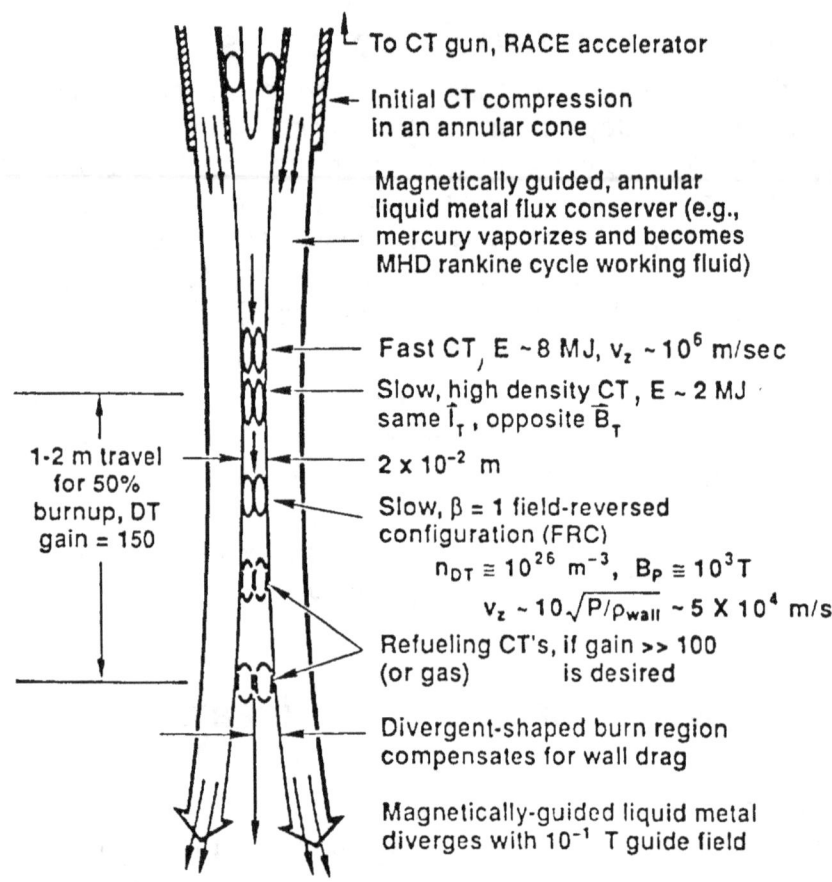

Fig. 7.13. Ring Accelerator Experiment (RACE).

Compact torus plasma accelerators such as RACE may be used as an alternate driver to the laser propulsion concept discussed in Section 2.0.

The gun is now in operation at the Livermore National Laboratory, being funded by internal Livermore research funding (Ham88 and Har88). Its advantages are light weight and simplicity. It requires a hard vacuum for operation. The gun is currently fired a single shot at a time but is believed to be capable of a 50 hz pulse rate. A back of the envelope calculation of its specific power indicates a 10 kW/kg level. The reactor would not be an expensive device on which to conduct fusion science experiments, costing on the order of $10M per year. The reactor scales reasonably for space fusion.

The disadvantages are that it is a relatively new concept and, therefore, has a very limited test background. The stable convergence of the plasma under fusion fuel confinement conditions and retention of hardware integrity under ignition are key concerns. A solution has been suggested (Ham88) to provide a liquid confinement wall (Fig. 7.13).

Stability of the plasma under the high compression is a matter to be concerned with as development proceeds, although this has been examined, and no reason found why in principle it would not work. As part of the fusion reaction, soft x-rays are produced, requiring some shielding. The efficiency has already been demonstrated to be greater than 40%, an order of magnitude greater than lasers. It will burn D-^{3}He, but requires a 50 MJ driver which is 10 times larger than that required for ICF. The means of fusion energy conversion to spacecraft propulsion and power will require analysis and experimentation. If the concept proves feasible for generating net power, it would be a very attractive reactor for space. Its linear topography is envisioned to readily convert to thrust or electrical power generation. It operates in a vacuum environment. Cooling requirements are anticipated to be low.

7.2.3.2 PLASMAK™-(CONCEPTUAL ONLY)

This confinement approach is quite different from any of the more traditional designs. It has the potential for a high specific power but lacks any test support at the present time. If proven, there could be a great opportunity for either space or aeronautical propulsion according to the inventor. To describe its design, an excerpt was taken from a proposal with the written consent of the owner.

> A PLASMAK magnetoplasmoid (PMK) to outside appearances is a highly compressible ball of plasma that is suspended in a thick gaseous atmosphere. Actually, it consists of a super hot magnetized plasma ring surrounded by an insulating vacuum poloidal field (Kernel), which is protected and cloaked within a Mantle of plasma, which interfaces with the surrounding dense gas (fluid) blanket. ... It is the Mantle that imparts uniqueness to the concept. It provides fully ionized plasma to seal the external vacuum field against impurity penetration, while at the same time providing an external confinement pressure for the whole PMK system.
>
> Fusion temperatures and densities are achieved through simple mechanical high pressure techniques which are used to heat the PMK. This is accomplished through a rapid increase in the pressure of the surrounding gas blanket. ... Consequently, ignition temperatures in the Kernel plasma fuel with aneutronic burn densities on the order of megawatts per cubic centimeter are achievable. The fusion energy is released by natural heat radiation from the burning Kernel fuel through the transparent Mantle and into the gas blanket, which becomes a hot, dense plasma which then can be used to operate a multimegawatt or gigawatt inductive MHD electric generator or a propulsion engine.

The concept is illustrated in Fig. 7.14. A number of reviews have been held. The unproven mantel physics is a key controversial element for this concept.

7.0 Fuel and Design Options for Space Fusion Reactors

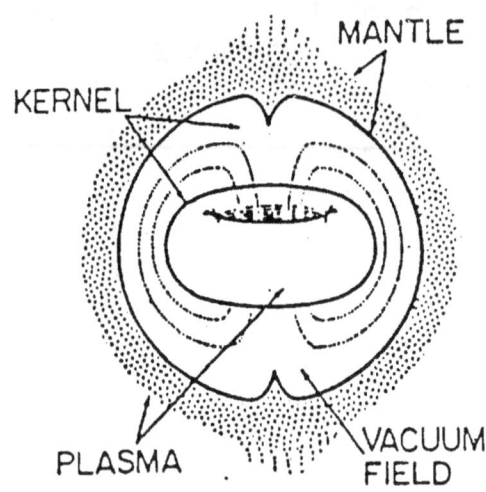

Fig. 7.14. PLASMAK.

Its development hinges upon the ability to generate and maintain the charged plasma sphere, similarly as in ball lightning—to inject fusionable fuels, to compress it mechanically, and to extract the energy to perform useful work. The concept is further described in Kol88.

7.2.3.3 MIGMA

Migma involves the use and generation of self-colliding beams to burn fuels to produce a pure aneutronic reaction. The Air Force had funded this work. Fig. 7.15 shows the reactor's design principle (Mag85).

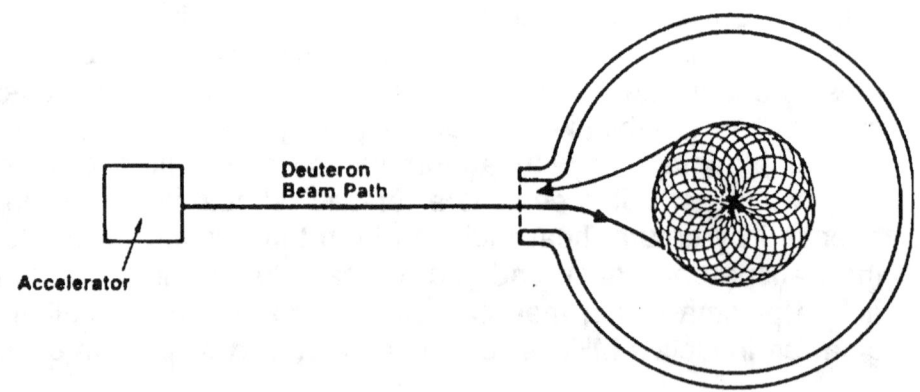

Fig. 7.15. MIGMA.

7-28

The concept's objective is to produce multiple particle orbits which intersect at the center producing head-on collisions of ions. The advantage of this concept is the concentration of high energy within a small space. It has a directed target of ions rather than to rely upon statistical (Maxwellian) collisions. On the other hand, that ion driven reaction entails recirculating power which must be maintained at a sufficiently low level to provide high system efficiency. This approach has high potential for space application by virtue of the directed beam which can inherently produce high specific power systems, provided the recirculation power can be held within reasonable limits.

7.2.3.4 MAGNETICALLY INSULATED INERTIAL CONFINED FUSION (MICF) (HASEGAWA/KAMMASH CONCEPT)

This approach combines features of MCF and ICF. It involves generation of the magnetic field directly from the plasma to insulate the plasma thermally from its surrounding surface.

The inverse of the ICF technique discussed earlier, this MICF scheme uses D-T coated on the inside of a spherical pellet wall (Fig. 7.16).

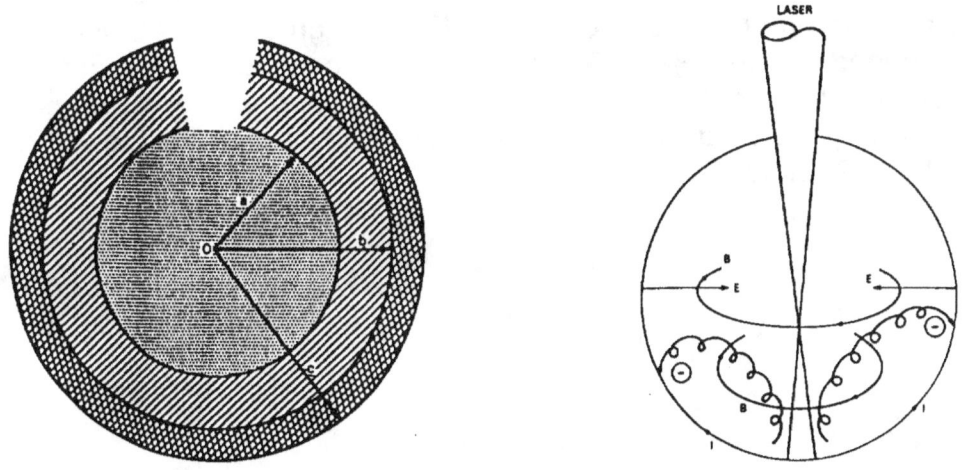

Fig. 7.16. Magnetically Insulated Inertial Confined Fusion (MICF).

A laser impinging upon the inner wall through a hole in the sphere provides the energy to create the plasma. The plasma is contained longer than ICF due to a metal wall constraint where the speed of sound is much higher than in gas as in the case of the ICF configuration. Plasma containment is via the metallic wall, and thermal isolation of the plasma from the adjacent structure is accomplished by the plasma generated magnetic field. The plasma burns and expands adiabatically into an expansion chamber and subsequently into a magnetic exhaust nozzle for propulsive thrust. Because the laser input is 3 MJ and is a single unit, this concept has some inherently attractive features. Refer to Has86, Kam87, and Kam88.

This concept is new and has not received close scrutiny. One concern is whether it will work since as the plasma builds up the laser cut off density is exceeded. The authors of the reference document believe this concern will be abated by the use of an alternate laser which is transparent to the plasma build-up. The Japanese are funding the MICF concept.

7.2.3.5 NEW INTIATIVES: MAGNETIC INERTIAL-ELECTROSTATIC AND MAGNETIC DIPOLE

Subsequent to the completion of this activity's review of confinement options, two new approaches have come to the forefront.

The Magnetic Inertial-electrostatic confinement approach is a new concept using an inertial-electrostatic spherical colliding beam fusion confinement design. It is "based on the use of magnetohydrodynamically stable quasi-spherical polyhedral magnetic fields to contain energetic electrons that are injected to form a negative potential well that is capable of ion confinement." (Bus91) This program, referred to as HEPS (High Energy Power System) is being funded by DARPA. The physics of the concept uses MHD stable magnetic fields to confine energetic electrons. Energetic ions are confined by the electron-generated potential well. This is illustrated by Fig 7.17a and b (Figs. 1 and 2, Bus91).

7.0 Fuel and Design Options for Space Fusion Reactors

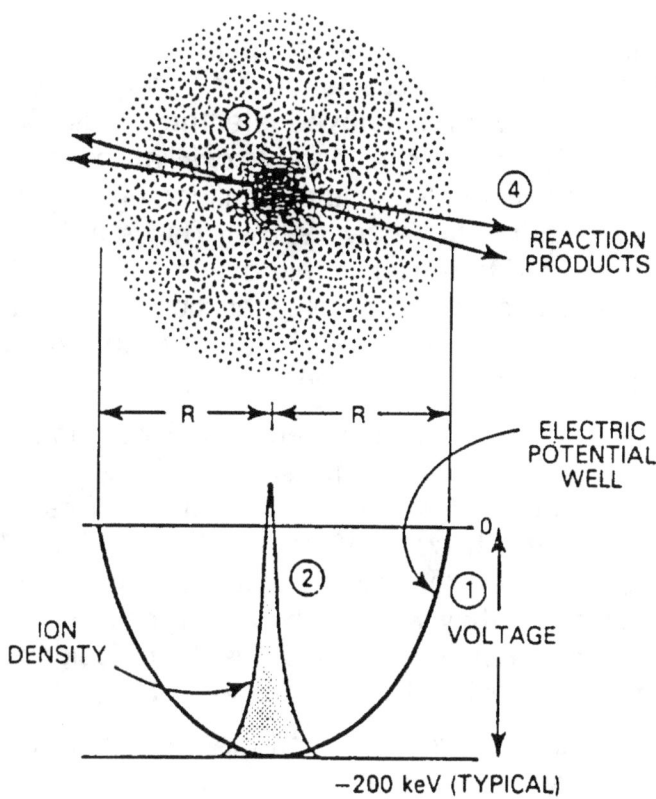

Fig. 7.17a. Inertial-electrostatic confinement: deep negative electric potential well ① traps positive ion fuels② in spherical radial oscillations ③ until they make fusion reactions ④.

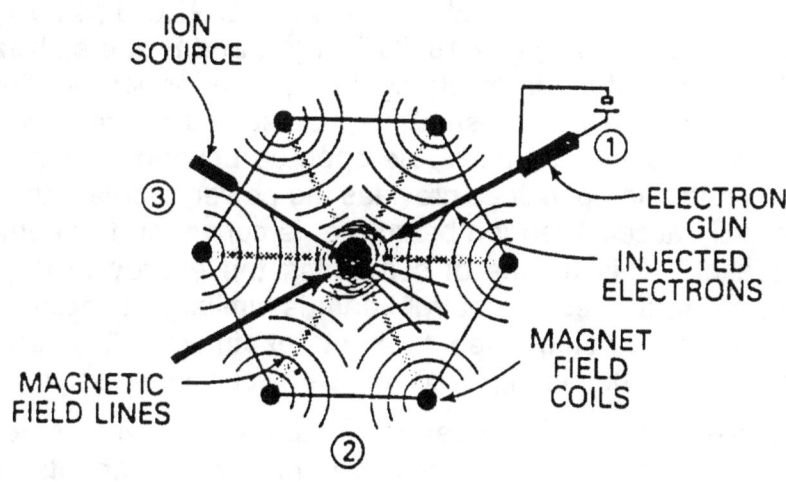

Fig. 7.17b. Inertial-electrostatic confinement: trapping well formed by energetic electron injection ① into cusps of polyhedral magnetic fields ② and ions fall inot well and remain until reacted ③.

The approach is interesting since it is an option to magnetic confinement of Maxwellian plasmas, the approach taken by nearly all of the fusion work performed. The NASA work employed electric fields; MIGMA is a non-Maxwellian approach; ICF is another non-Maxwellian concept. The reference cites the historical treatment of electrostatic confinement of plasmas, work which commenced with Langmuir in the early 1920's. This is a "focused" or "directed" approach which is referred to earlier in this document and intuitively seems to offer a chance of success above a "purely statistical" means to achieve fusion. The physics becomes complex, and the reader desiring a greater understanding should pursue the reference.

The other new development is a concept entitled "Fusion in a Magnetic Dipole." This confinement approach was suggested by Drs. Teller, Glass, Fowler, Hasegawa and Santarius and was presented at the "First International A.D. Sakharov Conference on Physics," Moscow, USSR, May 27-31, 1991. The work is directed to a space propulsion application of fusion energy. (Tel91) The authors concluded that they would achieve a specific power of 1 kW/kg with a system of this nature. With some improvements in design the possibility was suggested of perhaps increasing the specific power to 10 kW/kg.

The concept is well described in Tel91, and the description has been extracted here for ready reference.

II. REACTOR PROPULSION SCHEME

As a plasma confinement configuration, we choose the simple magnetic dipole shown in Fig. 7.18. Coil C (the dipole) carries a large current, of order 50 MA, and provides the strong field that confines the D-^{3}He plasma in an annulus about the coil, as shown in Fig. 7.19 Coils A, A´, and B (the stabilizer) provide a weaker field that levitates the dipole against gravity or acceleration, at a stable position between the coils. The stabilizer also serves as the "divertor," whereby the closed magnetic lines of the dipole open up beyond an X-point (field null). Heat diffusing onto the open lines provides the power to create thrust in the form of a magnetically accelerated ion beam that is converted to neutral atoms as it exists the rocket. This means of converting the energy of the magnetically confined plasma to a directed neutral beam is similar in principle to the neutral beam injectors now being used to heat tokamaks. The arrangement to accomplish this, sketched in Fig. 7.20, will be discussed in Section III.

We have chosen the dipole over the more extensively studied tokamak because of its greater simplicity, especially the divertor, and its higher specific power. Topologically the tokamak is also a "dipole," in that the toroidal plasma carries current producing the same (poloidal) field configuration as that shown in Fig. 7.18 but without the complication of a material internal ring to carry the current. However, the tokamak has other complications that appear to make it less suitable for space applications. Namely, whereas the metallic internal ring of the dipole configuration is rigid, current carried by the tokamak plasma causes violent instability that must be overcome by a much stronger toroidal field supplied by large coils interlinking the toroidal plasma. This has the virtue of creating high-shear closed magnetic flux surfaces to confine the hot core plasma. However, for space applications, there is the major disadvantage that a divertor coil to open up the flux surfaces to allow propellant to escape must compete with the strong toroidal field, whereas in the dipole the divertor field need only compete with the weak outer regions of the poloidal field of the dipole coil. Thus, though the tokamak has a divertor of sorts, it serves only to dump heat on the interior walls, inside the toroidal coil structure, and does not provide an escape path for propellant. In addition, the simpler dipole is expected to be much less massive than a tokamak of comparable power and therefore to produce greater specific power.

II.A. DIPOLE PROPERTIES

The dipole configuration has recently received renewed attention, as a candidate D-^{3}H3 reactor (4.5). As noted in these references, according to theory, supported by planetary and space observations, the dipole exhibits remarkable magnethydrodynamic (MHD) stability up to local values of the pressure parameter β exceeding unity.

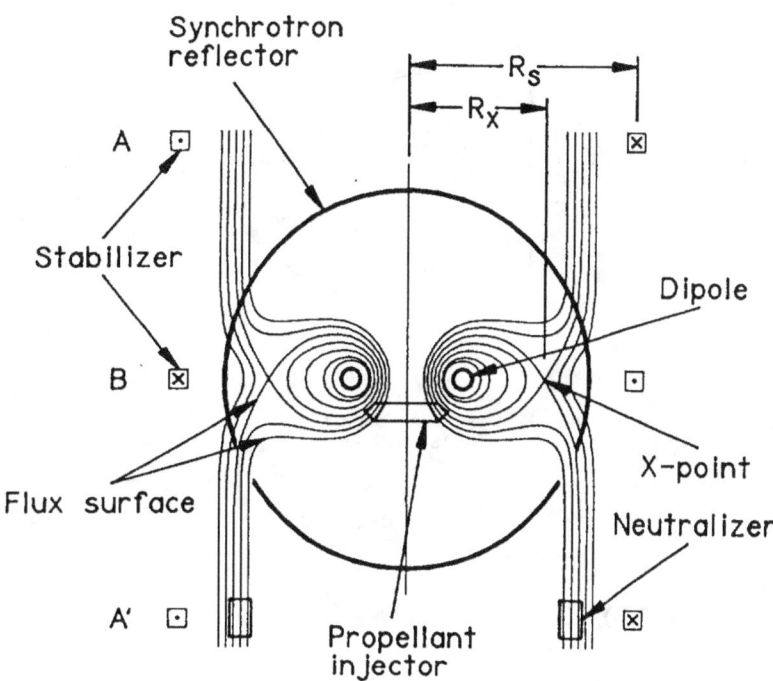

Fig. 7.18. Dipole Reactor Propulsion Scheme.

7.0 Fuel and Design Options for Space Fusion Reactors

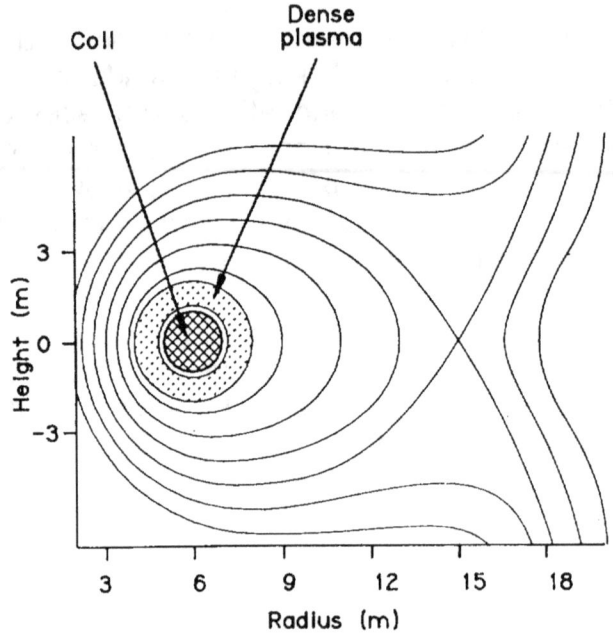

Fig. 7.19. Expanded cross-sectional view of magnetic flux surfaces and plasma.

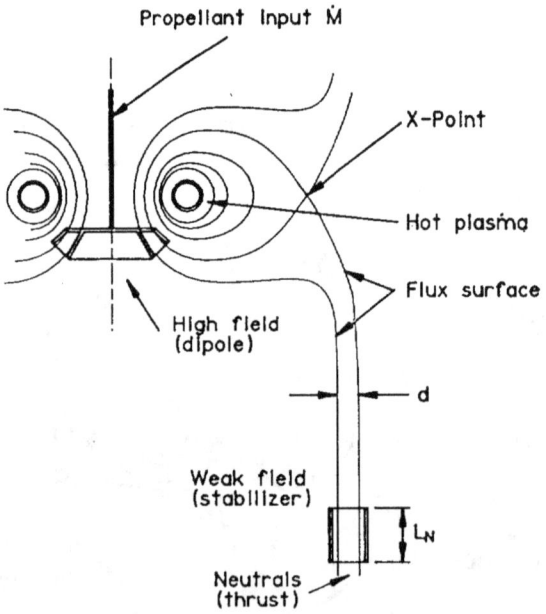

Fig. 7.20. Detail of propellant feed and thruster.

7.2.3.6 FURTHER READING

There are still other confinement approaches used or proposed. This document was not intended to provide an all-inclusion listing and description of devices proposed and tested. The judgments of many individuals were used to establish a representative cross section of key fusion experiments, at least sufficient to include a representation of the various fundamental approaches. Where later configurations have evolved, the early experiments which led up to the more matured experiments were not included. Additional details on some presented herein as well as a discussion of others not included in this document can be found in many reference documents. These include Gla60, Tel85, Che74, Mil76, and Rot86. Many details can be obtained from the American Nuclear Society's *Journal of Fusion Technology*. A good overview of fusion in a non technical book may be obtained from Bro82. I have also cited and updated concepts not found in any of the earlier texts.

7.3 EVALUATION OF SPACE APPLICATIONS OF FUSION ENERGY USING CURRENT FUSION ENERGY CONVERSION CONCEPTS

In the remaining part of this section the best developed and preferred fusion experimental approaches are considered for space propulsion and electrical power applications.

7.3.1 PROPULSION: FUSION ENGINE

The primary objective and interest in fusion energy for application to space is in development of a fusion rocket engine and secondarily in the generation of electrical power. Several of the reactor concepts discussed in Section 7.2 can be considered as potential options for space applications which should receive consideration for further design study and for experimental testing. Testing of space applicable fusion reactors is very important since, in the fusion technology particularly, many unexpected phenomena have been uncovered by testing, a situation not unlike our current chemical propulsion technology programs. The configurations for energy conversion systems for space propulsion will differ from that used to generate electrical power for utility companies. The fusion program's test results are **configuration sensitive**, and consequently new and different designs from the mainline program's tokamak experiment will produce new and different problems. This is a matter where the space and terrestrial programs diverge and is part of the rationale why NASA should become involved with space fusion testing, analysis, and experimentation.

The following discusses the DOE leading fusion design concepts and other proposed concepts as well. Comments relative to their applicability for space are provided with the goal of defining the direction where the space program should look as a starting point for fusion energy conversion.

7.3.1.1 MAGNETICALLY CONFINED FUSION (MCF)

Magnetic confinement is the initial concept by which efforts were made to produce controlled fusion as an energy source. The first major program, Project Sherwood, commenced in 1950. The DOE MCF systems have received the vast majority of the funding and therefore the most attention to theory, design, experimentation, and testing. With some exceptions, the MCF reactors have basically been low pressure, low β reactors; and they consequently exhibit low specific power characteristics.

As mentioned, the space fuel of choice is deuterium-helium-3. The most important question, then, is to decide the design nature of a reactor which can burn that fuel combination and meet the system performance parameters (Section 6.0) enabling NASA to meet anticipated mission requirements. This implies higher magnetic field strengths, low cyclotron radiation, low mass designs, and good plasma stability at higher temperatures.

Two basic MCF design configurations have received the most attention for terrestrial use: toroidal and linear. Keeping in mind that space reactor designs generally must produce a minimum specific power value of 1 kW/kg, preferably higher, one realizes that significant weight reductions are necessary in the current mainline MCF designs. The reactor's two major mass components are it's magnetic field load carrying and supporting structure plus its neutron shield. One advantage of the space operation is the readily available vacuum "tank," eliminating that terrestrial design requirement and the attendant mass to accommodate a "clean" vacuum facility.

There is interest in higher specific power reactors for the commercial power plants, too, although the motivation is more a profit related issue rather than preempted by a hard performance requirement. A study was performed at the request of the DOE on high power density fusion systems (Dav85). Some very positive suggestions were provided which supported the need, even for the terrestrial program which is not typically considered mass sensitive. Unfortunately, funding was not made available.

This study's conclusion after considering all of the MCF designs is that the clear initial choice for proceeding with the development of a possible NASA program for space fusion reactor design is the Field Reversed Configuration (FRC). Currently in an early state of development, albeit at a very modest level, work or it is being terminated in 1991. Details and discussion on the various designs' suitability for space follow. Considerable emphasis is placed on the reactor's β. Section 8.0 examines the preferred reactors in greater depth.

7.3.1.1.1 FIELD REVERSED CONFIGURATION (FRC)

The FRC (Fig. 7.3) presently appears as a concept potentially capable of meeting space required specific power levels. It combines some of the attractive features of both toroidal and linear systems. Although it is classified as a compact toroid, the linear topological nature of the external magnetic field lines would be conducive to the production of direct thrust. Several terrestrial FRC reactor designs have been performed. The machine in operation at Los Alamos was terminated in 1990 and the other at Spectra Technology in conjunction with the University of Washington is scheduled to be shut down in 1991 because of funding limitations. This work is considered to be applicable to the space program although no space version actually exists.

As mentioned, the attractiveness of this machine stems from its high β ($\leq$90%), good plasma confinement scheme, high power density, potential for steady state operation, and overall compact design. The confinement is provided by two end magnets and a reversed field which may be initiated and sustained by a number of methods. A toroidal current produces the confining magnetic lines of force along the poloidal axis (Fig. 7.3). One possibility to heat the fuel to ignition is by quickly compressing the plasma with a rapid ramping of the plasma current and by increasing the magnetic field. The fusion products heat the surrounding plasma, providing attractive reactor energy multiplication.

Thrust for a fusion engine is produced directly by a magnetic nozzle at one end, accomplished by a field imbalance. The use of the magnetic nozzle and plasma entrapment makes this concept attractive because the plasma remains physically away from the wall.

The performance predicted from the current work is presented in Table 7-5 (Cha89):

TABLE 7-5. FRC Predicted Performance.	
Fusion Power	0.5 GW
Plasma Volume	80 m^3
Ion gyroradius	0.01 m
Plasma radius	1.5 m
Ion Temperature	100 keV
Exit Velocity	10^7 meters/sec
Specific Impulse	10^6- 10^3 seconds
Thrust	0.4-50 kN
Propellant addition	0-0.8 kg/s
Specific Power	10 kW/kg

7.3.1.1.2 TOKAMAK

The mainline tokamak reactor, the best understood fusion experimental device, is designed to demonstrate controlled fusion for terrestrial electrical power generation. A tokamak suitable for space propulsion or in-flight power has not been designed, and it is likely that such designs would have difficulty achieving the required high specific powers. The current terrestrial design has a specific power of 10^{-4} kW/kg. Perhaps it could be considered for a lunar surface based power supply to produce laser driven or ablation propulsion systems as discussed in depth in Appendix B and Section 2.0.

7.3.1.1.3 TANDEM MIRROR

The tandem mirror, shown in Fig. 7.6, is the leading linear fusion reactor. It warrants further consideration. As discussed earlier, that view point is based upon preliminary conceptual designs considered by the University of Wisconsin (San88). The prime factors for continuation of its consideration are its capabilities for specific power greater than 1 kW/kg and for direct thrust conversion over a wide range of specific impulse values or, alternatively, the direct conversion of fusion energy to electrical power. A key difficulty in developing the tandem mirror for space is that, although a moderate experimental and theoretical data base is available, only a very small terrestrial research program presently exists for this concept. The funding constraints placed upon the DOE program allow it to continue with only one mainline activity, the tokamak being the one selected. That decision occurred in 1986; and, although some important tandem mirror experiments remain, a substantial effort would be required to develop a tandem mirror for space.

7.3.1.1.4 COMPACT TOROIDS

Compact toroids include the Spheromak and FRC. At this point it would be prudent to include the compact toroids as potential options since they exhibit relatively high betas. Their development significantly lags other experiments.

7.3.1.1.5 CONVERSION OF PLASMA TO THRUST

An investigation into the means for conversion of plasma into thrust was initiated by NASA subsequent to the initiation of this study. Plasma throat and nozzle analysis is being performed by MIT. That program assumes that plasma confinement under burn conditions can be demonstrated. Another thrust option, direct matter injection, high energy ion beams used to accelerate hydrogen propellant along magnetic tube of flux, was studied at NASA Lewis Research Center in the 1960's (Eng66).

7.0 Fuel and Design Options for Space Fusion Reactors

The Air Force recently completed a study concerning the "Characterization of Plasma Flow Through Magnetic nozzles." (Ger89) A wide variety of temperatures of plasmas was investigated ranging from 1 ev to 100 ev with some consideration given to temperatures as high as 1 keV. The report's abstract concludes with the statement that "... the use of plasmas for space vehicle propulsion is a natural and interesting application for plasmas that well deserves further study." Several points are raised pointing to the importance of further work on this subject. There is a question raised concerning the viability of plasma propulsion at high densities, where $n = 10^{17}$ to 10^{18} cm^{-3}, due to radiation losses during plasma transit through the nozzle. Another is the effects of plasma exiting the nozzle. These reports and analysis tasks are cited to illustrate the effect of the space application on the technology and the importance of a space fusion program. Clearly the technology needs much greater emphasis.

7.3.2 ELECTRICAL POWER

Two major separate flight applications have been identified for space-generated electrical power. Electrical power can be made available for the purpose of converting it to a propulsive devise such as an ion engine or for direct use for electrical power to accomplish flight operations such as science experiments, data transmissions, life support, materials processing, etc.

7.3.2.1 DIRECT CONVERTER

Design concept

The kinetic energy of plasma particles from one end of the reactor is converted directly into high DC voltage by the application of bias voltages to grids to recover electrons at one end and ions at the opposite end. The system is shown schematically in Fig. 7.21.

7.0 Fuel and Design Options for Space Fusion Reactors

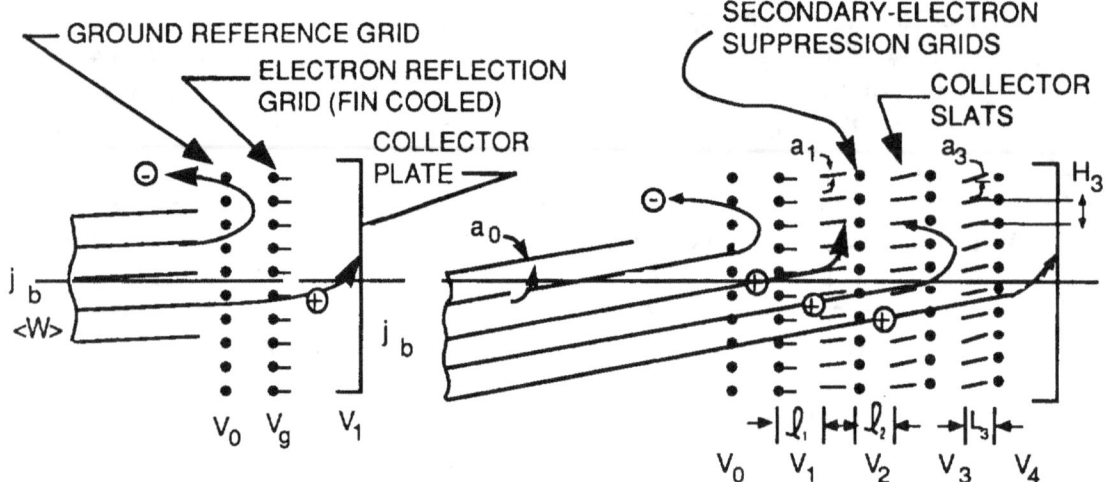

Fig. 7.21. Direct Converter (Mil76).

Cooling of the grids is required due to heating by the ion flux. It should be pointed out that this system produces high voltages, on the order of 1 MV. The techniques for the handling of these high voltages in a space environment, including system interactions with a fusion reactor and plasma exhaust, have only been addressed conceptually (Kul87).

Status

The objective of the direct conversion system is to increase power conversion mass efficiencies to avoid inherent inefficiencies attendant with the thermal conversion systems. The importance of mass minimization is illustrated by the performance advantage as shown throughout Section 2.0. Relative conversion efficiencies are illustrated by Fig. 2.53.

Charged particle products from the reaction make this approach possible, another reason for the D-^{3}He reaction preference. In a recent paper by Dr. Perkins, it is concluded that "ultrahigh efficiencies (above 60%) will require almost perfect mating of the components." (Per88) Testing on direct power converters has been accomplished at Lawrence Livermore National Laboratory (LLNL) at up to 100 kV with a net efficiency of 48% (Bar81). The best test result attained was 86.5 $\pm$1.5% (Bar77). High voltage breakdown was a concern expressed also by the researchers. A first cut was made at addressing high voltage breakdown and other operational issues for a space fusion reactor using direct conversion (San88). Tentative solutions were given.

7.3.2.2 SPACE APPLICATIONS

Only one magnetic fusion space power study was found to have been conducted in more recent times, SOAR. That preconceptual study was performed by the University of Wisconsin for the Air Force for a 1000 MW$_e$

fusion power plant reactor for space applications. The study results showed that a D-^{3}He fuel reactor could deliver a specific power of approximately 2 kilowatts of electricity for every kilogram of material orbited (Kul87). The order of magnitude of fusion systems is illustrated in Fig. 7.22 which compares both a 250 MW and 1000 MW reactor with the Shuttle Orbiter.

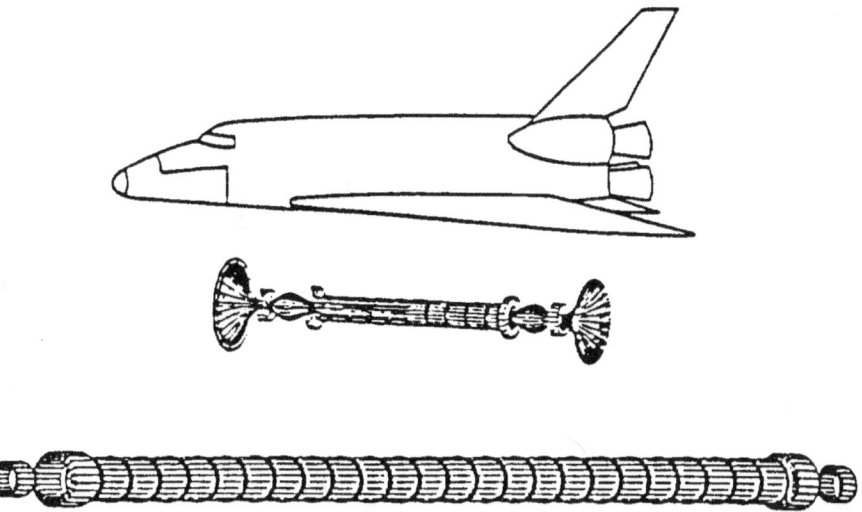

Fig. 7.22. Order of magnitude of fusion systems comparing both a 250 MW and 1000 MW reactor with the Shuttle Orbiter.

The systems were designed to yield that power for 600 seconds. Use was made of the highly efficient electrostatic conversion of energy to electricity. The optimum performance confinement scheme shown by their calculations is the tandem mirror design. While this particular design would not be applicable to the NASA space missions without modifications, at least for those missions considered herein, it gives an understanding of the advanced thinking and planning for space now under way outside of NASA and shows the feasibility of fusion power application, based upon preliminary calculations.

7.4 SUMMARY

The preferred fuel for space is deuterium-helium-3. There are many design concepts and experiments for demonstrating proof of principle. None have been given a through evaluation for the space application. The one preferred from this review is the Field Reversed Configuration. There are likely to be other options.

8.0 STATUS AND PERFORMANCE OF POTENTIAL SPACE FUSION REACTORS

The major objective of this study has been to examine high energy space science and exploration missions, mission requirements, the program impact on NASA for the implementation of those missions, and the planning required for their fulfillment. In particular it examines the means to achieve those high energy missions including the energy conversion approaches, their relative merits, and technical status. This section examines those fusion reactor confinement concepts presented in Section 7.0 having a sufficient data base to perform a reasonable extrapolation of the concept into the space reactor regime and emphasizes the status of the reactor designs having the greatest potential for space. It presents data relative to the critical fusion reactor parameters discussed in Section 4.0 and evaluates the chances of offering a successful space application, particularly with regard to meeting the vehicle system and mission requirements as presented in Sections 2.0 and 6.0.

8.1 TERRESTRIAL PROGRAM STATUS – GENERAL
BACKGROUND

Great progress has been made on a technology that can be considered as one of mankind's greatest technical challenges in the physical sciences.

The perception frequently encountered during this study, however, indicated a concern by individuals who are not acquainted with fusion technology that fusion is not a viable source of energy because it has been researched for a long period without producing net, controlled energy. That perception results more from a lack of understanding of the progress which has been made in the terrestrial fusion program rather than being based upon some fundamental physics or technical issue.

There are a number of key parameters to be considered when using progress yardsticks. Consider the ratio of the equivalent energy output per energy input (Fig. 8.1).

8.0 Status of Potential Space Fusion Reactors

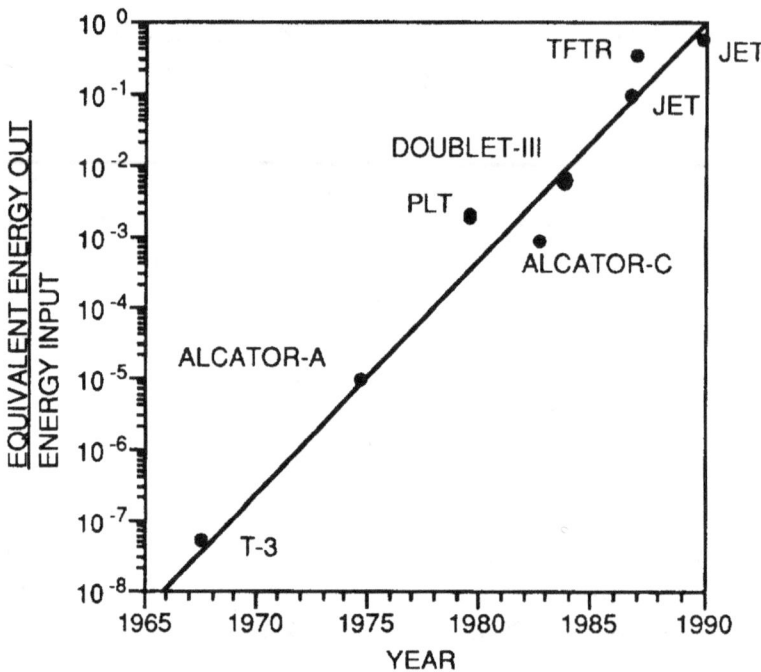

Fig. 8.1. Plasma experimental confinement progress as expressed by the ratio of the equivalent energy output per energy input (San88 – updated to 1990).

The magnitude of the technical difficulty with producing controlled fusion is demonstrated by a longer terrestrial fusion development period than initially expected, and the work remaining even after first demonstration should not be underestimated. Confidence that controlled fusion energy will be demonstrated is high because even more difficult problems than originally anticipated are being solved. During the early phase of this study, the projection was made that breakeven was expected within several years. The first demonstration holds great psychological importance. The production of net power will be of even greater significance as in the successful demonstration of any technology. Consider the difficulty with the achievement of the first manned flight, an event preceded by hundreds of years of failures, and then consider the progress in air travel once the technology had been demonstrated by the Wright brothers. The development of Bell's telephone illustrates the point too. Look at the advancements in electronics following the discovery of semiconducting materials. Fusion, once demonstrated, will most likely make great comparable strides. In fact, that is true with most technologies. Fusion would indeed be a technical anomaly if it should be unsuccessful in a rapid expansion after first demonstration!

The task of the fusion researcher is to produce a net energy output from the reaction. Almost 7 orders of magnitude improvement in the ratio of the equivalent D-T energy output to the energy input from the plasma have been accomplished since the late 1950's. Refer to Fig. 8.1 for a description of the progress. The Joint European Torus (JET) at the Culham Laboratory in England has reached a gain of 0.8. Thus, it is not too early for NASA to commence the pursuit of a space fusion development program. The progress achieved by DOE and the need in NASA for high energy propulsion systems make the present time appropriate to seize upon the benefits of the very modest investments made by the terrestrial program. The application of fusion energy to space missions is unique, requiring very directed attention by NASA.

8.2 COMPARATIVE RESEARCH MATURITY, FUSION REACTOR DESIGNS

For NASA the key question is "what is the work that remains to be accomplished for fusion energy to become viable for space?" To obtain a measure of fusion technology status let us then examine in this section the critical technical issues of the preferred concept(s) and the performance potential. A number of fundamental accomplishments have yet to be demonstrated, the most apparent of which is confinement of the plasma to produce a net energy yield. The capability to maintain a stable plasma under long duration steady state operation is still further away. Examples of critical issues recently examined for the SOAR, a linear reactor design, were: plasma physics, fueling, high voltage direct converter design, and the operational impact of low radioactivity (San88). Other reactor configurations and designs would have some of those issues plus others.

Major areas in need of investigation, some uncovered by this study, and shown to be necessary to support a space flight capability are:

- the development of high magnetic fields for space reactor designs,
- low specific mass magnets and light weight structures,
- plasma stability under the flight operational regime,
- efficient plasma confinement (β),
- minimal recirculation power,
- space start/restart power,
- the efficient conversion of plasma energy to controllable thrust power,
- ash removal,
- reactor flight control systems under net power operation,
- mass flow rate thrust augmentation,
- space fuel conservation in the exhaust plasma.

8.0 Status of Potential Space Fusion Reactors

The major focus of the terrestrial program has been on reactor plasma confinement approaches which is clearly the first topic of great interest in the establishment of the conclusions that are to be drawn from this study. There are two approaches toward plasma confinement as discussed in Section 4.0, MCF (magnetic) and ICF (inertial). Because the classified nature of ICF prevents further discussion in this report and since a study termed "VISTA," Section 2.0, was accomplished using ICF, this study analyzed MCF in greater depth.

The MCF information in this section depends heavily upon a recent Department of Energy study entitled the "Technical Planning Activity" (TPA), in which a committee of over 50 fusion researchers were assembled to examine the physics, technology, and systems aspects of the development of fusion power. The TPA produced terrestrial fusion reactor research plans to develop each of the concepts examined. The key publications for the TPA are: Bak87, Cal86, Abd85, and Dea85.

The TPA Plasma Science Group divided fusion plasma physics issues into seven main categories, given in Table 8-1 (Cal86).

TABLE 8-1. TPA Plasma Science Group Plasma Physics Issues.
Confinement Systems Issues:
Macroscopic Equilibrium and Dynamics
Transport
Wave-Plasma Interactions
Particle-Plasma Interactions
Composite Issues
Burning Plasma Issues:
Alpha Particle Effects
Burn Control and Ash Removal

Five classes of issues were considered to be primarily related to research in the major category shown as "Confinement Systems Issues," while other two classes of issues were associated with the next generation of experiments and the regime where the plasma is "burning," that is, where the fusion power produced exceeds the input power.

The term "Macroscopic Equilibrium and Dynamics" refers to how the plasma is held by the magnetic field. It is an important subject since it deals with the maximum plasma pressure that any given magnetic confinement scheme can achieve, that is, it determines the reactor's β. Issues included in this category are: forming and sustaining magnetic equilibria, minimizing magnetic disruptions, and in general, plasma kinematics when magnetically confined.

8.0 Status of Potential Space Fusion Reactors

The term, "Transport," deals with the heat and particle losses from a stable plasma, losses that are either normal to the magnetic field lines or parallel to them or combinations of both. Transport figures into the Lawson parameter, $n\tau_E$, the energy density and confinement time necessary to achieve plasma burning. As shown later, a value of $n\tau_E > 3 \times 10^{20}$ sec/m^3 at an ion temperature of 10 keV is required to burn D-T.

"Wave-Plasma Interactions" refers to plasma stability on the velocity-space scale, that is, degradation of plasma confinement due to amplification of short wavelength noise. In addition, it refers to heating the plasma for externally powered waves. This is an important option for raising the plasma's energy level to the temperature needed for burning. The reactor's current drive, presented in Table 8-5, is important for the operation of current driven machines like tokamaks, FRC'S, RFP's, etc.

"Particle-Plasma Interactions" refers to the interactions of the plasma with its environment, including neutral gas, atomic physics, plasma-material surfaces, impurity sensitivity, and neutral particle beams when used.

"Composite Issues" refer to the interrelationships and tradeoffs between the aforementioned plasma science issues. All of those plasma issues must be balanced from a systems trade perspective as no single issue dominates. It encompasses optimization of the reactor configuration, the plasma profiles, and the pulse length.

The major category "Burning Plasma Issues" refers to the plasma science issues, particularly with consideration of processes which are unique to the burning of plasmas. There are two major issues to address once breakeven has been demonstrated.

"Alpha Particle Effects" deal with the effects of charged particle reaction products on plasma equilibrium, stability, heating and transport, including alpha particle containment, heating effects, and effects of hot alphas.

"Burn Control and Ash Removal" deal with ignited plasmas in their approach to steady-state behavior – control of the plasma constituents and alpha particle removal without tritium or helium-3 removal.

The TPA study was primarily aimed at the D-T fuel cycle. Issues for the D-^{3}He fuel cycle would include proton effects in addition to alpha particle effects. Based upon a judgment of the present level of understanding for the configurations, the TPA Plasma Science Committee classified reactors into three categories:

(1) well-developed knowledge base,

(2) moderately developed knowledge base, and

(3) developing knowledge base.

8.0 Status of Potential Space Fusion Reactors

Table 8-2 summarizes the TPA's classification categories and the magnetic fusion experiment (MFE) program's status.

TABLE 8-2. Technical Planning Activity (TPA) magnetic confinement concept classification of reactor knowledge base.		
WELL DEVELOPED	MODERATELY DEVELOPED	LESS DEVELOPED
Tokamak	Advanced Tokamak	Field Reversed Configuration
	Tandem Mirror	Spheromak
	Stellarator	Elmo Bumpy Square
	Reversed Field Pinch	Dense Z Pinch

Table 8-3 below summarizes the TPA report contents with respect to the issues listed in Table 8-1. A judgment has been made regarding the level of understanding achieved by each concept for the issues, with a value of very good, good, medium, or low assigned. It must be emphasized that the table reflects this study activity's interpretation of the TPA's results, and it may not necessarily represent the official position of the DOE.

TABLE 8-3. Subjective level of understanding of plasma physics issues for the concepts examined in the technical planning activity.					
Reactor	Macroscopic Equilibrium and Dynamics	Transport	Wave-Plasma Interactions	Particle Plasma Interactions	Composite Issues
Tokamak	Very good	Low	Good	Medium	Low
Tandem Mirror	Good	Fair	Good	Medium	Fair
Stellarator	Good	Low	Medium	Medium	Low
RFP	Medium	Low	Medium	Low	Low
Spheromak	Medium	Low	Low	Low	Low
FRC	Medium	Low	Low	Low	Low
EBT/EBS	Medium	Medium	Medium	Medium	Medium

Another measure of the status of a concept is its performance with respect to values required for its reactor embodiment. Required parametric values, as projected for the burning of two fuel options, D-T and D-^{3}He, in the most applicable MCF confinement experiments evaluated during the TPA are presented in Table 8-4 as are ICF parameters. Refer to Table A-1, Appendix C.

8.0 Status of Potential Space Fusion Reactors

MICF is included in the table; but, a very recent approach, it is very much less developed.

TABLE 8-4. Characteristic plasma development requirements for generic fusion space reactors.

Parameter	ICF (Laser driver)		MICF (compact torus driver)		MCF (generic torus, R/A = 3.3)	
	DT	D-^{3}He	DT	D-^{3}He	DT	D-^{3}He
Fuel Ignition Temp., T_{ign}, (keV) (with DT spark plug)	1	2	8	10	8	10
Burn Temp., T_i, (keV)	15	60	15	60	15	60
Fuel Ion density, n_i, (cm^{-3})	6×10^{26}	2.4×10^{27}	10^{21}	6×10^{21}	6×10^{14}	2.4×10^{15}
Plasma pressure, p, (bar)	3×10^{13}	6×10^{14}	4.8×10^7	1.4×10^9	29	580
Magnetic Field, B, (T)						
• at β = 1	NA	NA	3.5×10^3	1.9×10^4	2.7	12
• at β = 0.06	NA	NA	NA	NA	11	49
f_c = $E_{charged}$/E_{fusion}	0.4	0.9	0.4	0.9	0.19	0.65
Auxiliary efficiency, η_a	0.1	0.1	0.5	0.5	0.5	0.5
Coupling efficiency, η_c	0.15	0.15	0.9	0.9	0.9	0.9
Fuel burn-up fraction, f_b	0.57	0.31	0.15	0.10	0.30	0.44
Fuel confinement time, τ_i, (s)	1.6×10^{-11}	2.2×10^{-11}	1.3×10^{-6}	5.3×10^{-7}	5	5
$n_i \tau_i$ product, (cm^{-3} s)	9.8×10^{15}	1.3×10^{16}	1.3×10^{15}	3.2×10^{15}	3×10^{15}	1.3×10^{16}
Energy/fuel confinement ratio	1	1	1	1	0.2	0.2
Plasma radius, a, (cm)	7.2×10^{-3}	4.8×10^{-3}	1.4	1.9	63	63
Plasma ignition energy, E_{ign} (MJ)	0.45	1.1	45	90	76	76
Driver energy, E_{ign}/η_c, (MJ)	3	7.3	50	100	84	84
Electrical input, $E_{ign}/(\eta_c \eta_a)$, (MJ)	30	73	100	200	168	168
Fusion gain, ($G_{ideal} \eta_a$)	250	111	56	220*	100	400*
Useful plasma output energy, ($G_{ideal} f_c E_{ign}$), (MJ)	300	730	1000	20,000*	1680	30,400*
G_{fom} = ($G_{ideal} \eta_c)f_c \eta_a$, charged MJ/electrical MJ	10	10	10	100*	10	180*

*Assumes an initial DT plasma re-fueled with D-^{3}He.

Table 8-5 (a), (b) summarizes the status and the test results achieved for the most important plasma physics parameters of the current MCF and ICF programs' experiments. The VISTA study assumed an ICF gain of 1,500, Section 2.0.

8.0 Status of Potential Space Fusion Reactors

TABLE 5 (a). MCF achieved parameter status.

Key fusion experiments	Ion Temperature, keV	Ion $n\tau_E$, m^3/s	Average β, %	Plasma Current, MA
Tokamak	30	2×10^{20}	7	6
Tandem Mirror	0.4	2×10^{16}	25	NA*
Stellarator	1	2×10^{19}	2	NA*
RFP	0.5	6×10^{16}	20	0.4
Spheromak	0.2	6×10^{15}	6	1
FRC	0.6	3×10^{17}	90	2
EBT/EBS	0.05	1×10^{15}	0.1	NA*

* NA-not applicable

TABLE 8-5 (b). Status of ICF reactor concept.

	Density, cm^{-3}	Time, seconds	Energy Input, MJ	Gain, Q
Status	10^{25}	10^{-10}	0.03	0.3

The Lawson criteria ($n t$) defines the net electricity breakeven condition value of $n t$ required at a given temperature T_i. Breakeven is the point at which the total fusion output, if it were converted to electricity and reinjected, the reactor would self-sustain burning. This provides an excellent first estimate of these parameters, although Lawson made certain assumptions such as 33% energy conversion efficiency and 100% efficient heating of the plasma by fusion products. Neutrons, as typical reaction products, are immediately lost from the plasma without a transfer of energy to the plasma. The charged fusion products, i.e., ions, are slowed by the background plasma, and their energy then serves to heat the plasma and any cold fuel input. The progress and status made towards meeting the Lawson criteria is shown in Fig. 8.2 (San89).

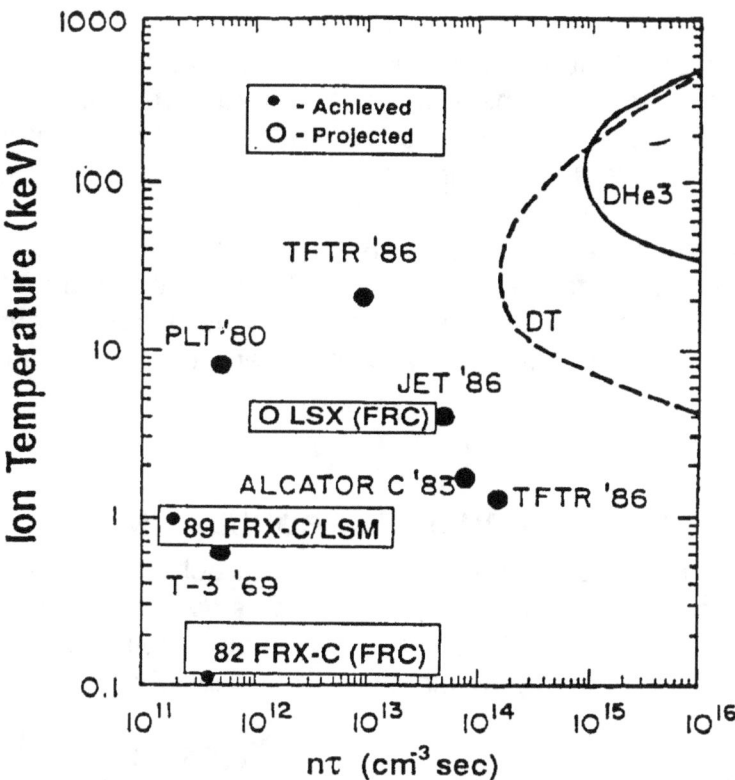

Fig. 8.2. Status of fusion experiments relative to meeting burning conditions (Lawson criteria status).

It shows that these parameters have each individually met the required limits for D-T ignition in the tokamak and that collectively they are rapidly approaching the confinement time, temperature, and density parameters required to achieve ignition. That progress is fundamental in understanding the positive conclusions concerning the ability of NASA to make use of fusion energy for tremendous benefits to space.

8.3 SPACE FUSION REACTORS (SFR)

For achieving the capability to accomplish the advanced space flight missions of the type that have been examined in this analysis we require that five key system parameters be met: low vehicle mass, high energy density, variable-high specific impulse performance, highly inherent system safety potential, and high potential for ultra-reliability over a long burn duration without maintenance. For fusion reactors, these transform into the following important parameters for space propulsion: high specific power, good plasma stability, high burn efficiency, operational simplicity, and ease of energy extraction. Many of the other desired traits presented in Section 6.0 will be designed into the system or achieved as a consequence of selecting a fuel which is as aneutronic as possible.

8.0 Status of Potential Space Fusion Reactors

High specific power will be attained only by high β machines. These will exhibit the minimum mass and size characteristics required for space. Of all concepts shown in Table 8-4, the FRC exhibits the highest β. It is therefore the reactor design on which the greatest possibility exists for achieving a SFR. The second option is the tandem mirror. The dipole (Tel91) is another. It should be stressed, however, that the conclusion about the FRC is based on general considerations and is tentative. It is not a well developed approach as shown in Tables 8-2 and 8-3. A final comparison can only be based on detailed designs for space, designs which need to be accomplished, and, of course, on testing of those designs.

8.4 ENERGY CONVERSION FOR PROPULSION

We can use the FRC as an example of the characteristics desired of fusion reactors for space. The FRC had been funded as two main experiments at Los Alamos (FRX, Fig 8.3) and at Spectra Technology, Bellevue, Washington, by the DOE at only a level of approximately $5M annually in 1989-90.

Fig. 8.3. FRX experiment at Los Alamos National Laboratory.

This is a very low level as discussed later. The Los Alamos program was cancelled in 1990; the Spectra Technology program is scheduled for termination in 1991.

8.0 Status of Potential Space Fusion Reactors

The reactor concept potentially of great benefit and interest to space unfortunately has a relatively low level of theoretical and experimental understanding, the consequence of a low program priority. Research maturity, the topic of Section 8.2, is our first interest as we consider the potential of the designs for meeting the space requirements as presented in Section 6.0. While the tokamak is better developed, Table 8-2, and is approaching a breakeven, as shown in Table 8-1, the FRC ion temperature and nt values have a long way to go. Reference to Table 8-3 will show a low level of understanding of the FRC reactor's physics.

Let us then develop further the reasons for the appeal of the FRC reactor to space. The space reactor parameters of interest as defined by Table 6-1 are presented below as Table 8-6 with a subjective qualitative evaluation of the capability of the FRC to meet those parameters. It can be assumed, unless otherwise quoted, that the data presented in this section of the report were obtained from the Eighth American Nuclear Society's Fusion Topical Meeting poster paper entitled "Fusion Space Propulsion with a Field Reversed Configuration" by Dr. Chapman and Dr. Miley, later published in *Fusion Technology* (Cha89). This is the only known reference showing details of the FRC for space.

8.0 Status of Potential Space Fusion Reactors

TABLE 8-6. FRC research status for meeting key space reactor requirements. All of the parameters' status are based upon very preliminary analyses and/or educated guesses. All require thorough analysis, design, and testing to validate that the parameters can be met.

SPACE REACTOR PARAMETER	FRC PERFORMANCE AND RESEARCH STATUS			
	Potential to meet	Does not meet	Unknown	Comments on FRC status
6.1.1 Specific Power	X			Limited study. Requires design.
6.1.2 Thrust				
(a) low: 1 N to 10 kN	X			Limited conceptual work.
(b) medium: 10 kN to 50 kN			X	Has not been addressed.
(c) high: 50 kN to 500 kN		X?		Requires a large plasma volume.
6.1.3 Specific Impulse	X			10^3 to 10^6.
6.1.4 Fuel Cycle	X			Can burn D-^{3}He. Appears capable.
6.1.5 Beta	X			90%.
6.1.6 Ignition	X			Needs design study. Requires testing.
6.1.7 Throttle capability	X			Needs design study. Requires testing.
6.1.8 Plasma Stability			X	Limited analysis done. Burn experiments required. The major issue.
6.1.9 Power Level	X			~1 GW.
6.1.10 Electrical power variability	X			Requires design study.
6.1.11 Dual Mode Operation	X			Requires design study.
6.1.12 Mass			X	Requires design study.
6.1.12 Efficiency			X	30% calculated.
6.1.13 Recirculation Power	X			None required.
6.1.15 Modes of Operation	X			Work will follow net power demonstration.
6.1 16 Low /no neutrons produced	X			<2%.
6.1.17 Failure Tolerance			X	Requires design study.
6.1.18 Space Environment	X			Requires design study.

The rationale for the optimism associated with the use of this design is exhibited by the large number of system parameters which it is believed this design will fulfill in meeting the space requirements. But the unknowns are of great importance in considering the capability of the FRC to succeed. D-^{3}He plasma stability at net power is of utmost importance. The large fraction of charged particles favor this confinement concept.

8.0 Status of Potential Space Fusion Reactors

The parameters of greatest interest to space propulsion are high specific power, high reactor power density, thrust, and variable, high specific impulse. One parameter to achieve specific power is high fusion plasma power density. For the FRC burning D-^{3}He the fusion power can vary from 0.5 MW/m^3 to 7 MW/m^3 as shown by Fig. 8.4. Thus, a 400 MW reactor could range in size from 60 to 800 m^3.

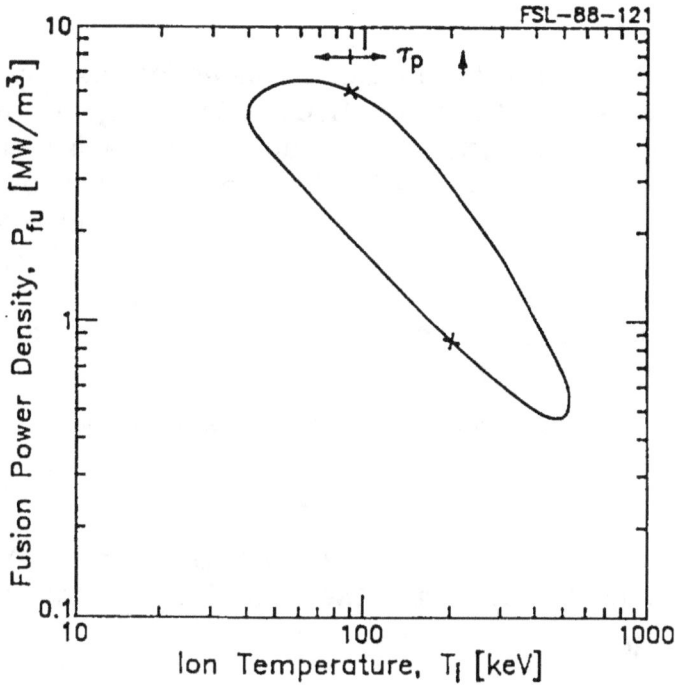

Fig. 8.4. FRC Power Density

The importance of β in the design of a compact reactor is clear from the fact that the reactor's power density scales as β^2. Consequently, a FRC β of 90% will have a power density roughly 160 times greater than a tokamak at 7% for a given magnetic field strength.

Dr. Tuszewski of Los Alamos National Laboratory summarized the rationale for the optimism for the FRC at the Eighth Topical Fusion Meeting.

> The FRC is ideal for use of the D-^{3}He fuel cycle. Its high plasma beta and power density allow substantial reactivity, little radiation losses, and most of the fusion power in the form of 14.7 MeV protons. These charged particles can be diverted in the FRC edge layer towards electrostatic direct converters, resulting in very high plant efficiencies. These attractive features are illustrated in Table 2, where the approximate parameters of a 1 GW FRC reactor are compared for a pulsed D-T system such as CTOR and for a conceptual steady-state D-^{3}He system. One observes that the 14

8.0 Status of Potential Space Fusion Reactors

MeV neutron production with D-^{3}He can be reduced by about a factor 100 compared to that of the D-T system. Another (possibly crucial) advantage of the D-^{3}He system is that gross FRC stability may be achieved at s ~ 10 with the help of high energy neutral beams, large-orbit protons, and possibly larger plasma elongations. This may not be the case for the D-T pulsed system at s ~ 30, in spite of the alpha particles. (Tus88)

In Section 4.0 the rationale is forwarded concerning the importance of using D-^{3}He as the reactor fuel pair, and in Section 6.0 the system requirements are presented. The FRC is a reactor capable of burning the desired fuel, D-^{3}He, as discussed below. Inherently, the FRC design readily allows for the direct conversion of plasma to thrust or electricity. Stability during testing has been satisfactory, apparently greater than predicted. Stability, however, remains a concern, and a large experiment (LSX) has been designed to address both stability and confinement. At the present time a detailed steady state reactor study has yet to be performed.

The FRC research issues which need to be considered are:

(a) gross stability

(b) confinement scaling with increases in S – a measure of the number of average ion gyroradii between the field null and the separatrix

(c) new FRC formation methods

(d) steady state operation (preferred).

The Large S Experiment (LSX) at Spectra Technology was designed to address these stability and confinement issues. It commenced operation in August 1990 and is scheduled for termination in 1991.

The Los Alamos experiment reported observations of internal tilt instabilities which is the first occurrence in what has otherwise been a grossly stable device. (Tus91) The authors conclude that "Additional stabilizing techniques will be required in future large-size FRC's." This work clearly illustrates the importance of testing and assists in forming the basis for a program strategy.

One recent study addressed the FRC's potential for propulsion: "Space Fusion Propulsion with a Field Reversed Configuration" by Drs. Chapman, Miley, Kernbichler, and Heindler (Cha89). The study examined burning D-^{3}He in a FRC. The analysis concluded, as presented in Table 8-6 that this concept offers many of the features that are important to space power, i.e., high power densities and a good confinement scheme. It appears more suitable to provide propulsion than the other experiments reviewed in the study as shown by Fig.

8.0 Status of Potential Space Fusion Reactors

8.5 (Cha89, Fig. 10). The concept was developed from a study of the FRC for commercial electrical power production (Mil78 and Mil79).

	Field Reversed	Tandem Mirror	Spherical Torus
Specific Impulse	○	○	○
Thrust (Power)	◐	○	○
Beta	○	◐	●
Power Density	○	◐	●
Thrust (Power)/Weight	○	◐	◐
Charged Particle Extraction	○	○	◐
Propellant Thermalization	○	◐	●

○ — Good ◐ — Average ● — Poor

Fig. 8.5. Comparison of Reactor Experiments for Propulsion.

The first parameter of interest is high power density. For space propulsion this implies a high percentage of charged particles generated by the reaction and a low percentage of neutrons. Fig. 8.6 shows the neutron production as a function of ion temperature.

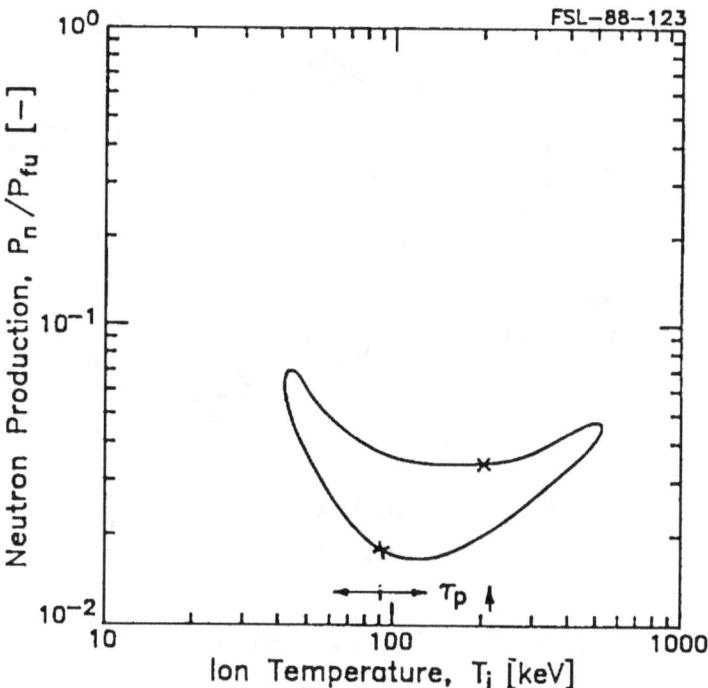

Fig. 8.6. FRC Neutron Power Fraction.

8.0 Status of Potential Space Fusion Reactors

Thus, it is seen that for reaction temperatures on the order of 60-200 keV, the neutron production will be below 2%. This subject is treated in greater depth by Dr. Kernbichler (Ker88):

> In the parameter domain of interest the fraction of fusion energy carried by neutrons can be reduced from 80% (D-T) to between a half (T recycle) and a third (no T recycle) of this value, if no ^{3}He is externally provided. With external supply this fraction can be brought down to several percent. Therefore, it seems justified to refer to D-^{3}He as a 'potentially neutron-free fuel.' Increase of the ^{3}He/D density ratio much beyond 2/1 does further reduce the neutron loss, but leads out of the ignition domain. It may also be worth noting that the shielding requirement does not reduce linearly, but roughly logarithmically with the neutron wall loading.

Particle confinement as a function of temperature is shown in Fig. 8.7.

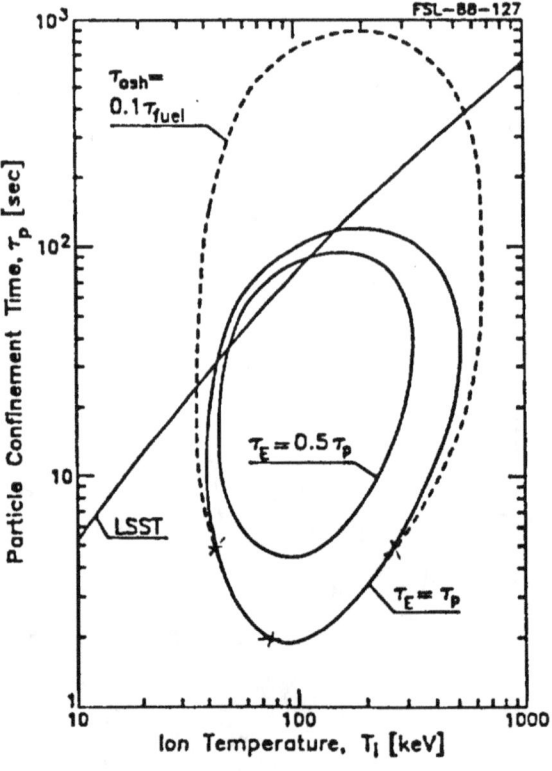

Fig. 8.7. FRC Confinement Time.

The burn efficiency improves with increases in τ as shown by Fig. 8.8. The figure quantifies the effects of confinement time on fuel burn efficiency.

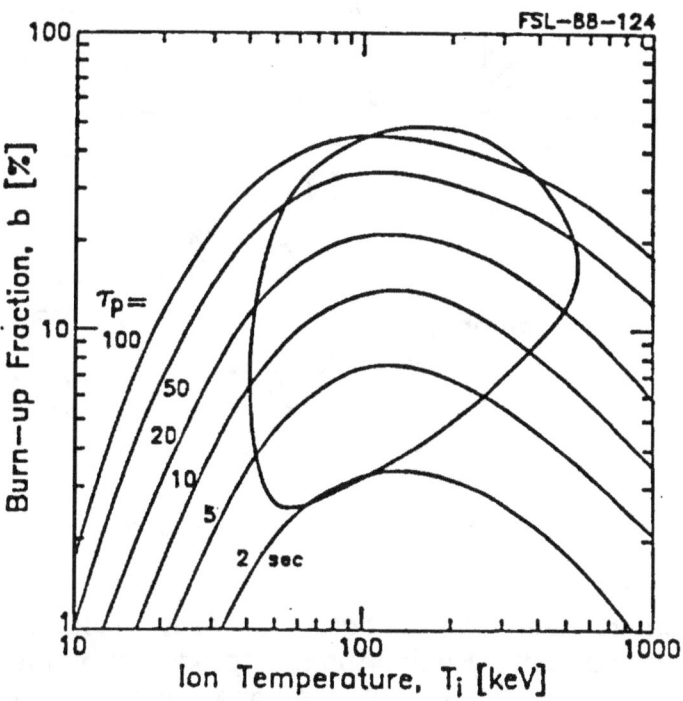

Fig. 8.8. FRC D-3He Fuel Burn-up.

Fuel (helium-3) loss is a concern that needs to be addressed.

For a space reactor, then, we would be looking for the FRC to deliver the following performance design parameters as shown by Table 8-7.

TABLE 8-7. Field Reversed Configuration (FRC) Space Reactor Parameters

Magnetic Field	5 T
Beta	0.76
Confinement Time	2 seconds
Electron Temperature	66 keV
Ion Temperature	86 keV
Mixture Ratio for ^{3}He/D	60/40
Specific Fuel Consumption Rate	5.2×10^{-7} kg/m^{3}s
Helium-3 Burn Efficiency	3%
Fusion Power	6.4 MW/m^3
Jet Power	29 %
Neutron power	1.9 %

The fusion engine concept is depicted in Fig. 8.9.

8.0 Status of Potential Space Fusion Reactors

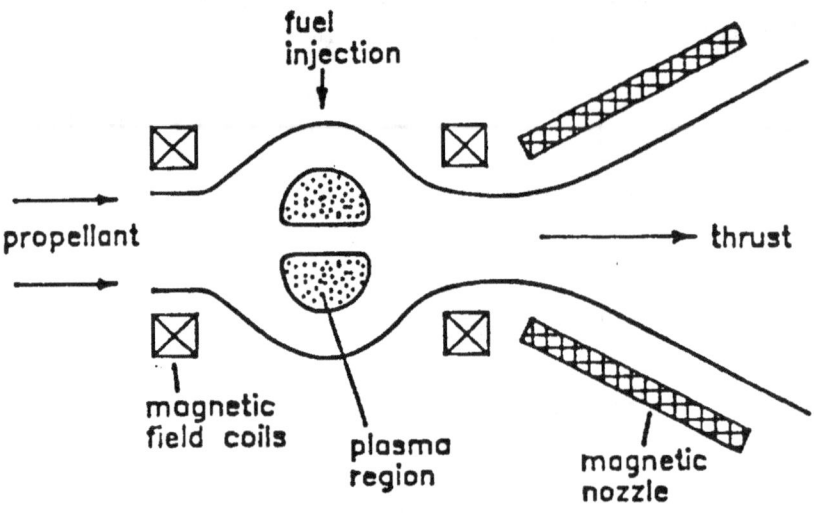

Fig. 8.9 FRC fusion engine concept.

The FRC, as can be seen from the figure, is ideally suited to propulsion by virtue of its external topography. Engine thrust is produced by the controlled release of a portion of the plasma directed by a magnetic nozzle. One advantage of this design, or any magnetic confinement reactor, is the absence of high wear moving parts, such as high speed turbines as used in the SSME (Space Shuttle Main Engine), and parts subjected to erosive wear, such as the nozzle of the SRM (Solid Rocket Motor). Thus, the reactor/thrust chamber inherently possesses features that are essential to the achievement of the long life time operational requirements of the space program. The reactor is supplied by fuel pellets which are injected into the plasma. Thrust and specific impulse are simultaneously controlled by the injection of propellant gas into the plasma scrape-off layer. The fuel injection system will probably contain moving parts, but this has not been studied. The thermalization of propellant is attained by heating from the plasma; the extent of thermalization is important to assure efficient use. Plasma thrust is produced and controlled by the release of plasma through a mass imbalance in the stabilizing external mirror magnets. Thrust produced solely from the plasma is low as a result of limitations imposed by magnetic confinement which currently permits plasma densities in the range of 10^{15} to 10^{16} ions per cm^3. A reactor of the power magnitude required by the manned programs would be characterized by the parameters as shown by Table 8-8 (CHA89, Table 2). This design is considered to provide a very high stability factor, where $S = r/3\rho_i = 50$. Stability is the main concern of the FRC.

8–18

8.0 Status of Potential Space Fusion Reactors

TABLE 8-8. High Power Design Parameters.

Total power	0.5 GW
Plasma Volume	80 m^3
Elongation Factor	6
Ion Gyro Radius	0.01 m
Plasma Radius	1.5 m
Stability Factor	50
Propellant Addition	0 - 0.8 kg/s
Specific Impulse	10^3 to 10^6 seconds
Thrust	0.4 to 50 kN

The propellant flow rate that is presented in Table 8-8 is based upon Fig. 8.10.

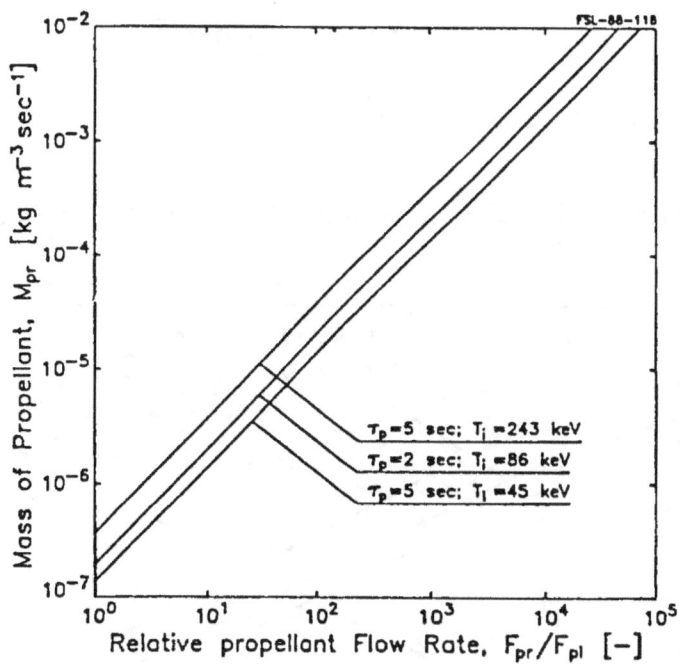

Fig. 8.10. FRC propellant mass flow rate in terms of propellant/plasma particle flow rates.

One of the key requirements for accomplishing high energy missions efficiently is to have the innate capability to vary the thrust and specific impulse. The thrust-impulse performance capability expected is shown by Fig. 8.11.

8.0 Status of Potential Space Fusion Reactors

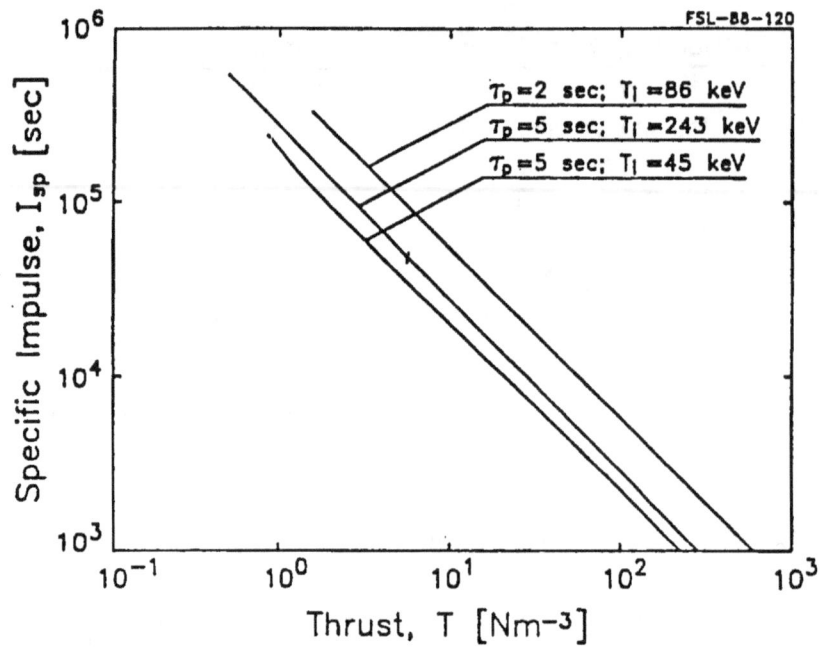

Fig. 8.11. FRC Variation of Thrust with Specific Impulse.

The variation of propellant flow rate with thrust and specific impulse is shown by Fig. 8.12.

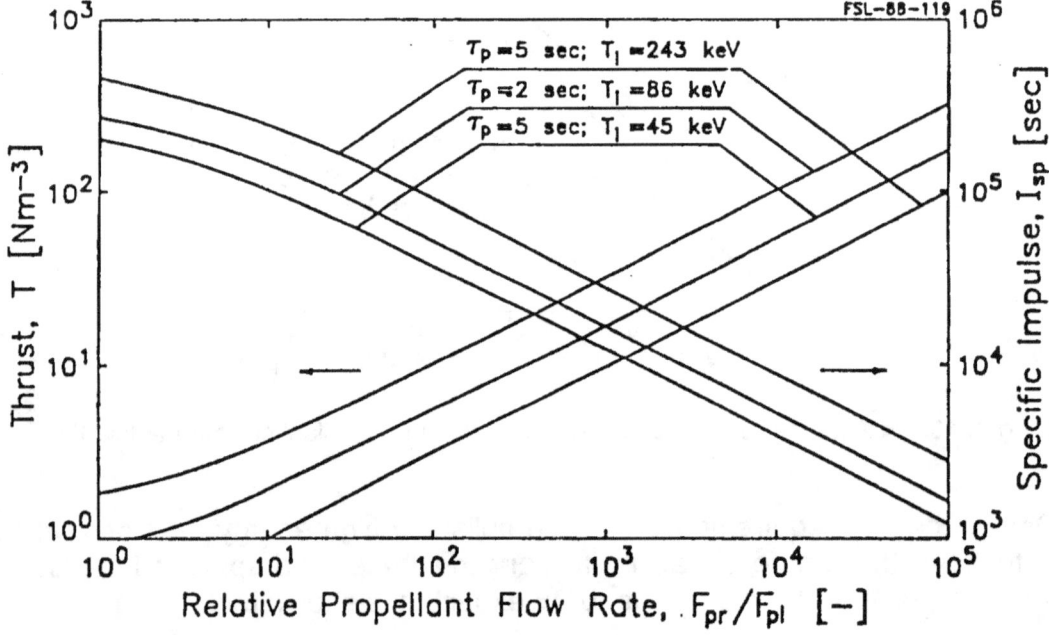

Fig. 8.12. FRC Thrust and Specific Impulse as Functions of Propellant/Plasma Particle Flow Rates.

8-20

8.0 Status of Potential Space Fusion Reactors

One question which arises is whether a need exists for thrust neutralization, the concern being based on ion engines. Those engines require a means of thrust neutralization to prevent a charged vehicle from developing as ions are ejected. In that event, the system would quickly terminate thrusting. In the fusion engine a plasma is produced which in the aggregate is neutrally charged. Both charged particles, ions and electrons, are magnetically ejected through the exhaust nozzle so that no neutralization is required. Any charge divergence will be automatically self corrected rapidly.

The very high temperature of the plasma and low particle mass produce the high specific impulses that are essential to attaining the next step in space propulsion. The high temperatures at which this fuel burns, approaching 100 keV (10^9K), are cooled by the injection of hydrogen gas in the nozzle, producing a higher engine thrust level. Based upon an extensively developed plasma model, Drs. Chapman and Miley estimated that thrust levels over the range of 0.1 to 40 kN are achievable. As indicated in "Mission Applications" Section 2.0, those are the ranges of great interest for space flight. Other studies indicate that specific powers on the order of 0.5 to 6 kW/kg are possible for various magnetic approaches although none has been studied for the FRC.

As discussed earlier, burning D-^{3}He is more difficult due to its higher ignition temperature in comparison with D-T. Lacking a strong experimental base for data, the authors used a model to define the parametric space available for burning D-^{3}He in this reactor concept. The power output expected from FRC reactors is on the order of 1 to 10 GW. This model is, as a side note, the same as the one used to establish the volume of plasma for aeronautical propulsion discussed in Section 2.2.11.

The FRC propulsion issues that need to be addressed are as follows:

****

- Limited volume: Its size is considered to be volume limited based upon stability considerations. The approach taken to achieve higher power levels is to provide a greater elongation factor. This consideration may be the ultimate limitation on the reactor size. Relativistic ions injected to orbit the plasma are anticipated to assist in the maintenance of plasma stability.

- Fuel efficiency: One important subject for investigation is the means to improve upon the fuel burn-up factor which is 3%. (Table 8-7)

- Reactor plasma efficiency: Thermalization efficiency of the propellants, ash, and reaction products is unknown and, if non-uniform and incomplete, will result in performance degradation.

- Steady state burning: Operation has only been in the pulse mode. Steady state is preferable.

****

8.0 Status of Potential Space Fusion Reactors

The only manner by which these issues can realistically be resolved is by full scale reactor designs and experiments at D-^{3}He burning parameters. The importance of testing has been illustrated by the FRC results to date.

Fusion reactors do not scale due to the physics involved. Currently in the conduct of MCF experiments the comparison of data with theory has resulted in good agreement with the exception of the understanding of transport. That is accomplished experimentally.

> Consequently, the approach suggested here is to proceed to a design that is considered ultimately capable of burning D-^{3}He.

While this approach may not be the preference of some, who prefer analysis and a multi-step phased program, it appears that the quickest and, therefore, most economical manner to obtain answers on the above is to perform a full scale design capable of reaching D-^{3}He burning parameters and to test it. It may be necessary to proceed first with neutral beam and start-up experiments. Answers to address the feasibility question are considered to be possible within 5 to 10 years. If this approach proves to be feasible, fusion energy for space could be available in 20 to 30 years, probably in time for the initial manned Mars trip, although this is clearly not recommended as the strategical approach for the first Manned Mars Mission. But if successful, the advantages of space fusion could become a reality. Spin polarization developments could aid in the design process by the reduction of neutrons.

The suggestion to proceed to a D-^{3}He burn configuration comprises a high risk approach, but the cost-to-benefit ratio to NASA for space is such that it is warranted. There is, looking at the opposite end of the spectrum, a large risk which results to NASA in the process of refraining from investigating the question of the FRC or other fusion energy concepts, for the abstention of fusion research presumes that we know the answer, i.e., that it will not work and therefore that FRC's should not be attempted – certainly an incorrect assumption.

> It is of utmost importance to the agency's future to assure its ability to carry out future missions by exploring key innovative high energy concepts.

8.5 ELECTRICAL POWER GENERATION

Two conversion systems have been researched more extensively: the periodic-focused (PF) collector and the Venetian Blind (VB) collector. These are treated extensively in (Mil76), and it is not the intent to repeat the material here other than to make the point that a number of concepts exist with various states of technology refinement. The application has been primarily for terrestrial application with only a small, 100K, study effort for space application funded by the Air Force SOAR study noted in the study.

Direct electrical power conversion has been extensively researched for the past 20 years at LLNL but has not been pursued since 1981. Additional efforts are required to enhance the efficiency of this preferred method of electrical power generation and to reduce heat load to the walls.

The venetian blind collector, Fig. 7.17, has been designed to convert plasma directly to electrical power via a set of biased plates.

Power conversion efficiency is gained by increases in the number of stages. Efficiencies up to 87% were reported. (Mil76)

8.6 SUMMARY

The FRC reactor concept offers great attraction to space from the standpoint of a high β, but it is among the least understood. Plasma stability has been a concern and during recent testing (1991) the FRC has experienced stability problems. The tokamak's program progress is optimism for fusion becoming demonstrated, but unfortunately it will have a specific power too low to be of interest to a space propulsion application. It will not burn the preferred fusion fuel, D-^{3}He.

Clearly a need exists to have available an energy source of the nature that fusion can theoretically provide, but this is neglected research topic. The point is frequently made in this report of the importance in initiating a space fusion program, and, further, that the lack of detailed studies have hampered the process of fully understanding the merits of fusion energy for space. Also, the importance of testing has been emphasized herein. It is this lack of emphasis on an analytical and test program that leads to large uncertainties, requiring assumptions without factual basis. Subsequent to the initiation of this study, one brief review was made to compare MCF to ICF (Sar88), but it encountered the same uncertainties.

The difficulty with all studies of this nature is that we simply do not have an understanding of what it takes to make either of these systems spaceborne. As one critical point, the ICF design approach must make assumptions concerning driver mass. At 5 MT, the start up power could likely be estimated low since lasers are very inefficient and since this is a pulse fired system. A low estimate could be used for the mass of the power supply, at 90 MT. Another critical point is that any study of an operational system must cover all aspects of a flight system, for example, the reliability factor has not been taken into consideration, i.e., how is redundancy achieved? Further, confinement options cannot be evaluated and extrapolated to flight system designs. For example, the above MCF study did not refer to the FRC and project its performance to net power.

8.0 Status of Potential Space Fusion Reactors

Until the reactor mass has been determined, studies can only make preliminary guesses at the performance potential of these propulsion systems. And the mass cannot be determined until critical fusion plasma confinement experiments which produce net power have been accomplished. An analogy which seems appropriate is made here with establishing and projecting a Boeing 747, a F-15, or the supersonic Concorde aircraft performances using the Wright brothers' Kitty Hawk as a model prior to the Kitty Hawk being designed and constructed. The same analogy is made with projecting electronic hardware performance, such as today's supercomputer technology, subsequent to the discovery of semiconducting properties of materials. The fact is that the gains were accomplished through active research and technological developmental programs by the developers and users of the technology.

9.0 ACCEPTABILITY: SAFETY, ENVIRONMENTAL, RELIABILITY, AND SPACE MAINTENANCE CONSIDERATIONS

Public acceptance of any radically new technology involving high energy will follow only after careful consideration and approval by the public on safety/environmental related matters. The new technology's impact on health and the environment is, therefore, among the top priority topics to address early. Acceptance of the technology for operational use by NASA is contingent upon two additional criteria, namely, the operational flight safety aspects and the inherent reliability that can be designed into the new system. This section treats all 3 subjects – safety, environment, and reliability – from the aspects of space fusion applications. In addition, a brief discussion on space maintainability issues is included.

PROGRAM ACCEPTABILITY FACTORS
• Safety
• Environment
• Reliability
• Maintainability

Studies and review papers were used to obtain insight into fusion safety and environmental issues as identified for the terrestrial application. A comparison of the same topical material as applicable to the space program is made in this section. Among the more recent documents to give the terrestrial fusion safety considerable attention are the ESECOM report, "Safety and Economic Comparison of Fusion Fuel Cycles," (Hol87); Kulcinski, "Apollo--An Advanced Fuel Fusion Power Reactor for the 21st Century," (Kul89); Emmert, "Apollo-L2, An Advanced Fuel Tokamak Reactor Utilizing Direct Conversion," (Emm90); Khater, "Activation and Safety Analyses for the D-^{3}He Fueled Tokamak Reactor Apollo," (Kha90), and an MIT safety and economic report (Bre87). (D-^{3}He was of secondary interest in the ESECOM and MIT reports.) In addition, a poster paper by Dr. Roth addressing the safety and environmental constraints on space applications, presented at the Vision-21 Symposium at Cleveland in April 1990, one of the most recent evaluations of the subject for impact both externally (public) and internally (NASA), was included in the review. (Rot89) The ESECOM report points out the safety and the environmental advantages of fusion over fission by the diminution of effects due to accidental releases and amelioration of the radioactive waste problem. The MIT report compares fusion with other energy sources and also compares the candidate fusion fuels.

9.0 Acceptability: Safety, Environmental, and Reliability Considerations

We first consider fusion's impact on public safety and the environment. Following those safety discussions, fusion's safety relevance to NASA as the user is presented – including system safety and operational safety. Appendix B offers some additional thoughts on this subject as well. The mission reliability expectations follow as well as a very brief discussion concerning space maintenance. Consequently by addressing all four topics, a development of the degree of acceptability of space fusion can be gained and, in addition, a better understanding of the need for the capability. A comparison between fusion and fission is also included in response to requests by reviewers of the report.

9.1 SAFETY

Safety in the space program comprises the system engineering aspects of the operations of a "system," both nominal and contingent, in the intended operational environment. Three components may be considered to comprise a system, namely, the hardware, software (operational instructions), and the environment.

The term "environment" enters into the space vehicle design equation from two distinctly separate aspects. The first environmental consideration is more recent and deals with the Earth's environment from the habitability perspective, i.e., does the device's operation adversely affect Earth's environment? It is a public safety concern. Avoidance of introducing unnatural contaminants into our environment is obviously important to all. That is the "safe" approach since nothing is changed. In a second and distinctly different vein, the flight operational environment constitutes a significant flight vehicle system factor to be taken into account from the standpoint of the space vehicle's <u>system safety</u> consideration. That is, a device must be designed to function satisfactorily in its operational environment. The demonstration of the system's ability to perform in its operational environment is paramount to a successful, safe operation. Problems concerning safety and product performance generally ensue where this fundamental principle is not fully addressed. This is, of course, an internal NASA concern.

In the event that the resolution of any major system safety issue cannot be achieved economically, the system will be rejected or its use severely restricted to those very narrow application's where the use is mandatory. There is a strong system design selection factor, too, in that the system must possess inherent properties or characteristics which permit the designer to implement a reasonable balance between the system's residual risks and the economics associated with the implementation of the system's hazard reduction methods. In other words, the device must possess inherent, fundamental operational principles such that the penalty for making the system acceptable from a system safety and environmental perspective will not be so great that the system has been rendered economically impractical, or alternatively, that its use could cause high risks to be taken. The bottom line is that the designer must be able to use the device to its performance potential.

In view of the dual safety aspects, i.e., to provide sound environmental protection measures and to implement a practical vehicle design using sound system safety principles, let us examine the impact of fusion on the flight vehicle and its mode of operation. What are the inherent system safety characteristics offered by fusion, both to the public and to flight safety, and what are the options available to the designer for control of hazards? How can space flight fusion system safety be achieved? What is the effect of space fusion on Earth's environment? Do these systems permit safe operational modes both from the standpoint of vehicle system safety and impacts to the environment? How does one gain confidence that the fusion engine system will perform satisfactory in its operational environment?

9.1.1 PUBLIC SAFETY AND ENVIRONMENTAL CONSIDERATIONS

Ultimately, the relative safety to mankind in general, or in other words, the environmental hazard, due the operation of any system is a measure of its waste products. Three questions address the impact of the system on the environment;

1. What is the environmental impact for bringing a system to the operational phase?
2. What is the consequence on the environment of the system's operation?
3. What is the environmental consequence of space operations without the system?

9.1.1.1 INHERENT ENVIRONMENTAL IMPACTS FROM SYSTEM DEVELOPMENT

This topic involves system usage considerations, the impacts only to bring the system on-line to a state of operational readiness. Fuels traditionally have the most significant impact since the mass is greater than any other system component. There are a number of important questions to address. What is the effect on the environment of manufacturing/producing the fuels? Are toxic substances required as a part of the vehicle manufacturing process or as part of the equipment used in the system itself? Does the manufacturing of the fusion fuels and the necessary ancillary production equipment require energy intensive processes? For example, the production of antiprotons in matter-antimatter requires an energy expenditure of ~1000:1 that can leave the Earth with a large residue of energy waste products unless solar energy is used.

9.0 Acceptability: Safety, Environmental, and Reliability Considerations

The production of fuel, an environmental concern for any energy process, is expected to be benign to the Earth's environment for D-^{3}He. Deuterium is extracted from seawater. If solar energy can be deployed for the extraction process, practically no impact is experienced. There is the usual mining impact to obtain helium-3, but none results to the Earth, only to the moon since the study recommends that total processing occur on the lunar surface rather than by processing materials returned to Earth.

From an environmental safety viewpoint, because tritium has a 12.3-year half-life, it must be continuously bred by Earth-based reactors. It requires the constant use of Earth-based reactors for fuel production. That constitutes an additional environmental hazard to counter and the attendant impact of additional energy and clean-up and disposal of the radioactive by-products, along with those attendant costs. An option is to breed tritium on the moon which could eliminate Earth environmental hazards. That could, incidentally, become a means to produce tritium safely for Earth utility power use.

For weapons related work, approximately 5 kg of tritium is produced annually at the Savannah River plant. To produce the 40 MT fuel mass for VISTA, either tritium fusion breeding, or alternatively the fabrication of ~4,000 Savannah River reactors, is required to provide a fuel production rate to meet a manned flight rate of one trip per year to Mars. The preference is avoidance, where possible, of reactor produced fuels which would contribute to the radioactive storage problem on Earth. Both fuel cycles, as a result of their use, produce radioactive materials to be disposed of in space, but the severity is reduced by those fuel cycles using reduced neutron production.

A similar concern exists for the use of fission energy for space propulsion and power. Since it is less efficient than fusion, there will be a greater waste disposal problem. At the present time, a national massive DOE clean-up program is being planned which is expected to cost tens of billions of dollars.

Consider the operations involved in placing space flight vehicles into their operational environment. Let us assume that these same high energy missions are carried out using chemical propulsion identical to the mix between the solid and liquid propellants as currently employed by the Shuttle. For the Manned Mars Mission alone, the propellant mass to be placed into orbit for chemical and probably for nuclear fission systems will be nearly an order of magnitude greater than that amount used by the high energy density systems. The MEM vehicle, a chemical propulsion system, required ~1,000 MT of propellants in LEO for accomplishing a 2-year trip, whereas fusion would consume 175 MT of propellants which transports a more massive payload for a quicker, 212-day mission trip (α_{p1} = 1 kW/kg). Then, consider the exponentially larger mass of propellants necessary to deliver that amount of propellant to orbit. The chemical propulsion system approach requires energy not only for the production of propellants but also for the production of numerous flight vehicles, the energy equivalent of 37 Shuttle launches per Manned Mars Mission. Chemical propulsion using Shuttle technology will contribute to pollutants in the

9.0 Acceptability: Safety, Environmental, and Reliability Considerations

atmosphere from the solid rocket boosters. When considered in the overall scheme of the total energy used and pollutants added, perhaps the percentage addition may be low, but there is a finite increase which is, of course, less desirable than none added, particularly if there is an option. The total quantity of energy consumed and the pollutants eventually added to the Earth's environment are impacts best avoided, if possible.

NEP has the same problems as NERVA (hydrogen propellant) but instead makes use of rare gasses for propellants. The rare gas quantities required of the magnitude to accomplish high energy missions of the nature considered herein are not readily available.

9.1.1.2 OPERATIONAL IMPACT

This topic encompasses the effects of operating the fusion system both in a nominal and contingent mode. In the case of energy release during nominal operations, we primarily need to consider the combustion products. From any of those fusion reactions shown in Section 4, we see that combustion products (ash) from fusion burning comprise alpha particles, protons, neutrons, helium-3, and tritium, plus any unburned deuterium, tritium, and helium-3. Of those, only tritium has a lasting environmental impact due to its 12.3-year half-life.

Tritium is a light element and a weak beta emitter and consequently does not possess the safety and health threat that fission reactors and RTG's do. Tritium released into the atmosphere presents the concern that some will form into tritiated water which could cause lung damage. Clearly then, as shown by Table 4-1, the operational hazards to counter from fusion energy are a function of fuel selection and reactor design. The combustion product hazard is enhanced or mitigated depending upon the particular fuel cycle used.

As mentioned in Section 7.0, one advantage of the D-^{3}He fuel cycle is the reduction of the neutron hazard. A fuel cycle of D-T yields 14.07 MeV neutrons and, for D-^{3}He, substantially reduced quantities of 2.45 MeV and 14.07 MeV neutrons. Unfortunately, the latter fuel cycle lacks the capability to eliminate neutrons altogether because of the D-D side reactions. Tritium will be present in the exhaust. The neutron-free fusion reactions are not energetically viable in terms of plasma confinement concepts being researched now.

Consider neutrons and charged particles. Neutrons have an impact on the local environment, surrounding spacecraft, and within the flight system itself. The local space environment will be minimally affected by the presence of neutrons, since they decay to a proton and electron within ~10 minutes causing no lingering impact to the Earth's environment. Ion plasma, if the fusion system is used for large scale operations, e.g., as could be the case for the Orbital Transfer Vehicle (OTV) scale operations, could perhaps have an effect upon on Earth's environment, particularly the very high upper atmospheric environment, i.e., ionosphere and magnetosphere. The Rot90 paper expresses a similar concern. Whether or not this is a significant impact affecting radio communications on Earth as a consequence of frequent operations would have to be studied in depth. Similarly, investigations are necessary to determine whether the deposition of charged particles and the effect that they might have on sensitive space instrumentation and science experiments are concerns.

The most significant safety and environmental problem resulting from fusion reaction products (ash) results from neutron activated elements. These are high energy neutrons which can activate any element. To reduce this environmental hazard to other spacecraft, either shielding or remote reactor operation are alternatives. Shielding is required for protection of the flight system and flight crew during manned flight. Heat from the neutron flux is another fusion by-product.

Radiation in the form of x-rays and γ-rays is also produced. The protection from neutron shielding is expected to suffice to reduce the radiation intensity to an acceptable level.

The conclusion drawn is that a strong environmental preference is expressed for those fuel cycles which burn nonradioactive fuels and whose ash comprise environmentally inert elements such as helium and hydrogen. From these, energy is extracted in terms of charged particles. One of the report's recommendations is to continue to study the means to minimize the neutron flux and to perform experiments that verify the reductions in emissions.

9.1.1.3 SPACE OPERATIONS WITHOUT FUSION

RTG's are of great concern having substantial potential environmental impact during launch to orbit and during operations while in low Earth orbit. Fission reactors are safety benign during launch, but there is concern during orbital operations from errors and failures that can result in an unintended reentry into Earth's atmosphere. Core meltdowns constitute a grave threat, particularly where the half life of many elements is sufficiently long to cause damage to Earth's environment.

The concerns with fission are an inadvertent entry into Earth's atmosphere, exhaust particles which are radioactive, contingency measures, and safe disposal. Ground testing will present radioactive debris hazards requiring environmental protection measures to be incorporated. A means is required to protect the public from the aforementioned hazards.

Flight operations without a high specific power fusion system results in the use of either chemical or nuclear fission to perform the mission. As discussed in Section 2.0, the Mars mission as defined and analyzed was not aided by the use of nuclear propulsion. For a manned Mars trip then, we are required to manufacture an additional 44×10^3 MT of propellant per trip. The advantage of the fusion system is in the vehicle's mass fraction. The Shuttle mass fraction, considering it as representative of a chemical manual system, is slightly better

9.0 Acceptability: Safety, Environmental, and Reliability Considerations

than 1%. An unmanned space system show would improvements over that performance. The manned Mars fusion vehicle would operate between ~25% to 70%, depending on the specific power. Hence, the result is that for current space vehicles which are less efficient, an expenditure of ~60 to 1 ratio of propellants to payload for launch to LEO is required.

9.1.2 SYSTEM SAFETY

System safety is, in part, the art of analyzing a vehicle's or system's sources of energy to determine the modes of energy release which can be to the detriment of either man or machine. Machines include the flight vehicle, or system, plus any interfacing equipment. System safety addresses the means for release of the energy, the magnitude of the hazard if an accident occurs, and methods of control to prevent occurrence. The controls to achieve system safety comprise those fundamentally inherent properties present within the system, as well as any added specific design features and operational procedures that allow the machine to perform satisfactorily for meeting mission objectives and providing for a safe operation.

Consider first the inherent safety properties of fusion. The sources of kinetic energy in a fusion reaction include: radiation, heat, neutrons, and charged particles. Energy in a MCF fusion reactor and within a fusion powered vehicle resides solely within the plasma (the kinetic energy source) plus that which is stored in the magnet (potential energy). Inherent safety of a fusion system is obtained by eliminating the explosion hazard. Neither an inadvertent mixing of the nucleons of the fuels nor chemically combining the fuels will result in a fire or explosion. That is due to the inert properties of the fusion fuels. In the latter case, hydrogen does not chemically react with helium. Safety in the former case is best illustrated by considering the large quantity of energy required to initiate the nuclear reaction as exhibited by the magnitude of the effort required to trigger the fusion reaction. Thus, there is no stored chemical energy in the fuels which can be accidentally released. That property is uniquely associated with nuclear energy, whether fusion or fission. There is some activation of the reactor by neutrons, resulting in afterheat. Safety of a terrestrial tokamak burning D-^{3}He was examined in the "Apollo-L2, An Advanced Fuel Tokamak Reactor Utilizing Direct Conversion" study:

> The main conclusion that one can draw from these calculations is that after a full reactor lifetime the Apollo-L2 structure can be disposed of as low level waste by shallow land burial. . . . The worst possible accident that can usually be envisioned for a fusion reactor with respect to controlling decay heat is to instantly lose the coolant while the plasma remains on. . . . The results show that two weeks after LOCA [loss of coolant accident], the maximum first wall temperature levels off at about 200 °C [3].

9.0 Acceptability: Safety, Environmental, and Reliability Considerations

Chemical propellants, by their nature, on the other hand, are intrinsically unsafe. It is their characteristic to react when contact is made at a very low energy level. There is considerable risk involved with their use, and extensive measures are required to prevent an accidental mixing of the fuel and oxidizer. The hazard is controlled by operational procedures and design approaches to maintain the machine within a defined safe operational mode. Hence the design is only as safe as the controls incorporated will permit.

Inherent with fission reactions, the entire vehicle's propulsion and power energy for release is stored within the small confines of the reactor core. The potential exists, therefore, for fission reactors to release their concentrated energy, resulting in meltdowns.

Let us then briefly examine the overall safety aspects of a fusion powered space vehicle, considering flight safety, top level system failure modes, and inherent hazards.

Flight safety:

High energy in the conduct of a Manned Mars Mission inherently aids safety. With fusion the energy level is sufficient to abort the mission for a return to Earth (RTE) abort mode. There is sufficient energy to transport additional spacecraft mass which allows for a greater number of safety devices to be designed into the spacecraft. There are additional capabilities for increasing the design margins of safety. The mission flight time is reduced, reducing the radiation hazards. Appendix B discusses radiation hazards and presents a space transportation propulsion system option. Additional points are made in Section 2.2.1.

Controlled fusion energy conversion machines cannot result in large accidental releases of energy. The fusion energy content is distributed within, and therefore limited to, the confines of the reactor plasma. While the kinetic energy of individual particles within the plasma is extraordinarily large, the total plasma heat content is quite small. Fusion is so efficient that only a small quantity of fuel is necessary to produce a substantial kinetic energy release to the propulsion system. If the plasma is uniformly space distributed, as in a space reactor, then in the event that the magnetic field is lost, the plasma's release of energy does not cause loss of the reactor, or an explosion, but merely results in rapid cooling of the plasma and immediate termination of the reaction. The total Q (heat content) in a plasma at any given moment is small. On the other hand, the loss of thermal control in a fission reactor results in a core meltdown in the worse case, causing grave consequences to the vehicle and to the local environment. Contingency measures in that situation are required to be placed into effect. For fusion, only a restart may be required; or, more typically, the magnet may be damaged.

9.0 Acceptability: Safety, Environmental, and Reliability Considerations

Failure modes:

While it is not characteristic of fusion reactors to fail violently as we have experienced with chemical systems, nevertheless failure of the reactor to perform is of great concern from a flight crew safety and mission success perspective. Plasma stability and sudden unexpected cooling of the plasma have been discussed. Two major failure modes are: catastrophic magnet failure and loss of coolant. For superconducting magnets, large stresses can alter the windings causing the magnet to revert to a normalicy. The magnet can overheat due to I^2R losses. Short circuits can be experienced. Loss of coolant can occur from line leaks, fittings, and components or from operational errors. The critical loss of coolant failure mode within a system is highly design dependent. Solutions for the avoidance of these latter two failure modes are well known. The magnet should be designed with sufficient strength to withstand the maximum current load with margin. Similarly, the maximum current load must be designed with a large structural margin.

Hazards:

One source of stored energy in a fusion vehicle is the reactor's start-up energy, presumed to be capacitors. Space start system options were not analyzed in this study. Another is the stored energy in the structural loading placed upon the structure by the magnetic fields and contained within the magnetic field. The use of capacitors for start-up energy should be taken as only illustrative of a typical energy storage system. Capacitors appear to have a good history of dependable performance and are not composed of moving parts, so the main safety concern is the usual accidental release of stored electrical energy concern, one with known solutions. The reactor and system design solutions have not been analyzed, however, illustrating another reason why a good MCF space fusion powered spacecraft system study should be undertaken. The reactor's starting energy level requirement is very fuel selection and to a lesser degree, plasma confinement configuration specific.

System safety of the fusion reactor, and its reliability for dependable flight performance, are functions of reactor plasma stability. Understanding of the reactor's plasma stability characteristics is critical to understanding the reactor's system safety. There is no theoretical subject of greater importance to fusion system safety.

Plasma stability characteristics are unique to the individual reactor designs. Of the five plasma physics categories discussed in Table 8-3, plasma physics theory is best developed for the tokamak, while considerable work remains for the FRC. Once again, this points to the need for a space fusion program to prioritize the technical areas of interest to space. The major system safety hazard with fusion plasma is a sudden unexpected change in the reactor physics causing plasma perturbations and reactor shutdown. It is crucial that

9.0 Acceptability: Safety, Environmental, and Reliability Considerations

experimental test reactors are built to a full scale and designed to flight operational parameters to accurately evaluate plasma stability. Safety margins must be designed into the mechanisms that assure stable plasma performance and that withstand disruptions. Expedited full scale reactor burn experiments are necessary to come to grips with this concern as quickly as possible. Establishment of scaling laws has not been as expeditious nor accurate as required for extrapolation to net power size reactors.

The other source of stored energy resides in the neutrons. In burning of fusion fuels it is high, sufficient to activate materials. The neutron energy level and flux are of such magnitude to require protection. The hazard is minimized by the selection of fusionable fuels emitting the minimal neutron flux. Even with D-^{3}He there will be some activation of the reactor, leading to afterheat. The design has to be made dual failure tolerant to the loss of coolant failure modes.

Beyond this preliminary hazard analysis, the current stage of fusion reactor systems makes a more in-depth critique speculative. A detailed design is required to permit further comment upon the system safety aspects.

The conclusion is that aneutronic fusion has inherent qualities that lends itself very favorably to a relatively safe application for space missions while exhibiting other desirable performance properties needed for application to high energy space missions. But much research is needed to cause the advantages to be verified and to be realized. A significant parameter to explore in the earliest phase of a research program is the life expectancy of any plasma stabilizing subsystems and its margin of safety.

9.1.3 OPERATIONS SAFETY

The most powerful safety defense measure at the hands of the designer is hazard elimination. With that goal in mind, this report has emphasized those approaches which eliminate hazards from a fundamental theoretical approach. The objective is to produce technology that enables future propulsion system designers to proceed with an inherently safer space transportation capability in the accomplishment of advanced, high energy space mission objectives. Where hazards cannot be eliminated, design controls must be implemented, resulting in a more complex, costly performance penalized flight system. In the operational phase, more procedures are added which exacerbates the integrated safety risks that programs undertake. As a consequence in either case, risks naturally have to be assumed. Hence, we need to seize upon those rare opportunities to design for hazard elimination whenever they exist. A major motivation for this report's emphasis on operations was to determine if a safer inherent design approach to future propulsion systems could be obtained. This commenced with the design selection of nonradioactive elements at the onset and on minimizing neutron activation of other materials.

9.0 Acceptability: Safety, Environmental, and Reliability Considerations

A radiation hazard to the public can clearly be eliminated at the onset by the selection of deuterium-helium-3 fuels over D-T. From a personnel safety viewpoint, tritium is a 19 keV beta emitter, one at a low energy level which can be stopped by a sheet of paper. The danger is primarily in breathing the gas, resulting in lung damage. In the outdoors, the hydrogen gas, being lighter than air, will rapidly rise to the upper reaches of the atmosphere where it is will safely decay or in some cases reach orbital escape velocities. The main concern is tritiated water. The MIT report reflects a similar view on favoring helium-3 (p. 53) "Overall, it appears that advanced fuel tokamaks present a lesser risk than D-T tokamaks."

Fusion fuels, when considered as potentially reactive chemicals, are inherently safe, that is, an accidental mixing of the reacting fusion fuels will not result in a fire or explosion unlike our current cryogenic and hypergolic propellant systems. Public safety is aided during launch as is flight safety. The difficulty with igniting these fuels is a fact to which personnel in the terrestrial program will readily attest. When mixed with an oxidizer, however, either deuterium or tritium will be as chemically reactive as hydrogen. The use of helium-3, an inert element, in place of tritium will reduce the probability of a fire or explosion as well as its magnitude simply due to the reduced quantity of flammable fluids present. With D-^{3}He as the fuel cycle of choice, fire is a concern only during atmospheric operations due to the absence of an on-board oxidizer. That makes it inherently safer to the flight crew and vehicle during space operations than lox-hydrogen systems. While the objective of the enriched helium-3 in the D-^{3}He mixture is a reduced neutron flux, an additional safety benefit is gained from a reduction of deuterium mass. Obviously, for fire safety reasons, one would prefer the ^{3}He-^{3}He reactions. Unfortunately that fuel cycle's reactivity is quite low.

The D-T reaction produces 14 Mev neutrons, requiring massive shielding to protect the spacecraft's occupants. Special restricted operations will have to be placed into effect during a burn near Space Station Freedom to protect its occupants from neutrons. Unshielded, Dr. Roth estimates a "safe" distance of 16,000 km for D-T which compares to 2,900 km for D-^{3}He for a 200 MW reactor (Rot90). The neutron flux from the D-^{3}He reaction is substantially reduced in comparison with D-T, but some reduced protective measures are still to be required.

With D-T, the first wall material will become highly activated by the neutron flux of ~3-5 MW/m^2. The use of the deuterium - helium-3 fuel cycle lowers that to 0.1 MW/m^2. The ensuing activated elements are numerous, depending upon the type of materials selected for reactor and spacecraft construction. Because the level of radiation accrued will also be a function of burn time, it is conceivable that operational restrictions will become a function of the vehicle's flight time as the radioactive level increases. That will be the situation with either fuel, but the rate and level of radioactivity increase are considered to be many times greater using D-T than D-^{3}He, exacerbating the likelihood of a quicker and more severe radioactive hazard situation arising with D-T powered space vehicles.

Operations in any endeavor comprise the major contributor to accidents. Flight operational hazards are created by an activated first wall since their removal and replacement for launch vehicle refurbishment in a space environment can be anticipated to become part of the planned maintenance activity. In the case of either fuel, robotics will be required for repairs, and shielding is needed for protection of the crew. First wall disposal, after removal from the reactor, adds further to the operational hazards, risks, and costs, ones which are best avoided, where possible.

Not only does the flight operational performance characteristics favor the use of reduced neutrons, but the the fusion vehicle system design is simplified as well. Flight spacecraft system structure can either be eliminated or its mass reduced that otherwise would be necessary to absorb the neutron flux. Heat control design problems and heat rejection mass are reduced. With D-^{3}He the heat rejection requirement from the decay of tritium is eliminated. The on-orbit maintenance necessitating first wall removal is either at a significantly reduced time scale, or eliminated. Shielding of other orbital facilities or operational restrictions are substantially reduced. The impact on contingency planning is reduced, or at least simplified, such that, as a minimum, contingencies for accidental releases of tritium are eliminated from consideration.

Helium-3 reacts with deuterium at higher energy levels than tritium. Radiation losses will require considerable attention. As we proceed into developmental research with D-^{3}He, plasma stability as a consequence of expanding the operational temperature regime over and above the lower pressure reactors will be a subject of significant research endeavors. One key D-^{3}He developmental goal is to provide the design means for the reactor accepting a greater tolerance to plasma instabilities, i.e., the reactor must be designed to contain 14.68 MeV protons with operational margins. Verification that the reactor can be operated with a significant tolerance to values from the reactor's design point and examinations of system sensitivity to operational divergences will be necessary for assuring a reactor's operational system safety. The higher temperatures will undoubtedly increase the reactor's sensitivity to contamination such that the purity quality of the fuels and propellants will be an important consideration in achieving a reliable operational vehicle. A drawback to the D-^{3}He reaction is the fact that it still contains 1% to 5% neutrons. Whereas the low flux reduces the neutron environment, a given quantity of mass is still necessary to absorb neutrons. There are trade-offs, but the balance is clearly in favor of D-^{3}He. Emphasis must be placed on research for an aneutronic reactor design.

Any payload launched without radioactive substances will have a significant advantage in terms of less protective mass required. The costs to acquire launch approval will be considerably lessened. There is a strong safety preference to the use of space-based vehicles which are non-radioactive.

9.0 Acceptability: Safety, Environmental, and Reliability Considerations

Some techniques to reduce the number of neutrons in the D-^{3}He have been examined as discussed in this report (Section 4.0). More in-depth studies and verification experiments to investigate neutron reduction techniques are worthwhile. One such important study would be to establish the function of neutron flux on a flight reactor's and vehicle's design mass. That data would be useful in conducting trade studies to establish the vehicle design leverage of neutron reduction. From that information, planning can be accomplished to provide guidance for the level of funding for neutron reduction research. Hence, additional analytical efforts to numerically establish and demonstrate the performance and cost benefits and system tradeoffs of a lower neutron flux D-^{3}He fuel cycle play an important role in space fusion research.

Fission reactors present inherent safety hazards. The reactors' radioactivity level will increase significantly with use. Unfortunately, the probability of need for maintenance attention to the system will also naturally increase with system operational time. EVA repairs on an activated reactor will not be possible. Instead, robotics will be required or the reactor abandoned. In fission, heavy radioactive particles will be produced in the exhaust which will settle on the surface of the moon, or Mars, or will be deposited into space including LEO. These elements typically have long half lives and high energies. There will be a significant mass penalty for shielding of the crew since the particles are at a higher energy level. There is a hazard, too, in the result of a strike to a reactor by a meteorite of sufficient size and velocity causing damage that results in the release and spread of radioactive materials in space. Unlike ground contingency operations, there is no way practical technique to clean up the radioactive debris in space. Perhaps these are concerns that after detailed study may be found to be insignificant, that is, the probability of a damaging strike may be acceptably low, but the option to eliminate these hazards is preferred.

9.2 SYSTEM RELIABILITY

If man is to settle Mars in an economical, safe manner, the performance requirements for the Manned Mars Missions will necessitate new design approaches and operational standards which contrast significantly from current expectations of rocket engine performance. Instead of burn times of seconds, as expected for solid propellant motors or minutes in burn durations for liquid engines, we now must examine what it takes to run a fusion engine for a continuous burn duration of two months. Based upon the proposed design using magnetic fields, magnetic nozzles, and the lack of highly stressed moving parts like turbo pumps, that approach appears reasonable. The obvious primary concern is the damage that the vehicle will sustain from the accumulated neutron exposure over those operational periods.

Beyond the Manned Mars Missions, to perform the more distant high energy missions – particularly those beyond the solar system – reliability requirements

9.0 Acceptability: Safety, Environmental, and Reliability Considerations

of a new dimension will be placed upon space propulsion and power systems. A rendezvous mission to the nearest star takes on the order of 200 to 300 years using systems with very high specific power characteristics for the missions as considered herein. The fourth stage reactor will be expected to function for approximately 50 years after being dormant for nearly two hundred years. The first, second, and third propulsion stages are each respectively expected to provide the electrical power for the annual or biannual science data transmissions while those stages are respectively operating in a propulsive mode. The option is for the fourth stage to be cycled once or twice annually for the data transmissions, that mode requiring a large number of start cycles being placed upon the machine over a 250-year interval. The fourth stage reactor will be shutdown at the completion of a 50-year burn period. Then it will again be called upon to transmit science data twice annually for another 10 to 20 years after completion of the thrust mode.

To accomplish such a mission, new standards for flight system design reliability must be set. On-board self diagnostics and corrective measures will be required to avoid system divergences. This will set new standards for diagnostic tools for artificial intelligence by flight vehicle systems and for self corrective tools to remedy undesirable situations before they occur. Once again, if the shielding can be adequately accomplished within the vehicle's specific power limitations, the magnetic confinement of fusion plasmas and thrust conversion and vector control appears to provide a reasonable technical approach. Reliability is achieved by the inherent capability of a "solid state" propulsion system design, redundancy of flow control devices, and predictive diagnostics to prevent deviations leading to system malfunctions or failure.

For fission systems, mission durations of the stellar nature are not considered feasible due to half-life limitations. The question of the means to achieve fission system reliability for the more reasonable Manned Mars Mission has not been addressed for a flight system. Let us examine some of the high reliability approaches and issues.

Fission reactors, in which gas flows over a core, will have an erosion concern. Erosion is a life limiting concern as well as a safety concern, the latter resulting as a consequence of ejected radioactive particles. NERVA type systems are high thrust systems requiring higher propellant flow rates and rotating hardware. For hardware of that nature, large design margins will be important where space repair and maintenance are not anticipated to be cost effective. Redundancy of the high wear system equipment is also considered to be important to achieve system reliability.

Fission reactors require the use of multiple control rods and actuators. Therefore, from a controls design perspective, these are more complex than magnets. The chance for failure is greater in a dynamic system than in a static device. Operation of the fission reactor is required in flight to cool the reactor mass. That will increase the number of control cycles in comparison to designs that avoid the requirement. The fact that such operational requirements are imposed upon a flight system design shows that that design approach is a reliability life limiting concept. It will be important to provide a wide design and operational factor of safety with fission reactors and to avoid pushing the fission engine close to the design limit to achieve high specific power capabilities.

Design trades involving fission need to incorporate a mass allowance for achieving a safe, reliable vehicle, just as cost trades must account for space maintenance, operations, contingency planning, reusability, and life cycle expectations as well as the R&D and manufacturing costs. That is, a complete systems approach must be taken. Ultimately, for a manned application, the safest design approach is a redundant reactor with a test verified-high degree of margin system that is tolerant to multiple operational errors during flight use.

9.3 SPACE MAINTENANCE AND LIFE CYCLE APPROACHES

For missions within the solar system, there will be the economic necessity to reuse these massive engine systems with a minimum of orbital maintenance. The fusion energy source offers unique features that meet these challenges – non radioactive fuels, charged particle energy, liquid refueling, low neutron flux to name several. Activated materials will be a cause for restricted maintenance operations.

The entire fusion vehicle is anticipated to operate as a "solid state propulsion system" to the greatest extent possible, the exception being the transfer of fuels to the reactor. Through asymmetric field strengths in the engine's magnetic nozzle, thrust vectoring can be simply accomplished. Inherently reliable magnets provide the net propulsive mode. Unlike chemical systems, large highly stressed pressure vessels are not required. High speed moving parts are eliminated. In so far as the propulsion systems are concerned, surface erosion of nozzle throats or other components over which mass flow occurs is avoided by the magnetic field confinement approach.

So while the magnets impose a mass penalty, there are tradeoffs offering benefits for other mass and design complexity savings and in the provision of inherent system reliability and low maintainability.

Maintenance of a fission reactor system will probably not be possible or, at least, will be a very expensive proposition. An extensive use of robotics will be required to replace man who cannot accept the high levels of radiation. Robotic servicing devices will also become activated and require either disposal, space storage at a safe distance, or storage in a shielded location. "The biological dose rate after shutdown at the back of the shield is excessive implying that all remote maintenance is needed." (Kha89).

Fission reactor reuse is less of a possibility than fusion where refueling can be achieved simply by the transfer of cryogenic fuels. The fission reactor cannot be so simply refueled, if at all, in space; so life cycle costs can be anticipated to be higher due to a limited reuse capability.

9.0 Acceptability: Safety, Environmental, and Reliability Considerations

9.4 SUMMARY

High specific power–high specific impulse space flight vehicle systems reduce the number of Earth surface-to-LEO flights, thereby reducing the environmental stress and lowering the Earth's energy strain. Fusion's attractiveness stems from its inherent characteristics of high performance and safety when D-^{3}He is used. The large amount of energy required to initiate the reaction is the feature which makes that high degree of safety assurance possible. High specific power and high specific impulse systems all inherently have the capability to make flight operations safer and environmentally more benign by simply reducing the number of launches required from Earth's surface.

These nuclear systems are safer than chemical propulsion from the viewpoint of their lack of chemical reactivity with each other. Burning of D-^{3}He eliminates radioactive fuel hazards related to servicing and handling operations.

If tritium is used, it is a weak beta emitter, having a 12.3-year half-life. D-^{3}He fusion energy is safer inherently than fission, not only from the absence of radioactive or low radioactive fuels, but also from the standpoint of not being capable of excursions. All of fusion's energy is stored within the plasma. It simply shuts down if a problem occurs and any damage is self contained. With fission, all of the vehicle's energy is stored within the reactor's core, a relatively highly concentrated energy source.

The fusion reactor's first wall maintenance requirements are simplified by the selection of deuterium and ^{3}He which has a substantially reduced neutron flux. Refueling in orbit or another planetary body is simply accomplished by a fluid transfer operation.

Environmentally it is a very sound approach. If accidental releases occur, the helium presents no hazard; and tritium is expected to rapidly disperse into the upper atmosphere where the breathing hazard, the only radioactive fuel hazard, will not typically pose a threat. The main concerns are tritiated water and local entrapment. Production of the deuterium and helium-3 fuels is not expected to have an impact on the Earth's environment, unlike chemical propulsion and fission. The production of the large masses of tritium, on the other hand, as required by the VISTA concept would be expected to have a significant impact. The space fusion fuel cycle's "ash" is comprised of all environmentally sound products with the exception of tritium.

The charged particle ash from burning the fuels offer an inherently sound approach for a "solid state propulsion system," a necessity for the long trips. For D-^{3}He the main, fundamental hazards are cryogenic fluid temperatures, deuterium fires when used in the presence of an oxidizer, and a reduced neutron flux which activate first wall materials. These are controllable.

10.0 SPACE PROGRAM OPERATIONAL ECONOMICS AND PROGRAM IMPLEMENTATION

10.1 PURPOSE AND APPROACH OF THE ECONOMIC-PROGRAM IMPLEMENTATION EVALUATION

Earlier we examined the mission enabling capability of fusion energy and its technical status, approach, and feasibility. The primary follow-up question is whether or not the current management structure will accomplish the program for space. Hence, we examine the factors which will influence the program:

```
***KEY MANAGEMENT FACTORS***
SPACE FUSION VIABILITY
AND IMPLEMENTATION STRATEGY
─────────────────────────────
-MISSION ECONOMIC CONSIDERATIONS
  •PROGRAM PRIORITIES
-ORGANIZATION OBJECTIVES
  • AGENCY COMMITMENT
```

In this section then, we consider the relative economies resulting to NASA with and without fusion energy. Since economics constitutes a significant part of any strategical planning activity and since it is perhaps the most overriding factor in all endeavors undertaken, the consideration of economics is of utmost importance in the development of the proper space fusion program strategy. It determines the priority of the fusion program among competing resources. The purpose of this section, therefore, is primarily to consider the economic benefits to NASA from the presence of a space fusion energy conversion capability. The conclusion reached is that the economical gain from fusion energy flight systems is enormous. Whether or not the actual achievement of a space fusion capability occurs will be a function of NASA's commitment to fusion. That commitment will be exhibited by the funding level to accomplish the challenging technical considerations which have been discussed in Sections 6-8.

In order to assure that the economic advantages are realized once the commitment is made, the next topic to address is the organizational structure most likely to achieve the desired results. Consequently, it is necessary to proceed one step beyond the purely economic matters to develop the means for the implementation of a program. A new NASA space fusion program would provide a program management system to focus upon the space fusion goal. This need is based upon the reality that program funding levels are directly linked to organizations, their functions, and mission objectives, all of which

10.0 Space Program Operational Economics and Program Implementation

factor into spending priorities. Hence, the current governmental organizational structures and the likelihood of it producing a space fusion capability in a timely manner is also an essential consideration in the development of a space fusion energy strategy.

This section, therefore, incorporates an evaluation of organizational roles and the influence of those roles upon the objectives mandated for the respective organizations, NASA and DOE, in the execution of their assigned responsibilities. What are the products and results expected from the two agencies? The approach taken here is to develop a funding strategy by using the technique of examining the influences of fusion energy on space flight operations. Looking upon fusion research as an investment which pays tremendously in terms of operational dividends as discussed in Section 2, we can establish a funding yardstick for the level of fusion research for space applications. Thus, the logic for the strategy and the recommendations from the viewpoint of managerial considerations is discussed and rationalized.

Organizations like NASA and DOE are instituted and constructed generally along the very specific line needed to meet the objective of producing focused results to resolve particular problems. The problems upon which the agency focuses are accomplished by very specific program goals. Thus, in deriving a recommended space fusion program for NASA, one must take into account, as a prime consideration, the uniqueness of its missions. Further, if we compare the NASA missions with those of the DOE to examine their similarities and differences, this process provides guidance in the development of a program strategy for fusion energy for space.

NASA MISSION	DOE MISSION
• FOCUSED SPACE RESEARCH	MULTIFACETED PROGRAM:
• FOCUSED AERONAUTICAL RESEARCH	
• DEVELOPMENT PROGRAMS	• DEFENSE
• FLIGHT OPERATIONS	• CIVILIAN
	• RESEARCH
	• DEVELOPMENT
	• UTILITY COMPANY
	-OPERATIONAL RESPONSIBILITY
	-PROFIT MOTIVE

The end objective of this section, then, is to concentrate upon the relative impact that fusion energy can make upon the <u>NASA mission</u> in the application of fusion

10.0 Space Program Operational Economics and Program Implementation

energy. Mission differences do require and result in major variations in program funding levels. In examining mission differences, one must consider all of the various elements of which programs are comprised in order to optimize system or mission costs. There are two major components in the final mission cost equation for organizations like NASA and DOE, namely, the developmental investment costs and the operational costs.

With regard to these two components we note a significant, and fundamental difference. DOE's mission is energy for defense and utility electrical power needs. Fusion research comprises only a very small effort in the total DOE budget. For terrestrial fusion energy conversion, the ultimate goal of DOE is to <u>develop</u> fusion energy as a source of power for the generation of commercial electrical energy. It is the function of the public utility companies to <u>operate</u> the fusion system profitably. For NASA there is a profound difference. It has both a <u>developmental</u> role as well as an <u>operational</u> one. Instead of profit motives, we operate on the principle of whether or not the fusion device is mission enabling or mission enhancing to meet program goals for space exploration or for science. That distinction – profit versus space mission achievement – creates a fundamental difference in organizational perspectives and goals.

The flight operational costs can be divided into two categories: initial deployment costs and day-to-day operational costs.

FLIGHT OPERATIONAL COSTS	
INITIAL DEPLOYMENT	DAY-TO-DAY

This analysis considers first the effect of fusion energy on the initial space deployment operational costs. Day-to-day operational cost considerations were then estimated from analyses of space mission operational costs. Then, from an understanding of the total mission operational economic trades, a figure of merit was derived for the other component of the cost equation, namely, the level of reasonable developmental investments for space fusion experiments and a program leading to the implementation of a flight fusion powered spacecraft. Flight operational costs for NASA are a major consideration in undertaking a program of this nature since in contrast with the terrestrial program, NASA, unlike DOE, in the end is responsible for the flight <u>operational</u> aspects of its missions. In DOE, demonstration of first principles and cost competitiveness with other sources of energy is the goal. Consequently, there is a major difference between the two agencies as exhibited by the execution of flight missions versus the production of commercial power, as discussed below.

10.0 Space Program Operational Economics and Program Implementation

10.1.1 AGENCY COSTING PHILOSOPHY

Clearly one major goal of all agencies, or any organization for that matter, is to successfully accomplish its mission with minimal costs incurred. And the fundamental goal of the programs within organizations is to accomplish the mission objectives within the allotted cost and time constraints. The total program cost is comprised of the sum of *all* of its elements. To minimize the total program cost, however, the acceptable costs for any specific program component, or hardware, can vary from agency to agency and from program to program, depending upon mission objectives. Thus, to assure that the minimum *total* program costs are attained, some programs may well consider a higher cost of its components to be an appropriate expenditure over that spent for the same hardware used in other programs. This difference can become even more pronounced in different organizations that have different missions.

10.1.2 DIFFERENCES IN ORGANIZATIONAL PHILOSOPHIES

As an illustration of this key point, in the space program we willingly pay a premium price for high reliability electronic piece parts. The high reliability electronics piece parts serve as an insurance policy. Whereas the space electronic circuitry may be similar to that used in various terrestrial consumer devices, the space application's operational environment is more extreme and the economic consequence of failure far greater than that for consumer products. Additional rigorous testing of space hardware is required to eliminate potential defects or marginally performing hardware. The philosophy adopted for space programs is to provide a high degree of assurance that an electronic piece part will not fail during the systems verification checks while in launch count down or while in flight. The least expensive impact of a failure is at the initial time of hardware possession, namely, component acceptance testing. The cost of failures increases as the vehicle progresses along the launch processing flow path. The costs from failure dramatically escalate as the time of launch approaches. Those additional quality assurance costs to achieve high reliability, when summed over the entire program, can more than pay for the investment costs in the acquisition of quality parts. Those added reliability costs would be dwarfed by one launch delay. The cost of failure in orbit is even much greater. Manufacturers of televisions, radios, computers and the like, on the other hand, in their mission to produce the least expensive product, would consider the additional space reliability measures incompatible with their mission objective of meeting severe international consumer product competition.

Similarly, we pay more than an order of magnitude cost premium for rechargeable space qualified batteries. The rationale is obvious. A $500M to $2,000M spacecraft, plus the additional transportation cost to low Earth orbit (LEO) of more than $300M and the impact of mission failure, all trade very favorably toward the additional insurance which is provided by a small expenditure of funds for achieving higher battery reliability for flight energy storage systems which cost $200K to $400K. But for ground applications of rechargeable cells, one would be more likely to assume the risk of battery failure and would not spend those additional funds simply because the impact of its failure to ground operations is negligible in most situations and not worth the space program's cost trade.

10.1.3 IMPACT OF AGENCY MISSION VARIATIONS

Consider the differences in the two agencies' missions, particularly with regard to the operational products, the degree to which the technology is enabling versus enhancing, and the requirements imposed upon the technology. With fusion, the mission cost trade between programs in agencies like NASA and DOE is different due to mission differences. The terrestrial fusion program – having the goal of providing a means for the generation of utility power – must meet the mission performance objective of producing an energy conversion system which is cost compatible with the other current energy sources that include coal, crude oil, fission, or hydraulic power and to do so within the required safety and environmental constraints. Otherwise, it will not succeed. As shown later these alternate energy sources are competitors with fusion. Thus, for utility power purposes, the final number in the DOE fusion cost equation is the anticipated relative cost of electrical energy delivered to the network grids for utility company use. The electric utility company's product concludes with the production of safe, economical power for use by industrial and individual consumers. No further performance requirements reside for that program. Thus, the terrestrial fusion program funding level is considered to be appropriately keyed to advance at a rate compatible with the terrestrial energy cost projections.

There is one other major mission consideration. For for some space missions, fusion is an enabling technology, a fundamental difference in the missions of the two agencies. For terrestrial applications, fusion energy is only a complement to the existing energy supplies. Fusion is not "mission enabling" nor operationally more economical than current energy supplies. Obviously the DOE is not concerned with budgeting its office to operate utility company fusion reactors. That results in a difference not only in technical requirements but in agency priorities as well.

Technical requirements differ. The differences between the mission requirements and the implementing hardware designs for terrestrial fusion application and the space fusion applications are anticipated to be as pronounced as the difference between a diesel engine designed for a diesel locomotive and a turbine designed for commercial aircraft, or as the fission reactor for use at an electrical utility company compares with the space nuclear power plant, i.e., the SNAP type space power plant, although each of those pairs respectively share common fuels. One must take into account the differences in the physical environments where the respective reactors will operate. Water for cooling terrestrial reactors is readily available on Earth,

10.0 Space Program Operational Economics and Program Implementation

whereas NASA relies upon radiation cooling in space. The space reactor has a ready supply of vacuum available whereas it is naturally denied to the Earth bound reactors. Maintenance is a different problem altogether for the two applications where, at best, space may be considered relatively inaccessible in many situations and absolutely so in other cases.

In the <u>space program</u>, the cost of the energy to accomplish a mission is a significant component in the cost equation. But in space, in contrast to the terrestrial application, the fusion produced power is only one component of a mission. It is a means to a space program's objective, not the end purpose. That implies that in the space program, we would readily accept a higher cost of energy as a wise investment, provided that a positive **net benefit** to the <u>total</u> space system mission results. That is a point on which this section concentrates.

It is shown that, where space missions durations are significantly shortened by fusion energy, or where the number of Earth orbital launches are reduced, the opportunity exists for NASA to reap substantial economic advantages. The economic advantage of fusion to space currently contrasts with the terrestrial program's economic incentives to develop fusion. Fusion, as currently envisioned, is not economically compatible with currently used energy sources in today's international energy economics.

As shown later, the energy leveraging power of fusion energy, when converted to space economies, is enormous as compared to chemical energy's costs for performing a lesser mission. That is a critical point for NASA's consideration since its budget does include a very significant level of funding for operational costs.

Paradoxically the United States' space fusion development is unfortunately presently governed at a rate determined by the mission of the terrestrial fusion program's goal to develop economical utility electrical power. If fusion were an engineered capability today, because of its currently conceived higher cost of electricity (COE), those power plants would sit idle due to fusion's present inability to compete economically with other energy sources. But if it were available for space today, fusion would be playing a major role in space exploration.

10.1.4 SYNOPSIS OF FUSION COSTS ON SPACE MISSIONS

For NASA's missions, the cost of fusion energy is only one part of the system equation, the mission being ultimately to conduct science and to move man and cargo through space to conduct space science and exploration missions. For space, fusion energy is the means to the end, not the end in itself. Today, if a fusion reactor existed which has the properties ascribed to it in the text of this report, NASA could very well accept a higher energy cost for a new source of energy unlike the power companies, provided that the new energy source permits either mission enabling or mission enhancing technological advancements. *From the considerations noted in this report, it appears that, with the availability of this new fusion technology, a new mission enabling capability is provided, space mission costs reduced, safety enhanced, more data intensive science missions permitted, and more data returned from science missions. If space fusion were developed now, man could conduct space exploration missions on a scale not otherwise permitted.*

10.2 REACTOR AND FLIGHT VEHICLE OPERATIONS

Let us now evaluate some of the key cost elements which are unique to space operations. Operational cost advantages for NASA are a function of the operational manpower level, wear out rates, and recurring costs traded against speed of task accomplishment and quantity of work performed – an operational efficiency factor. These will have to be factored into a space fusion system design and development program. Each are discussed below.

10.2.1 OPERATIONAL MANPOWER LEVEL

Consider first that an advanced degree of autonomy must be achieved for space fusion systems because of the great distances and speeds involved. The most stringent mission objectives, those for stellar flight fusion systems, which demand the delivery of unerring or, alternatively, self correctable fault tolerant systems that endure for lengthy, multicentury periods of time, beyond 4 light year distances, such as with an Alpha Centauri flight. The importance of artificial intelligence (AI) was alluded to earlier regarding the innovative in-situ conduct of science missions. In the context of the vehicle system's operational aspects for missions, AI is equally important for operating and controlling the spacecraft's systems, thereby accomplishing the vehicle's mission objectives. The spacecraft must possess the intelligence to fly itself. If key failures occur, then self detection, diagnostics, and repairs must be accomplished by the spacecraft, in situ. Included in this category are guidance, navigation, and control of the spacecraft and of the propulsion system.

The capability for the spacecraft to fly with minimal human assistance is important for the planetary missions, particularly where the Earth based response reaction time is on the order of 8 to 10 hours for the round trip data transportation time at Neptune's and Pluto's distances. Because two thirds of the vehicle's flight time is in a thrusting mode, the vehicle's system design must accommodate and correct errant spacecraft behavior without waiting for the lengthy Earth response. That is a suggested difference in operational philosophy and system requirements between constant acceleration, long thrust duration fusion vehicles and the current non-thrusting, low velocity spacecraft missions which tend to be more closely monitored.

10.2.2 REALISM OF AUTOGENOUS SPACE FUSION REACTOR

Is an Autogenous Space Fusion reactor a concept that can be considered? Let us attempt to examine further the extent of what is involved in the accomplishment of the autonomous system objective, at least in principle.

To better understand the magnitude of flight support personnel needed for autonomous flight systems, let us examine the level of manpower involved with the flight operational aspects of terrestrial fission reactor operations and compare it with the level used for fusion "reactor" experiments. There, a staff of approximately 300 to 400 persons is used for a continuous 24-hour operation throughout the year. All plant activities are included in that number, many of which are not applicable to a space fusion reactor.

Contrast that with the large fusion experiments that involve approximately 100 persons for a large machine. The "Dense Z Pinch Experiment" at Los Alamos requires less than 10 people; the estimated cost is on the order of $300K. The magnitude of the autonomous space fusion reactor, as being suggested in this study, is such that those relevant tasks which are now performed by these ground reactor and test personnel, will have to be performed by intelligent flight systems rather than using a trained staff. Obviously these are not identical comparisons, but they indicate an encouraging trend. Can the trend be such that the number of support staff diminishes to zero? That accomplishment may be determined primarily by the wear-out rate.

10.2.3 WEAR OUT RATES AND COST FACTORS FOR FLIGHT OPERATIONS

There exist, consequently, strong rationale and motivation for the development of a concept coined here as a "solid state propulsion system," that is, one without any moving parts. Perhaps that is not achievable, at least in the absolute sense, if for no other reason than due to the requirement to control the flow of fluids. Nevertheless that at least should be the goal to serve as a space fusion program technical focus. With MCF, the likelihood of successfully achieving that goal is much greater than with any other system. ICF systems, it is concluded, will have to be developed using other means to achieve a high state of operational reliability, accomplished traditionally through the application of hardware redundancy or by providing design margin. That conclusion was drawn from the basic design concept which require active flight operations, such as the loading of pellets with fuel, pellet ejection devices, and very precise ignition triggering mechanisms.

The parameters to be measured for mission operations and flight control are not fully defined, since the flight reactor has yet to be built. Nor are they totally undefined either. We know that some fundamental system parameters of importance for reactor control in flight will include the thrust level, the electrical power demand and the generated values, plasma temperature, magnetic field data, system temperatures, restart system status, and fluid flow controls, to name some of the obvious. None of those, based upon current knowledge, appear to involve the high frequency, millisecond control system responses, such as is necessary, for example, with the Space Shuttle Main Engines' turbopumps. The software code for the accomplishment of the current control and housekeeping tasks for fusion experiments has not been complex, actually, a manual task for the tokamak, until recently. Those measurements include plasma position detection and the prevention techniques to maintain the plasma in a location away from the wall. The development of the research codes for understanding plasma stability and transport is another matter, involving Crays for research analysis and modeling, and that, of course, is not a requirement for flight operations.

Whether or not the goals we set forth here can ultimately be achieved must receive extensive review and analysis, much of which awaits the determination of the means to achieve the safe and reliable production of net fusion power. At least from the initial considerations given to the subject here, the required level of autonomy does appear to be reasonably promising.

10.2.4 RECURRING COSTS

For purposes of this document, recurring costs consist of the costs of flight consumables – primarily fuels – plus the costs to deliver the fuels to low Earth orbit, the major cost factor in the implementation of high energy missions. Scheduled maintenance is another recurring cost.

The cost of the energy product consumed is relatively small. Thus the trade here is to decrease the quantity of mass placed into orbit by either lengthening the space flight times or by an increase in the propulsion system specific power and impulse. In the case of fusion energy, the high specific power and impulse performance offer the cost advantage of drastically reducing the number of launches to low Earth orbit to accomplish greater mission objectives than any other currently developed energy sources. The recurring fusion fuel cost for performing the solar system missions is small since the quantity of fuel consumed is small. Scheduled maintenance is minimized with the proper fuel selection.

10.3 COMPARATIVE AGENCY ECONOMIC ISSUES

Let us then consider the economic importance of fusion to the DOE and NASA in terms of setting program priorities.

10.3.1 FUNDING LEVEL FOR TERRESTRIAL FUSION

Estimates of COE (cost of electricity) are periodically made by the terrestrial fusion program to establish figures of merit for fusion reactor power plants, particularly to establish fusion's cost effectiveness in relation to current power plants. The ESECOM study was a recent attempt to better define those costs. The community generally considers the tokamak COE to be in the range of 50 to 100 mil/kW·h. Current utility fission energy costs approximately 40 mil/kW·h and approximately 35 mil/kW·h using crude oil. Until the production costs of commercially available energy rises as the crude oil supply decreases, there will be relatively little incentive on the part of Congress to fund the development of a terrestrial fusion program much beyond the current rate, one which at least maintains the professional capability. The annual United States' spending rate trend for MCF, plotted since program initiation in 1954, is shown in Fig. 10.1 (anom87).

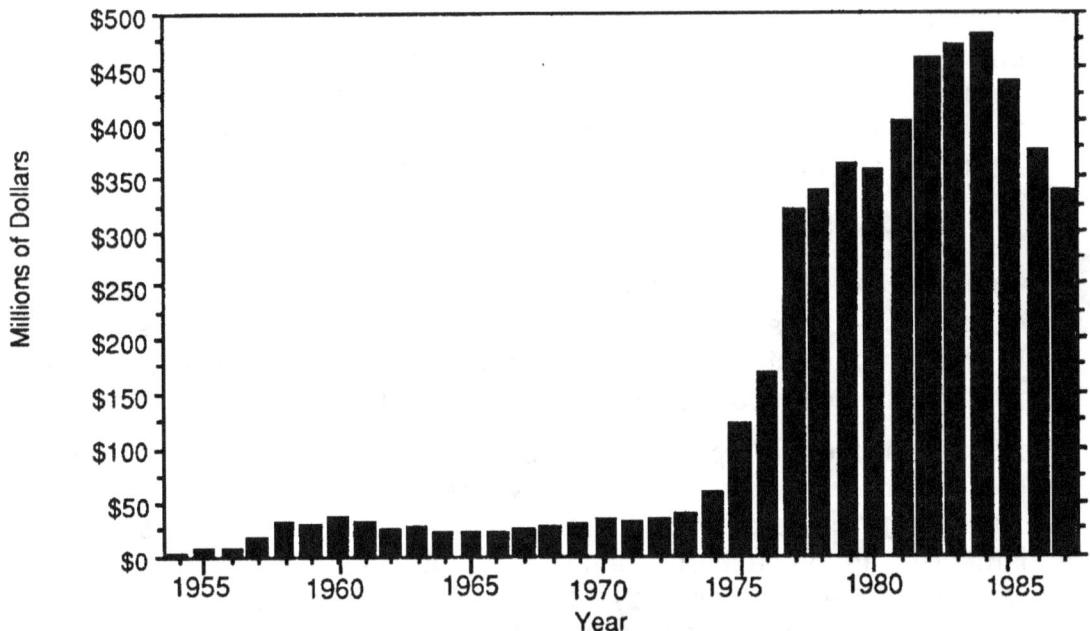

Fig. 10.1. Historical Magnetic Fusion R & D Funding, 1951-87 (in current dollars).

The spending surge began during the energy bind of the 70's. In more recent years it has been decreasing, and in 1991 it is at ~$275M. Shown in real year dollars the value is considerably reduced, probably at approximately half of the later years values.

10.3.2 FUNDING LEVEL FOR SPACE

The United States space fusion technology research lacks any funding. Discussions with Soviet fusion personnel did not reveal that they have considered it. So worldwide, the amount for space is probably at zero, although the Japanese appear to have a program under consideration. Even at NASA where the gains are much greater, at least in the relative near term, the agency only saw fit to fund it at $1M (Appendix A).

10.3.3 PROBABILITY OF APPLICABILITY OF TERRESTRIAL FUSION FOR SPACE BENEFIT AND TIMELINESS

Is the DOE program likely to yield direct benefits for the space program any time in the near future? To answer that, consider that the 1990 year funding of ~$325M is adequate to maintain the operation of the one primary experiment at Princeton plus several million dollars for a low level funding of several alternate experiments. The experimental reactor of interest to space is the FRC which was funded at a total level of $5M at two facilities in 1990. The program is being terminated in 1991 due to a declining fusion budget in DOE, as are all alternate experiments that could have potential space use. Compare the complexity of the task between the development of fusion with the simple chemical rocket that Dr. Goddard set out to develop. His experiments commenced in ~1910 under his funding. In 1916 he requested $5K for the continuation of the experiments. The total requested cost was $11K. In comparison with the gross national products between that time ($48B) and today, there is very little difference between the relative funding level that the chemical rocket received and which the very complex FRC fusion reactor is currently receiving.

To place the national fusion research and development funding level into perspective, consider Fig. 10.2 (anom87) which shows the relative program funding levels between the years 1980 to 1987 for solar, fusion, fission, and fossil energy sources, plus the amount allocated for conservation.

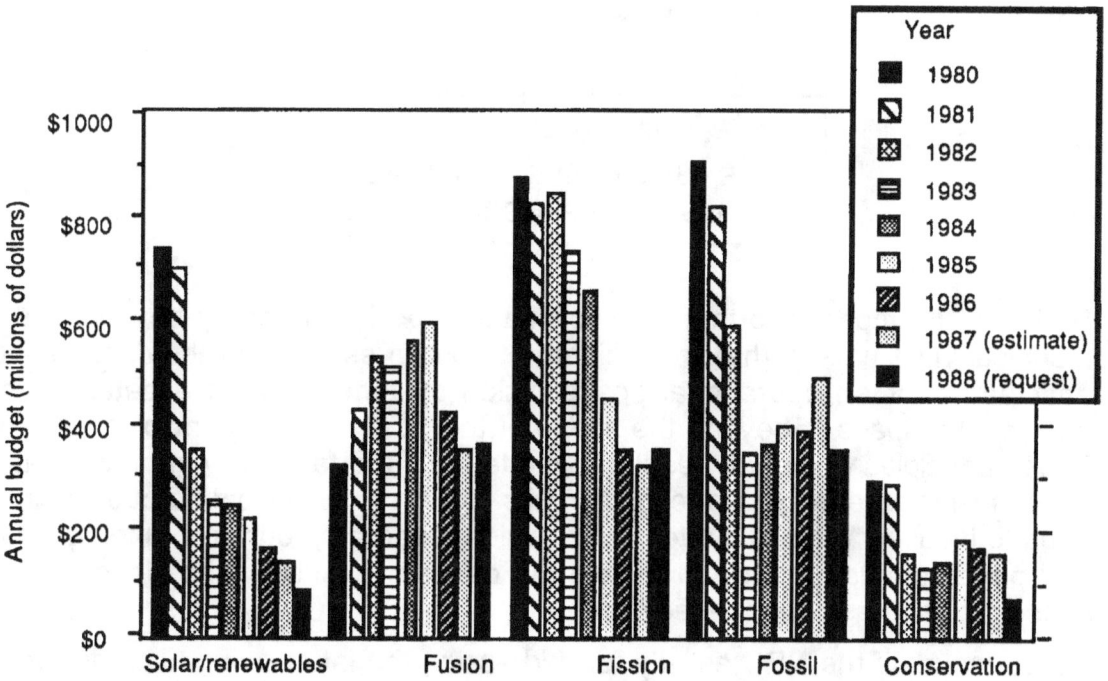

Fig. 10.2. Annual Appropriations of DOE Civilian R & D Programs (in current dollars).

Fusion, the most technologically demanding of all, and the least developed for practical use, is funded at an integrated level over that time interval below fission which has been demonstrated since 1942, below fossil fuels which have been demonstrated for well over a century, not much higher than solar energy. And this is for the time when fusion was funded at its highest levels!

If we compare fusion research with the funding expended over the years by NASA for chemical propulsion research and development, then one can expect that under the current plan, spin off benefits for NASA's use cannot begin to be addressed for many decades, if ever, without some miraculous breakthrough. Yet we have immediate applications for it! The funding priority is a direct function of the agency mission.

A grant was made to the National Research Council by the DOE to "..... conduct a study of the priority and pace of magnetic fusion research and development in the context of long - term policy and the factors that should enter into that policy formulation." The results of the study, chaired by I. L. White, are published in the National Academy Press report entitled "Pacing the U.S. Magnetic Fusion Program," 1989. There is direct relevance to the space fusion energy application as discussed below.

10.3.4 ECONOMIC ADVANTAGES OF FUSION FOR SPACE

The NRC study dealt with fusion energy for commercial electrical power production. Let us return momentarily to the space program.

> RATIONALE :
> SPACE FUSION DEVELOPMENTAL
> RESEARCH PROGRAM

Probably the most important question addressed in this study beyond that of technical viability is, "What is considered to be an estimate of the cost savings anticipated for an operational space fusion system?" While a detailed cost analysis for space is beyond the scope of this study, a few principles emerge. These principles and study results illustrate the importance of not only initiating space fusion research but of expediting its progress. As follow-up questions we need to inquire, "What is the terrestrial fusion research program's funding status and projected plans, and will it yield the desired results for the United States space program as now needed?"

Let us focus on the NRC activity for addressing the latter point. The answer is contained in the "Executive Summary" of the NRC report, which addresses the United States fusion program status. The statement made serves to illustrate

10.0 Space Program Operational Economics and Program Implementation

the point that the space fusion program should proceed under space program management:

> The committee believes that current domestic funding levels are inadequate to meet even the near-term objectives of the plan. (p 1)
>
> The committee further finds (Chapter 2) that the United States is no longer the dominant leader in international research on magnetic fusion. The United States lost its leadership position in magnetic fusion energy to the European Community largely because of a 50 percent decrease in funding (measured in constant dollars) since 1977. As a consequence, the United States program has also narrowed. (p 3)

As budget problems increase, inadequate funding can be anticipated to exacerbated, perhaps to the elimination of funding for all alternate concepts.[1]

The NRC report summarizes, unknowingly, the importance of NASA initiating a space fusion program to focus on a space application, provided that a flight mission need and economic payoff can be shown. The earlier sections show that fusion energy is a tremendous mission enabling and mission enhancing technology. The need is shown to be real. The purpose of this section is, then, to better quantify the economic potential for the manned and unmanned space science missions.

---

[1] Unfortunately this prediction made in 1988 appears to be correct. "Faced with a Congressional cut of $50 million from the magnetic fusion budget (see our November newsletter), the Department of Energy Office of Fusion Energy has decided to terminate essentially all of its experimental programs aimed at developing a more attractive magnetic fusion reactor concept. Instead, it will protect the budget of its "conventional" tokamak program." (FpaDec 90)

10.0 Space Program Operational Economics and Program Implementation

10.4 MANNED MARS MISSIONS

Mission operational costs for the comparative purposes of this report are defined to comprise the launch costs plus those day-to-day operational costs required to accomplish equivalent mission objectives. To provide, then, an indication of the magnitude of cost benefits, consider first the launch costs today if we were now to proceed with a Manned Mars Mission. Later in this section the planetary space science fusion mission operational costs are evaluated. The Shuttle flight operational costs, during a typical 5-day mission, are not available, so an estimate was made of those values with assumptions clearly identified. Let us assume that either a chemical, fusion, or fission propulsion capability presently exists which can technically meet system requirements, and therefore either system is available as an energy option to accomplish this mission. Nuclear Electric Propulsion (NEP) is not considered an option since it is not sufficiently energy intense for these missions. To examine the relative mission operational costs, the focus of the discussion in this section, we need to account for the costs to launch the spacecraft, its propellants, and propulsion system into low Earth orbit (LEO) plus those costs necessary to perform the daily manned mission flight operations.

RELATIVE MISSION OPERATIONAL COSTS

-COSTS TO PLACE SPACE VEHICLES IN LEO

-FLIGHT OPERATIONAL COSTS

10.4.1 OPERATIONAL COSTS TO LEO AND COSTS IN ORBIT

To transport the mass of the Martian spacecraft into LEO, the Shuttle is assumed to serve as the launch vehicle for reference purposes. As discussed later, other vehicles would probably be used, but this program is current and as such, provides us with a well defined data base as opposed to projected capabilities. The cost of a single Shuttle launch of a payload to low Earth orbit at the time of this report is $317M, a number quoted as representing the averaged mission cost through 1993. The cost is stated to decrease to $190M as the expected average for the following three years. The Shuttle's upper payload mass limit is 27 MT. That normalizes to a launch cost of $9.3M/MT to LEO.

10.4.2 LAUNCH SYSTEM OPERATIONAL COSTS

In the Mars Excursion Module (MEM) study (Can69), the initial vehicle mass in Earth orbit was 1,000 MT using the following: chemical energy propulsion for Earth orbit escape, aerobraking energy dissipation at Mars' orbit for capture, chemical energy propulsion for escape from Mars, and aerobrake energy dissipation capture at Earth. The 50 MT MEM carried a crew of 4 to the Martian surface from a Martian orbit for a 30-day surface exploration period. To place that quantity of mass into Earth orbit using the Shuttle requires a minimum of 37 trips. At today's launch cost, that is $11,840M for the launch operational costs to orbit the Mars vehicle into LEO alone, plus the additional energy costs for the consumables including the Mars trip's fuel and oxidizer, a minimal number by comparison to launch costs. The flight time for one of the mission's windows to Mars using aerobraking was 160 days; for the return trip, it was 230 days.

10.4.3 FUSION POWERED LAUNCH OPERATIONAL COSTS

The flight operations cost scenario below assumes that a specific power of 1 kW/kg fusion propulsion system is available and is on orbit for multiple mission use. The fusion launch cost is now factored into the Manned Mars Mission cost equation. This mission, which was the one analyzed in Section 2.0, sends a 133 MT payload to Mars and returns a 61 MT payload to Earth. The mass equivalent of the MEM in this fusion mission example is 72 MT (133-61 MT). A summary of the costs to conduct a Manned Mars Mission, based on today's cost and performance references, is shown in Table 10-1. Assumptions that were used in the generation of this table are discussed in the text below.

TABLE 10-1. Cost comparisons ($M) between fusion and chemical propulsion for a Manned Mars Mission (MEM, Can69).

MISSION PARAMETER	CHEMICAL PROPULSION VEHICLE	α_{p1} FUSION VEHICLE	α_{p10} FUSION VEHICLE
ROUND TRIP TIME, DAYS	390	215	182
INITIAL VEHICLE MASS, MT	1,000	402	185
PROPULSION SYSTEM MASS, MT	NA	94	23
PROPELLANT MASS, MT	NA	175	30
OUTBOUND/INBOUND PAYLOAD MASS, MT	NA	133/61	133/61
TOTAL SHUTTLE LAUNCH COSTS, $M	11,840	3,520	1,920
PROPELLANT LAUNCH COSTS, $M	NA	1,920	320
FLIGHT OPERATIONS COSTS, $M	182	99	<90
NUMBER OF SHUTTLE LAUNCHES	37	6	1

NA - not available

Consider a fusion powered Manned Mars Mission which requires 212 days for the round trip flight time, exclusive of the Mars visit time. While faster fusion powered round trip times, such as 160 days are possible, the 212 day flight time substantially reduces the initial fusion vehicle's mass placed in Earth orbit to the recommended level of ~400 MT. That is significantly less than the 1,041 MT initial vehicle mass required for the 160-day trip while the 400 MT mass allows a reasonably fast trip.

For the 400 MT initial vehicle mass mission: the propulsion system mass is 94 MT; the payload comprises 133 MT; and the propulsive energy mass for transporting a vehicle to Mars and returning it to Earth LEO for this mission is 175 MT.

10.4.4 FUSION PROPULSION SYSTEM POWER PLANT COSTS TO LEO

This study assumes that a reusable fusion propulsion system has been stationed in LEO. The operational costs to place it there are small on a per mission basis. Three Shuttle launches are required to place the 94 MT fusion vehicle mass into Earth orbit. Assume a reuse life of 20 missions. The single Manned Mars Mission portion of the operational cost to place the fusion vehicle into LEO is then $37M.

10.4.5 FUSION FUEL COSTS

Because the specific impulse requirement is low, 12,000 seconds mission average, all of the propellant mass — except for 9.15 kg — will be a diluent, assumed to be liquid hydrogen. The D-^{3}He energy figure of merit is 19 MW-year per kilogram of ^{3}He. The power required to propel the Mars vehicle to and from Mars within 6 months, as discussed in Section 2.0, is 145 MW or 72.5 W-years, assuming a 1 kW/kg propulsion system and a Mars vehicle sized to 133 MT out-bound and 61 MT in-bound payload mass. If we assume a 70% efficient reactor, the mass of helium-3, required is 5.45 kg for a 50-50 mixture of fuel. The helium-3 is brought from the moon directly to LEO. The remaining 3.7 kg deuterium mass is transmitted by the Shuttle to LEO from an Earth based facility. The total mass transported to LEO for a 1 kW/kg specific power fusion propulsion system to perform the Manned Mars Mission is 175 MT propellant mass plus the 133 MT payload mass. For fusion propulsion, the 212-day mission uses 7 Shuttle launches, or $2,200M as <u>launch operational costs to orbit</u>, using today's Shuttle costs and flight performance as a comparative baseline. While other larger launch vehicles could reduce the number of flights, there are the developmental costs to be included plus the additional equipment and operational costs incurred by the larger vehicle. But if there are any savings, those would also be of similar percentage benefit to the fusion system.

10.4.6 FLIGHT OPERATIONAL COSTS

In addition to the launch costs, an approximation of the daily manned space fusion operational costs to fly the vehicle to and from Mars is required to complete the operations cost comparison. Current operational cost information was not available so estimates were made which are clearly identified for updating if more definitive cost information becomes available.

10.0 Space Program Operational Economics and Program Implementation

Consider, for example, that a staff of 500 persons, contractor and civil servants, support the flight at a cost of $120,000 per person per year. That amounts to a total cost for personnel services of $60,000,000 per year; the daily cost is $250K/day for salaries of direct flight support personnel. That number is assumed to double when the costs for equipment and all the support services, facilities, etc., are taken into account. Because future space operations for a Manned Mars Mission can be expected to be supported by the Space Station Freedom (SSF), an arbitrarily selected cost of $50M per year for a Mars mission dedicated by SSF support was used. (That mission support role and cost for the Space Station is not yet defined.)

From purely the flight time savings factor, the impact of fusion on mission flight operational costs is not as great as for the launch savings, simply because the total operational costs are not as large by comparison. Safety and other science enhancement advantages of the larger mass permitted by fusion were not factored into the operational cost advantage, although these are major program gain considerations which need to be factored into the mission's cost equation. The difference in flight time for missions using the chemical powered vehicles missions in comparison with fusion is 390 − 212 = 178 days. Let us assume that the

> **TIME SAVED = 178 DAYS**

then the following costs are representative of fusion flight operations for a Manned Mars mission (Table 10-2):

TABLE 10-2. Operational cost summary for a manned Mars mission.

• Mars flight control personnel operational costs	$60M
• Support cost for flight operations (assumed equal to the flight support personnel costs)	$60M
• Space Station flight support for mission (arbitrarily assumed value)	$50M
• Total operating costs	$170M
Costs per day	$0.5M

10.4.7 CHEMICAL PROPULSION FLIGHT OPERATIONS VERSUS FUSION FLIGHT OPERATIONS

For the MEM chemical propulsion mission, the operational cost to conduct the mission is $182M as calculated using the above assumptions; for fusion,

$106M. Thus, the operational cost to launch the Mars mission's flight systems plus mission flight support using chemical propulsion is:

$$\text{CHEMICAL}.....\$11{,}840M + \$182M = \$12{,}022M$$

as compared to the same for fusion:

$$\text{FUSION}.....\$2{,}219M + \$106M + \$37M = \$2{,}362M.$$

For normalized mission comparisons, the actual difference between these two missions is much greater than the $9,659M amount indicates. The MEM mission analysis had transported a manned cargo vehicle (i.e., the Apollo Command Module analog) of an unspecified mass plus the MEM, the Martian lander, and the ascent vehicle, which returns only 136 kg of soil sample to Earth. The MEM's total mass was only 50 MT. The MEM equivalent for this study was 72 MT. Unfortunately the contractor did not identify the payload mass for the trans-Martian/Earth crew quarters and Earth Atmospheric reentry module. But it is believed that if we compare the performance of the two Earth-Mars vehicle masses using equivalent payload masses, then the chemical system's operational cost would dramatically increase when compared to the fusion vehicle's 133 MT outbound and 61 MT return payload masses.

Other significant advantages of high specific power systems which were not factored into the cost equation must also be accounted in comparative evaluations of space energy systems. One design objective for fusion spacecraft is to provide a system compatible with fully automated control such that the spacecraft and flight crew will have maximum control over the flight system operation. With the additional on-board payload mass capability of the fusion vehicle, greater space autonomy can be designed into the spacecraft. Hence, the spacecraft can be designed to permit the flight crew to more readily accommodate extensive in-flight spacecraft maintenance functions as well as other functions that otherwise would be accomplished on the ground, thereby providing for enhanced in-flight autonomy using the more massive fusion powered vehicle. Therefore, in determining the operational costs, an allowance for additional ground flight control personnel was not included for in-flight monitoring of the fusion propulsion system status. A cost credit should be given instead to reflect the autonomy.

Larger fusion powered spacecraft can transport additional flight crew, permitting the accomplishment of more science per "unit mission." More massive remote science payloads can be designed to remain on the Martian surface (or moons) to serve as Martian science outposts. Most importantly, a significantly more massive Martian soil sample can be returned to Earth, representing a larger

10.0 Space Program Operational Economics and Program Implementation

variety of geologic conditions on the Martian surface. A more favorable science mass returned per mission cost value is achieved. These added advantages have not been factored into the cost benefit equation.

10.4.7.1 ALTERNATE LAUNCH VEHICLES

Examination of the means to reduce the costs to orbit have been on-going since the mid-1960's. The space station had originally been designed ~1970, but prior to its construction, a cheaper means to transport it to orbit and to travel back and forth was considered necessary rather than to use the throw-away Saturn V or S-1B vehicles – a very costly approach. Thus, prior to proceeding, a new manned reusable launch vehicle was considered necessary to carry out the mission of NASA. A manned vehicle was deemed necessary in order to deploy and retrieve payloads and in general important for a reusable vehicle's operational approach to space flight. The flight of man, while offering mission flexibility, results in a mass penalty. If we were to redesign the Shuttle as an expendable launch vehicle and use the current Orbiter's mass as orbital payload, making allowance for retention of the main engines, then the payload mass fraction increases from ~1+% to ~5%. That represents ~4-fold improvement in the payload delivery capability.

It must be remembered that the original rationale for the Shuttle design was to reduce the costs to orbit of the expensive large expendable launch vehicles by designing the reusable manned vehicle. The Shuttle was designed to replace expendables, whether of the Saturn class or smaller, as part of the space transportation infrastructure. So any projected cost reductions from new large vehicles have to be closely examined in view of the history. Going to space is expensive – the rationale for this study's emphasis on reducing the requirements to low Earth orbit through higher performance space propulsion, whatever the source of energy, whether fission or fusion. Many serious efforts were made to solicit and use "airline-like" procedures to reduce Shuttle costs. Advanced technologies and cost reduction approaches to existing technologies were used. Clearly, there will be one significant advantage to the large launch vehicles, namely, fewer launches are required for the Mars mission, making it possible to more quickly to assemble the large payload mass in low Earth orbit. The Shuttle at its demonstrated launch rate would have taken approximately 9 years.

10.4.8 ADVANTAGES OF HIGH SPECIFIC POWER

If a 10 kW/kg specific power reactor can be developed, the same 212-day mission cost for the launch of consumables would decline to the expenditure of a single Shuttle launch. The Shuttle's payload capability is limited to 27 MT. Sufficient fusion propellant, i.e., 23 MT, could be launched into LEO to perform the 212-day mission. The total launch cost for the mission's Δv energy in that case is only $357M (which includes the prorated allowance for the launch of the propulsion system), a significant reduction from the $2,200M value for a 1 kW/kg propulsion system (when referenced to the Shuttle launch costs). The flight operations cost is estimated at $90M for the half year trip time. *Hence, we are now looking at a total launch and mission operational cost of only $440M, plus the 5 launches required to launch the 133 MT payload, a constant cost factor regardless of the space propulsion system.* Payload efficiency is high, and operational costs are minimized.

There are other very significant cost factors which impact the overall program costs in the design and development of payloads which are real, but difficult to quantify, that also benefit from high specific power propulsion system. Because of the greater energy, higher margins of safety and reduced redesigns to save weight will result in substantial cost savings to the program. Refer to the discussions in Section 2.0.

As iterated throughout this study report, the research investment in increasing the propulsion system's specific power is the best up-front research funding possible. It is high risk but extremely high gain research technology.

10.4.9 OPTIONS FOR THE FUTURE

It is clear that we will need to find a new means of doing business in space, or the goal of exploration of the solar system and the universe will not be achieved, solely on economic grounds. Transmitting payloads to Earth orbit using the current technology systems will never be cheap due to the physics and chemistry involved. To reduce launch costs using the chemical propulsion systems to transport payloads to LEO, a new flight operational approach is mandatory. We must reduce launch demands placed on those LEO bound chemical propulsion systems. A higher probability of achieving public support for frequent missions to Mars can be anticipated where the trip costs to place the Martian vehicle and consumables into LEO are significantly reduced. We can anticipate round trip missions to be required once or twice annually in support of settlement. The use of high specific power space propulsion systems, like fusion, is a fundamental change in approach of the nature needed for future missions. The development of these systems, however, is not anticipated to be quick.

> Because the potential for real cost benefits is so great, it is important to NASA and its future that we proceed now to commence fusion experiments and related test programs to seize upon that potential.

10.4.10 FLIGHT CREW SAFETY ADVANTAGES

While it is not possible to place accurate cost values on flight crew safety, some positive, non quantitative statements are obvious. As stated before, the integrated radiation exposure from galactic cosmic ray background exposure is reduced as a direct consequence of the shorter flight times. Refer to Appendix B for a discussion of additional details. The heavier payload capability makes the weight penalty for shielding for protection from solar flares more readily achievable with less mission performance penalty than with chemical propulsion. Further, we have the advantage of shorter trips which lessen the probability of exposure to solar flares. Whereas aerobraking maneuvers are required for the chemical system, the preference is for a totally propulsive retro maneuver in lieu of aerobraking. The preference is based upon one of response time. With aerobraking there are no abort options. Acceptable navigational errors or errors in aerodynamic properties can be expected to be in the domain of a very narrow band width. Without a precursor unmanned demonstration of a full scale craft, the risk resulting from uncertainties will be high. With propulsive braking, flight operational options exist to continually change a course of action. The environmental exposures (drag loads, heating, heating rate, etc.) on an entry vehicle using retrograde propulsive braking are less severe. Hence, given the option, the preference is to conduct propulsive retrograde maneuvers in lieu of aerobraking. From this perspective the nuclear fission system will be safer, without requiring a precursor demonstration mission it will probably be cheaper too; refer to Fig. 2.4.

In addition to flight crew safety, the Earth's safety is enhanced because the environmental impact is reduced using D-^{3}He fusion. Furthermore, less of Earth's energy is used since fewer Shuttle missions are required.

10.4.11 MISSION DEVELOPMENTAL COSTS

Obviously, any program developmental costs which are associated with hardware made unnecessary by fusion will add to NASA's savings. The greatest is the elimination of the requirement to qualify man for 1 to 2 years of space travel. The current plan for Manned Mars Missions is in a direction toward missions to the Moon and extended durations in LEO first as precursor missions. That program can be expected to take ~15 years to complete. The fusion powered Mars mission can be accomplished in 3 months of flight time. The US astronaut has already been qualified for longer than that time by virtue of the Skylab Program. If fusion were available now, we could proceed without those alternate program activities that require the qualification of man for longer space flight times. The savings in that case is obviously enormous. Another example is the heat shield. A current experiment, the Aeroassist Flight Experiment, for obtaining design parameters would not be necessary – if a fusion energy capability currently existed. That total mission cost will be greater than $600M. Further, mass will not be as critical, permitting the use of lower cost hardware. Offsetting such savings are the costs to develop AI, for example, and the costs to obtain ^{3}He from the Moon. Note, too, that both of those have cost savings advantages – reduced flight operational costs as discussed in the case of AI operational efficiencies and from the sale of ^{3}He for Earth based fusion powered reactors.

10.4.12 FISSION OPERATIONAL COSTS

Fission propulsion of the NERVA high thrust class could under some circumstances reduce chemical propulsion operational costs of the MEM payload class by reductions in the number of Shuttle launches. To prove this, we only need to refer to Fig. 2.4. The mass of propellant saved over that of chemical propulsion-aerobraking approach was 250 MT – 9 Shuttle flights or $1.7B. That approach used a hybrid propulsion mix that was comprised of: nuclear propulsion for Earth orbit escape, aerobrake capture braking at Mars' orbit, and chemical propulsion Mars orbit escape, and Earth aerobraking. An all fission NERVA type nuclear system increased the operational costs by $1B since it requires an additional 100 MT of propellant to be delivered to Earth orbit. For more massive payloads, the higher specific impulse of fission is expected provide more significant benefits over chemical systems. The MEM Rockwell study team used a lower value of specific impulse, 800 seconds, to allow some conservation in 1967-68 for testing which by 1972 ultimately demonstrated 850 seconds. Higher values up to 1000 seconds are being looked at now, but some degree of caution must be exercised before we can rely on these numbers and implement irrevocable planning prior to demonstration.

10.5 SCIENCE MISSIONS

In Section 2.0, this study examined a single science payload mass (20 MT) for a variety of missions. That single payload mass provided a normalized study evaluation process for the comparative evaluations of a constant and consistent set of input data for multiple science missions, namely, the performance of sample return vehicles transporting 20 MT payloads to its science target(s) and returning 10 MT payloads to Earth. The mass of the initial vehicle and propellant loaded varied depending upon target mass differences and distances, as well as trip times. A follow-up study would be to conduct a similar analysis which considers mission parameters for alternate payload masses, both heavier and lighter which would better define the capabilities as well as the lower limits of fusion. Different payload return targets other than those returned to Earth should also be studied as an option. For example, these mission analyses would include evaluations of the transport of large science laboratory masses to a Martian colony from other locations in the solar system. Mars could serve as a major science outpost base, provided that in-situ production of propellants become a reality.

10.0 Space Program Operational Economics and Program Implementation

> For science missions using fusion energy, we can decrease flight times and save on operational costs while gaining an enormous return on science data.

10.5.1 EXAMPLES OF SCIENCE RETURN IMPROVEMENTS AND SAVINGS USING FUSION ENERGY

In the Galileo Program a 0.44 MT science payload is flown to Jupiter over a 6-year mission flight time. If a fusion propulsion power system yielding 1 kW/kg were available today, we could send to Jupiter a 20 MT total science payload and, unlike Galileo, be able to return a 10 MT payload with a soil sample from Europa in 1.56 years of flight time. That reduces operational costs by $450M since those mission operational costs are $~100M annually. The developmental and fabrication costs for a typical spacecraft of a Magellan class are approximately equal to the flight operational costs saved by using fusion energy for one Galileo class mission. Or contrary to our situation with chemical systems, the added operational flexibility of enlarged launch windows is now available. As an operational option and attractive feature, the higher performance capability in space reduces the launch pressures and schedule impacts that have been faced in launching planetary spacecraft in the 1980's and earlier. A missed launch opportunity costs NASA significantly to carry over the program staff for the period of time until the next launch opportunity arises, one year or greater depending upon the mission. That can amount to $200M, or greater, depending upon the time and nature of the flight program being delayed.

It should be pointed out that the achievement of this greatly enhanced payload capability requires additional propellant mass. The Galileo spacecraft and propulsion system are orbited using one Shuttle flight. For the fusion spacecraft, 243 MT of propellants are required to perform the 1.56-year mission, using a specific power propulsion system having a 1 kW/kg level of performance. If we could develop a 10 kW/kg system, then the launch costs to LEO would be comparable to a chemical system -- but with one to two orders of magnitude gain in payload delivery mass plus a sample return and a reduction in total flight time by a factor of four.

Another mission advantage for the fusion powered spacecraft is that multiple payload deployment missions are possible. The overall mission concept proposed here is to load fully one large fusion vehicle and conduct multiple target missions. A cost comparison between a single objective mission like Galileo and multiple missions like the one proposed here is beyond the scope of this work. It is proposed as a concept for further consideration as discussed below.

10.0 Space Program Operational Economics and Program Implementation

In the manner which we now perform space science missions, a substantial payload science mass inefficiency results. That inefficiency is apparent from an examination of the mass of current technology science spacecraft in relation to the actual science payload carried on-board. That inefficiency is illustrated in Table 10-3 below (Yea85)

TABLE 10-3. Galileo vehicle mass apportionments.

Orbiter mass	1.138 MT	100 %
Usable propulsion mass	0.938 MT	83 %
Orbiter science instruments mass	0.103 MT	9 %
Residual Orbiter mass	0.103 MT	9 %
Probe mass *	0.335 MT	
Total science instruments mass	0.44 MT	30 %

* The entire probe is listed as "science" mass although it actually contains only 0.028 MT for science instruments. The remainder is the structure, heat shield, power, telemetry, etc. necessary to support Galileo science. The Probe's science mass efficiency would obviously not be altered by the presence of fusion energy.

The mass of the Jovian Probe is unaffected by the propulsion delivery system except for the allowance of a more massive overall payload probe which could be flown on a high specific power system.

Significant mission economic advantages result from multiple targeting missions. The Multiple Asteroid Mission discussed in Section 2.0 illustrates that only a very small additional time is needed to increase the number of targets from 3 to 6 when the distance between asteroids is as great as one AU per target. To extrapolate that data to the planetary moons, it will not be unreasonable to expect more extensive (greater mass) science coverage for all of the outer planets in less flight time and with the added benefits of greater science instrument mass fractions, defined as space science instrumentation mass to the total spacecraft mass. The objective is to increase the science mass fraction substantially beyond the typical 30% that is now possible using chemical propulsion powered missions like Galileo. Additional benefits from greater mass samples returned to Earth has not been factored in the cost trade. Efficiency of multiple targeting is obtained by the economy of reduced integrated solar gravity, gained by multiple targeting per singular mission (and perhaps by phasing too, if proper timing is selected). This Multiple Planetary Solar System Mission (MPSSM) of the outer planets could be performed in less than 10 years with samples back at Earth. The important point is that efficiency

10.0 Space Program Operational Economics and Program Implementation

is gained by providing payload mass for science instruments in lieu of mass for on-board spacecraft propulsion.

A new, alternate mission approach is to employ the fusion powered space vehicle as a mobile launch platform from which spacecraft are launched to their individual targets, but involving on-board chemical propulsion. Whether it is cost optimal to target directly to the planets solely, relying upon the fusion spacecraft, or whether it is better to target to an optimal trajectory taking the fusion powered vehicle along an arbitrary trajectory in space to optimally deploy science payloads using chemical propulsion maneuvering systems to transition individual science payloads to their respective target planets, is the subject of a complex science objective-trajectory-mission cost analysis study. It is presented here as a mission study concept consideration.

10.6 LOOKING FORWARD

The fusion mission scenarios developed in this report show very favorable cost trades in favor of the high specific power systems for flying high energy missions. With decreases in payload masses, the advantage of fusion decreases because, at the present time, we consider the design of fusion reactors as devices of large inert mass which are not known to scale downward.

Some consideration was given to the topic of smaller fusion devices in this study. Research in this small reactor area is not anticipated, at least not until the the distant future. MIGMA – colliding ion beams – is an approach. The small, several megawatt size reactors are not believed to be of great interest to the large utility power companies which appear to favor the large centralized utility power plants, typically in the low gigawatt range. The several megawatt fusion reactor is a more difficult technical feat to achieve and is a subject for advanced fusion research for space. If that development can be achieved, fusion can serve an even wider space role.

10.0 Space Program Operational Economics and Program Implementation

10.7 SUMMARY OF OPERATIONAL ADVANTAGES AND SUGGESTED DEVELOPMENTAL COST INFERENCES

> Serious attention must be given to high specific power energy sources. Whether or not fusion is developed, the importance to NASA of high specific power energy conversion devices and large payload mass fractions is very clear. Because of the great dividends, NASA should invest significantly in this research field. It should be funded at a level sufficient to accomplish meaningful test and experimental results, commensurate with the potential gains to space science and space exploration.

For this very preliminary study evaluation phase, any specific cost numbers can be challenged. The assumptions used above are straight forward, well defined, and point toward a significant trend of cost savings and safety. High specific power "detunes" payload mass variation sensitivities. That is, there is less of a cost penalty for exceeding payload weight allotments. There is the option of developing heavy lift chemical propulsion launch vehicles to reduce the number of Shuttle launches. Any mission requirements, which can be attributed to the laws of physics, however, are undiminished by repackaging the same energy source. We can gain some cost reductions, but using the low performance chemical systems, the total energy consumption for space missions will always be great as will be the performance penalty. The developmental/operational implementation costs for such new heavy lift chemical propulsion systems can be expected to be expensive, requiring billions of dollars, perhaps on the order of $10B. That is now under study. Furthermore, benefits for the chemical system using more massive launch vehicles will translate into similar savings for fusion powered spacecraft.

The cost to just maintain and use the current chemical propulsion technology, which NASA now spends, is estimated at approximately $500M per year. This approach, without looking forward to high specific power systems, is a vicious circle leading only to a low propulsion capability and a less forward looking NASA than is possible with the use of high specific power energy devices. To iterate the theme of this report, additional mission studies are not needed to verify the advantages of fusion. The results have been known for some time. Needed instead are the fusion relevant analytical computations, experiments, and test verifications, or, in other words, activities that yield fusion hardware and produce meaningful, tangible technical results in the field of high specific power energy conversion.

High specific power flight systems for space propulsion and power applications are essential for NASA's future. <u>Space exploration and science programs will become energy limited in the not very distant future, making the realistic achievement of the high energy demanding programs in a timely or cost</u>

10.0 Space Program Operational Economics and Program Implementation

<u>effective manner impossible or with a sufficiently low payload mass capability to make them unattractive.</u> Yet the development time for fusion is long. **A technology void between fusion technology capabilities and NASA's mission requirements can reasonably be anticipated.** The operational cost savings between a Manned Mars Mission using the chemical propulsion systems of today's technology compared with one having a high specific power of 1 kW/kg is on the order of $9B per mission, or a savings of about $2B per round trip flight ticket per person, for a four person flight based on today's costs. In addition, the mass of returned Martian soil sample using a chemical energy propulsion system is substantially less than with fusion or any equivalent high specific power propulsion system.

> A highly focused program to develop high specific power/high specific impulse space propulsion technology must be initiated by NASA to permit the accomplishment of anticipated future space missions.

That decision would allow NASA to set fusion program priorities along with the fusion system performance capabilities to meet those mission requirements. **Considerable time will be required to develop these systems, and it is therefore important that NASA commence fusion research <u>now</u> in order to realize the cost savings at the earliest date possible.** With the low level of funding allocated for fusion research, the DOE program cannot be expected to be of direct benefit to the space program for many decades, if ever. As a 10% investment on the future, a minimum of $50M to $100M annual budgeting should be allocated for a new start, the objective of which is to develop a fusion propulsion system of a minimum of 1 kW/kg and high specific impulse of 5×10^3 to 10^6 seconds. That 10% value ($50M) to develop a superior energy system represents an investment of 10% of the current funding level which is now designed to only maintain the current chemical propulsion technology. The $100M represents the amortization of 10% of the **savings** from the difference in operational costs for just **one** Manned Mars Mission between chemical propulsion technology and systems having a specific power of 1 kW/kg. Amortized against the future science gains and additional exploration programs – and even the future of NASA's space missions, it is a very small investment. Fusion energy appears as the most energy realistic source for achieving that level of specific power. The question is "What will that level of resources provide NASA?" – a topic of Sections 13 and 14.

In summary, then, fusion provides leveraged power. The costing leveraging power of high specific power propulsion systems is very clear. In round numbers, a fusion powered manned spacecraft, as envisioned in this report, would save over $9B in operational costs for one Manned Mars Mission

10-27

10.0 Space Program Operational Economics and Program Implementation

alone, considerably more for an equivalent performance mission! In addition, six planetary sample return missions were considered in Section 2.0. **The savings for all of these missions can amount to many tens of billions of dollars savings, plus it provides the opportunity for a more rapid, more extensive science return capability.**

It is important that a space fusion research program be structured to address the unique space related parameters if a more timely application of fusion energy is to be realized for space programs. As shown, of paramount importance is the propulsion system's specific power. The terrestrial program is not as mass sensitive, at least not critically to the extent that NASA's flight programs are affected. Any major developments for high specific power systems will have to originate from NASA. Actually, the desirability of and the need to accomplish more research on terrestrial reactor compactness, energy conversion and systems trades were recognized by the Senior Committee on Environmental, Safety, and Economic Aspects of Magnetic Fusion Energy (ESECOM) (Hol88, p 51):

> The design characteristics offering the most important potential benefits for fusion COE [cost of electricity] are as follows: 1. compactness (including but not limited to high mass density), which reduces the capital cost of the fusion power core, reduces, as a result, the sensitivity of COE to plasma performance, and may ease maintenance, 2. high level of safety assurance..., 3. advanced energy conversion systems...

But funding was not provided to accomplish the recommended tasks.

One good example of terrestrial and space program reactor design requirement difference pertains to relative emphasis on reductions in the reactor is specific power. A committee was charged in 1984 by the DOE to consider high power-density reactors. The report on the committee's findings and recommendations stated that

> ....increased emphasis should be given to improving the mass power density of fusion systems, aiming at a minimum target of 100 kWe/tonne. (Dav85)

That target is still an order of magnitude below the required power density for space. That is the same level of difference also presented in the National Research Council's Committee on Advanced Fusion Power. (Mil87) Another example is the space start requirement, particularly where there is a limited energy storage capability limit. On the ground, network grids are available for restart power. As a third example, space requires a minimal radiation

10.0 Space Program Operational Economics and Program Implementation

production to reduce vehicle mass and for operational simplicity. To present a fourth difference, we are concerned with the vacuum effects on the high voltages produced in the energy conversion; terrestrial applications are not. Other factors include heat rejection approaches which obviously differ in the two applications. Ash disposal and space vacuum availability are two other differences.

The conclusion is that fusion will pay for itself many times over, but a space fusion program needs to be initiated to accomplish the space operation, requirements, and mission priorities. That conclusion is proven by the DOE plan to delete the alternate experiments commencing with the FY91 budget.

11.0 BROAD ISSUES

This section presents the major issues that arose during the study. Rationale is provided to address each.

11.1 PHILOSOPHY ON ADVANCED SPACE RESEARCH, PARTICULARLY ON LONG DURATION STELLAR OR OORT CLOUD MISSIONS

With missions enduring for over 300 years, one will ask whether such endeavors should ever really be given serious attention. After all, 300 years ago, calculus had just been invented. The United States was not even a country, only a colony. Where will mankind be 300 years from now? Will there still be a civilization? Surely by then, we will have new theories and new technologies that will permit the future generation spacecraft to pass up the old. How can we expect to have equipment endure so long and still perform reliably? How can we expect to underwrite a mission that will not reach its destination until so far into the future with so many uncertainties? What will the gain in man's knowledge be as a result of a stellar mission that may not already be achieved by alternate technologies? Will it be too expensive?

These and many other questions will certainly be pondered extensively before ever undertaking a venture of this nature. The stellar mission will undoubtedly be the most expensive space mission that man is likely to consider for the sake of science, at least for quite some time. Three hundred years is a very long mission flight time!

Let us address the above items. A stellar mission with science as the objective should certainly be a welcome pursuit as a national goal, that objective being the gain of knowledge rather than funding destined to man's mutual destruction or otherwise to his detriment. Although no stellar mission cost evaluations were performed here, it would not be surprising to be in excess of 10 billion dollars, maybe as high as $100B, depending on how the mission is costed. Even if it is as high as $100B, when averaged out over the life of the mission, the $30M annual costs for the science returned from mission costs is trivial, probably orders of magnitude less than the amount spent on intoxicants, tobacco, and drugs, etc. It must be remembered that this is a long term, "active" science mission. Considering our current spacecraft, we pay typically $750M for the one or two year missions. On a per year basis, the stellar mission can be expected to be at least an order of magnitude less expensive.

The science gained from the technology developed for the stellar mission would benefit all. Because it can light the imagination of every thinking individual person on Earth, the program could be made an international one without burdening the financial budget of a single nation. A cooperative major joint scientific venture solely for the benefit of man by providing knowledge about his

11.0 Broad Issues

galaxy would provide a spiritual unification of unprecedented magnitude. Even if that spiritual unification and international understandings were the sole accomplishment of the mission, then it would have achieved for a comparatively small sum that which large defense budgets and centuries of war have been unable to accomplish.

The stellar mission was structured to provide "real time" flight data, using very high quality telescopes, on a biannual basis. It would be a short time before the stellar spacecraft would be beyond the bounds of the solar system, providing astronomy of interstellar quality – i.e., not subjected to natural influences of the solar system's physical dynamics. The science would commence being available early, thereby providing a rapid payoff, not restricted to the benefit of a fifteenth generation. Thus, it would be for today's scientist, as well as those for many generations of succeeding scientists – a gift from the older generations to the young, yet unborn.

Stellar exploration is the next step beyond solar system exploration. It is indeed a large one. The distance, times, mass, velocities, and power levels associated with just a simple visit to our next door stellar neighbor are impressive. To arrive there quicker, the energy requirements become staggering, just numbers with which to deal.

Our technological knowledge has become so impressive in the past 300 years, why not wait for later technologies to be theorized, researched, and developed, as for example the next hundred years, and arrive before this first fusion powered machine? That may be the situation, and the same logic can be employed to postpone the trip 50 years from now, 100 years, or however long one may wish to rationalize the delay. But the same situation is valid with most items that one purchases in our personal lives. How many of us turned down the purchase of an automobile at a given time because a better product can be anticipated to arrive later? How many avoided the purchase of 78 rpm records, the mono long playing records, or stereo, awaiting for the development of compact discs? Who refrained from black and white TV to attend only to the purchase of color? The list is lengthy.

The goal of the stellar mission is for man to seek knowledge about his universe. If man is destined to inhabit other solar systems, the time required just to obtain suitable targets is extraordinarily lengthy. Let us postulate a sufficient improvement in Earth-vicinity sensing systems, such as interferometry, and that they even become advanced beyond our expectations, and that planetary structures to discern habitable planets are discovered within the realm of those postulated missions in Section 2. Immediately following the initial excitement of the discovery we can anticipate the question of life. We will wish to explore and study any biological phenomenon present. Thus, there will be much interest in knowing the physical details of those bodies.

The only way reasonable will be to obtain a microscopic view of the planets by a remote sensing spacecraft. If analysis of the data yields positive results, then

11.0 Broad Issues

a program will be initiated to permit habitation by man. If negative, then the answer to the question of life elsewhere is such that we need to consider the species of life here on Earth even more precious than ever.

Even without discovering and arriving at a habitable planet this program will have great value. Because data are to be returned to Earth throughout the stellar mission, unique in-situ interstellar science is to be retrieved over a continuum. If a mission catastrophe were to be experienced after a period of time such as 100 or 200 years, the frontiers of space science exploration will have been enormously expanded.

Based upon rationale given below, these missions will never be quick. A start on preparations as early as possible is, therefore, advisable for a well planned cost optimized program. Certainly a significant time scale is essential in providing the capability prior to commitment of an unmanned stellar rendezvous mission. A precursor unmanned program will be necessary before any manned flight can be contemplated.

The possibility that future new energy conversion theories would invalidate the current theoretical transportation limitations and offer breakthrough technical improvements is perhaps in the minds of some individuals the primary reason why such a mission should not be attempted. It is not realistic to claim that we will not become more learned or that we cannot do better. But it is instructive to examine the assumptions of this study, science history, and our state of knowledge today to make some suppositions about the future.

The energies involved in the computation of this study's trajectories did not consider any relativistic effects because the velocities are too slow. For a new technology spacecraft to arrive there any quicker than the "Alpha Centauri Explorer" – our contemporary 2000 year fusion starcraft – the Δv of the new vehicle will probably have to be increased at least an order of magnitude to pass it. Dr. Goddard commenced work on liquid rocket propulsion in 1908. Thus, the time to develop our current chemical propulsion technology from its original inception was over 50 years to the first manned space application of that energy conversion system. Man has been involved with the conduct of research to harness fusion energy for over thirty years for peaceful applications. It is difficult to project the nature of the enormous new energy source that could provide the amount of energy needed to pass up the Alpha Centauri Explorer – our original starship in flight.

The total energy consumption to do better than fusion is enormous. Basically what we look for is higher specific energy. The proton-antiproton reaction, the next step up in specific energy, provides a 1 to 2 order of magnitude improvement over fusion. But the production of the quantity of mass for matter-antimatter (mirror matter) power will probably consume a century or greater, following the demonstration of that technology. The amount of energy consumed to produce the mirror matter is currently in the ratio of 10,000 to 1. Even upon proof of principle of practical energy conversion using this reaction,

11.0 Broad Issues

we require high specific power flight systems which may be very difficult. There is a limitation imposed by physics too. Note in the performance curves in Section 2 that the higher specific power propulsion systems require higher specific impulse, requirement values that ultimately exceed the speed of light (3×10^8 m/s). Hence, the performance of very high specific power systems will be constrained by the laws of physics. The exhaust velocity is determined by:

$$v_e = g\ Isp \approx 10^7\ m/s.$$

With reference to changes in theory that would allow an object to exceed the speed of light, it is noted that the relativistic equations of motion are being confirmed in particle accelerators and other experiments. The spacecraft's energy requirements increase as

$$1 \Big/ \sqrt{\left(1 - v^2/c^2\right)}$$

and becomes infinite at the speed of light. Traversing a space atmosphere having a density of 1 proton per cc at a speed in excess of 0.9c is estimated to be the upper limit of material technology due to the effects of drag from interstellar gas heating. The energy requirements to get to that value are even greater than when the drag loads are taken into account.

It must be remembered that theory of relativity produced "restrictive" or "limiting" knowledge compared to the unrestricted velocities permitted by Newtonian mechanics, a reason stated as rationale why a current science theory may not remain invariant with time since later theories may "invalidate current" theories. Actually, with reference to the original literature, one will note that Newton did not incorrectly state the laws of motion when referenced to revisions by relativity. His exact statement for the second law of motion translates from Latin as: "The rate of alteration of motion is ever proportional to the motive force impressed . . ." He then elaborates upon the definition of motion: "The motion, quantity of motion, or momentum, of a body has been defined as the product of the mass of the body and its velocity." (*The Mathematical Principles of Natural Philosophy*, I. Newton, 1686). Stated mathematically:

$$F \propto \frac{d}{dt}(mv) = m\frac{dv}{dt} + v\frac{dm}{dt} = m\frac{dv}{dt},$$

only provided that dm/dt = 0, the 1686 assumption. Relativity proved mass not to be invariant with velocity and thereby changed our concept of mechanics for

11.0 Broad Issues

the general case. Newtonian mechanics was therefore refined, but refined in an upper limiting sense, i.e., "restrictive," not vice versa in an expansive sense. Quantum mechanics is also a restricting science. Heisenberg's Uncertainty Principle is another example. Based upon those trends, it is not likely that the current principles will be invalidated but that they will become more "restrictive." The source of the vast improvements in new capabilities provided to mankind by science has resulted from the utilization of <u>current</u> theory to enhance technology rather than relying upon the development of new, fundamental theories.

The sources of energy are well defined, but perhaps a better understanding of the binding forces of nucleons or a better grasp of the nature of gravity will be mastered for mankind's benefit. The challenges of the matter-antimatter reaction technology and its attendant costs, plus the system and the safety problems show that it can not become a viable space energy source for a very long period of time. The best real hope for space transportation improvements in the near term lay in the technology of fusion and the ability to increase the space vehicle's specific power. The probability of that happening is a lot greater than the speculation that new fundamental principles may be derived which invalidate the current theories already having a good demonstrated empirical basis. We should, therefore, not preordain abstention as rationale for excluding space fusion research.

The fusion reliability requirement for 200 to 300-year missions is a very real challenge. But space is an excellent location for longevity in many respects, particularly regarding abstinence from an oxidation environment and from thermal and mechanical fatigue induced stresses. The Voyager spacecraft provides good evidence. Vehicle staging does provide a good approach to reduce the system's operational time to a more reasonable value. The major life limiting concerns will be material age life degradation related, particularly if a solid state propulsion system becomes viable and thermal control. The use of an aneutronic fuel assumes an even greater importance to space fusion in this context. That assumes that a suitable high power density plasma can be developed, one with a sufficiently low neutron flux that is compatible with the flight vehicle, its thermal environment, and reactor system materials for the life times needed for missions. A trend of extended life does exist. I doubt that anyone involved with the design of the B-52 would ever dreamed that it would still be actively used 40 years later. We can make a similar statement about the DC-3, but instead ~60 years is its use history.

11.2 ROLES OF NASA AND DOE IN THE DEVELOPMENT OF SPACE FUSION REACTORS

As widely recognized, magnetic fusion energy development has been unclassified and is pursued on a world-wide basis without the ownership of any single nation or organization. All of the activity is devoted toward resolving the

11.0 Broad Issues

energy problem for electrical power production. DOE has been diligently pursuing its charter, namely, a commercial terrestrial power plant capability, powered by fusion energy. NASA's charter is space and aeronautics; NASA internally provides for its energy capability including some of the largest and most powerful machines – the Shuttle's Main Engines (SSME) or the solid rocket motors (SRM), the Saturn V launch vehicle, etc. These energy conversion devices are developed internally by NASA, not by an outside office. The authority to do so is mandated by the original Space Act of 1958: "(2) The improvement of the usefulness, performance, speed, safety, and efficiency of aeronautical and space vehicles; . . . " (Anom58).

A strong *internal technical capability* of any developmental research organization like NASA is essential for proper management decisions relating to technical matters. That technical capability comprises the best "system safety" organization that can be established because it assures an inherent understanding of the technology in use. A consuming organization of any nature optimizes its resources and priorities to best service its needs. Under the present day situation, if NASA decides to use fusion energy, its ability to implement fusion energy is predicated upon the development of fusion for commercial application. NASA no longer has any expertise in fusion technology. But if a new space fusion technology developmental strategy is adopted by NASA, it will establish priorities for space application, and most importantly it will internally possess the technical expertise needed to use the technology.

Significant differences exist between the NASA and DOE missions and therefore between the program priorities of the two agencies.[1] In the DOE, research on fusion energy development must necessarily compete with all forms of energy alternatives for Earth-based power applications without attaching any particular importance to space energy needs and NASA's program priorities. The physics of space programs require light weight, high specific power, high specific impulse propulsion systems designs that are radiation cooled. Hence, the space program fuel of preference is deuterium and helium-3, the physics of which is more difficult to demonstrate net energy production than the mainline terrestrial program's fuel, D-T. NASA has the requirement for a propulsion system capability and preferably a reactor design that will serve a dual function of electrical power generation and propulsion. NASA's desired output capacity varies from small - tens of megawatts (or less) to gigawatts or even terawatts if sufficiently high specific power systems can be built, as opposed to a commercial power plant's gigawatt size requirement. The space operational environment of vacuum and zero gravity could provide

---

[1] The concerns expressed in this section on the importance of NASA initiating a space fusion program are illustrated by the elimination of the space important alternate experiment programs (Fpa90). The DOE program priorities do not reflect the importance of continuing experiments like the FRC which are applicable to NASA's space mission.

design options for space not available to the terrestrial power plant designer. The commercial power plant designer in turn has design options which are not available to the spacecraft designer. These fundamental differences will result in different approaches to research emphasis and thus in the fundamental reactor configuration. Further, what is not widely realized outside of the fusion community is that the resolution of the reactor physics in one design will not necessarily resolve the physics in another reactor. Precedence has been set for a NASA fusion program as discused in Appendix A. DARPA has recently become separately involved with the funding of a fusion experiment. The Air Force has funded MIGMA.

11.3 ENVIRONMENTAL IMPACT TO EXTENSIVE LUNAR MINING

This is a question raised at the NASA Lunar Helium-3 Workshop which is being pursued as an activity from the Workshop. The most significant concerned raised there was the creation of dust clouds affecting future lunar observatories and the effect on astronomy. Ref. Bil89 addresses these matters further.

11.4 INVOLVED PARTIES – NATIONAL OR INTERNATIONAL – IN THE DETERMINATION OF THE ENVIRONMENTAL IMPACTS

It can be expected that in the not too distant future, as the Earth becomes more fully recognized as a "system" rather than an isolated independent series of land and water masses, the international aspects and environmental considerations can be anticipated to increase. An international environmental community can hardly object to the flight of deuterium or helium-3, but questions and concern will be raised over tritium or any radioactive fuel. This is illustrated by a Sierra Club policy paper written to oppose tritium. There has been opposition to burning tritium in the Princeton TFTR.

11.5 PRIORITY OF NET POWER OR THE DEVELOPMENT OF AN ENGINEERING MODEL FOR THE D-T EXPERIMENTS BEING A REQUISITE FOR THE INITIATION OF THE ADVANCED FUELS VERSUS THE BENEFIT TO BE GAINED BY COMMENCING THE ADVANCED FUEL TESTING NOW

Actually testing using the advanced fuels has a wider experience basis than D-T which as of this date has not been placed into a reactor. To allow the earliest identification of problems, it would be appropriate to commence testing on an experiment as early as one is aware that it holds potential future benefits, which is now the case.

11.0 Broad Issues

11.6 REASONABLE PAYLOAD MASS FOR THE ADVANCED SCIENCE MISSIONS

The 20 MT outbound/10 MT return value assumed in this study represents an arbitrary educated guess for a desired payload mass to conduct desired future science missions. The real value should be determined from a group of space scientists to consider. Any new performance capability made available, however, is always put to a full capacity application with a need typically existing for even more capability. The use of a high specific power and impulse system desensitizes the payload mass's impact on vehicle size and flight time such that significant increases or decreases will not substantially alter performance characteristics.

11.7 TIMELINESS OF THE STELLAR MISSION

The next logical step for space science exploration beyond the solar system is stellar exploration. The length of mission duration is a factor which favors the earlier development of the capability. Because the developmental time will be lengthy, an early, relatively low funding level program would be the optimal fiscal, technical, and managerial approach. A well thought out program can be initiated early to optimally plan a program in an orderly fashion. Whatever development time is involved, the flight time is an additive, delaying the science gains that much longer. Thus, it is not too early to commence planning for a stellar mission now.

12.0 CONCLUSIONS REGARDING HIGH ENERGY SPACE MISSIONS AND THE FEASIBILITY OF SPACE FUSION ENERGY

This section summarizes the study's key conclusions. Additional points and conclusions reached in the study are discussed within the text.

12.1   A new way of doing business in space is absolutely essential for the future of the space program, particularly where high energy is required. The greatest economic and safety dividend from fusion energy resides in the <u>manned planetary missions</u>. The expense associated with the energy costs to launch repeatedly manned payloads to Mars for settlement will become too large to perform any more than a token level of space exploration. The best method to ensure positive results for preventing an "energy shortage" with our future space missions is to initiate a space fusion program now. While there can be no "guarantees" that this effort will be successful, the results from this study's analysis and investigations appear encouraging.

12.2   Based upon projected space mission needs, upon the progress made in the terrestrial fusion program, and upon the terrestrial fusion program priorities, it is <u>timely</u> for NASA to initiate a program to develop fusion energy for space. Any time scale that is projected for fusion's availability for space has a high degree of uncertainty. The commencement of a space fusion program now is important because there can be no guarantee that the development will be quick to accomplish. But without the program there is a guarantee that it will not happen.

12.3   The goal of any program is to successfully accomplish its mission objectives with <u>minimal risk</u>. To successfully accomplish the high energy missions of the nature considered herein will place very stringent demands upon system performance. <u>Highly reliable performance over very long durations</u> is ultimately required as we expand the space frontiers, a subject which must be stressed in the research of this technology. The importance of understanding the system's design and manufacturing process mechanisms and their influence on life performance is of utmost importance to successfully meeting the long lifetimes since these systems cannot be life test demonstrated for hundreds of years and flown subsequently. Of all the known sources of energy, including the most recent type considered to

12.0 Conclusions

be theoretically possible, fusion energy has the most attractive properties that will be consistent with meeting the requirements of long life, high reliability, high performance, and high level of safety that is ultimately considered to be necessary. Fusion, then, appears to offer the minimal risk approach in initiating a program to perform these missions. On the other hand, there is a substantial level of technical risk with the timely development of this hardware.

12.4 If space fusion energy were available for space now, NASA could make very <u>cost effective</u> use of fusion energy and take advantage of its safety features.

12.5 The development of fusion energy for space is critical to the <u>United States leadership</u> role in space. To not accomplish the research is to abdicate to whomever desires to assume that leadership role. The prime contender is Japan. They have proven their technical capability and have accumulated the wealth by which to finance the work. They are initiating an active space program and have a good strong fusion program funded. It would provide an advanced, new international image, one showing a strong research capability.

12.6 A new class of <u>high energy space missions</u>, which would substantially enhance man's knowledge of his solar system and cis-interstellar space, can be accomplished if space fusion energy can be suitably engineered. The science and exploration return realized is a result of the use of more massive payloads, faster missions, more operational flexibility, and enhanced safety.

12.7 Great <u>mission benefit</u> is to be gained from the large energy release available from fusion, but fusion energy is a consideration which has been neglected as an ingredient in NASA's planning of advanced missions. The omission is attributed, to a large degree, to the belief that fusion will not be developed until a very long time into the future. Consequently, no high energy science mission requirement has been developed. Because no such high energy mission requirement has been established, the less energy intense alternative energy technologies are being pursued instead. The scenario is reminiscent and suggestive of a Pygmalion effect, i.e., the consequence of a perception, whether real or not, can achieve the perceived vision as a result.

12.0 Conclusions

12.8 The <u>future economics</u> of NASA's missions will be severely impacted, either positively or negatively, respectively, by the presence or by the absence of high specific power propulsion systems. The current NASA investment at an averaged $200 - 300K level over the past two years on high specific power systems is not reflective of its true importance to NASA's space missions nor of its economic, mission, and safety vitality for NASA. Regardless of whether fusion is developed or not, the importance of the role of high specific power propulsion devices to our future must be recognized with the commitment at an appropriate level of resources reflective of its merit. If a fusion system, or any other high specific power energy conversion device, were developed today, it would be used to <u>great economic and mission objective benefits</u>. The current operational cost savings between using a Shuttle launched chemical energy system and a fusion system for just one Manned Mars Mission would be over $8B. Fusion, however, is not expected to be available for the first manned Mars flight if the mission occurs within the next 20 years. The loss of those initial mission savings is the economic penalty that is to be paid by the absence of a high specific power system, along with the loss of the additional <u>safety advantage</u> that would have resulted by virtue of quicker trip times, greater payload masses, and other safety features. Future science and exploration missions will require the availability of high specific power. At the current rate of spending, fusion will perpetually remain an advanced concept. Spending must be at a level commensurate with its true importance and the anticipated developmental time.

12.9 The <u>safety</u> of manned interplanetary missions would be substantially aided by the faster trip times. The most significant benefit is the relief from the integrated exposure to cosmic rays. Significantly reduced launches would enhance flight and ground safety operations. Another benefit is the capability to respond more quickly to contingent situations over long distances. A trip to Mars could be aborted in transit to return to Earth. The selection of D-^{3}He over D-T renders inherently greater safety benefits for the Earth's population and simpler system designs to achieve a high level of safety. Less severe operational constraints are also expected to result in space from its use.

12.10 The fusion developmental historical trend has been demonstrated as overly optimistic in terms of <u>projected dates for the development</u> of fusion reactors because the design solutions have proven more difficult than anticipated. The fact that steady progress is being made in spite of the greater than anticipated obstacles gives confidence to the eventual

12.0 Conclusions

outcome of controlled fusion. For terrestrial use, the question of its economic viability logically looms for the relatively near term application which retards terrestrial fusion funding and priorities, but that economic concern is not applicable to space missions where the performance advantage is so significant that major cost savings are made possible over the other energy sources.

12.11 The conclusion that fusion's <u>feasibility</u> exists only at some point far into the future is not necessarily subscribed to from the results of this study. Practically all emphasis in the terrestrial program has been placed upon "statistically" (Maxwellian) driven fusion reactor concepts as the best, most practical approach to cause the fusion of nucleons. Perhaps a more "focused" approach could be better suited for producing success. That suggests that design approaches like Migma and ICF may have merit, provided, of course, that the focusing process does not consume an inordinately large quantity of recirculating power and preferably that tritium not be burned.

12.12 The developmental <u>costs</u> to achieve fusion energy conversion are compensated by the anticipated reductions in space operations for accomplishing more ambitious missions. Fusion produces an inherently high energy yield process and is therefore operationally more efficient. More payloads per mission can be carried per spacecraft. A reduction in the number of chemical propulsion atmospheric launches results from the higher performance fusion powered propulsion system. The reduction of chemical propulsion launches makes fusion more kind to Earth's <u>environment</u> and to its energy consumption.

12.13 For space to become operationally established for the benefit of <u>commercial endeavors</u>, a large payload mass fraction space vehicle design is essential. A reactor designed to a high specific power makes possible large payload mass fractions, from 20% to 70%, depending upon the flight duration selected and the reactor's specific power capability.

12.14 The <u>space system requirements</u> differ from those in the terrestrial fusion program which DOE pursues for the benefit of commercial electric power production. As a result, it is reasonable to anticipate that NASA's space reactor requirements are sufficiently different to cause an entirely different reactor design to evolve, and that is indicated by the results contained herein. The tokamak reactor, if the mass/power is extrapolated in a straight line from the present mainline experiments, could never be a suitable space flight reactor due to its low power output for its intrinsically large mass, although it could be used as a stationary fusion powered electrical power station operating on the lunar or Martian surface. Differently designed reactors could very well be required for the different applications – space versus terrestrial. The operational environments and the modes of operation demand it. That the physics learned from one reactor may not necessarily be applicable to another is a product of the results of the various experiments to date.

12.15 The space fusion reactor designed for space power and for fusion engines to provide space propulsion will probably not be a direct, simple straight forward spin-off from the terrestrial program where the reactors will be designed for the generation of commercial electrical power. Ground power units typically differ from space and aeronautical power units. Also, space and aeronautical power units differ. There could be important synergism between the two programs.

12.16 For NASA to take advantage of fusion energy for space, a dedicated civilian space fusion program is required. NASA has a requirement for high specific power systems. Success in the terrestrial program does not guarantee a high specific power flight system being available for space use. The advantage of leveraging the past terrestrial fusion program experiments is substantial, and NASA can seize upon the advantage of the extensive work accomplished since the mid-1950's. Precedence has been set for a NASA space fusion activity by the Lewis Research Center's fusion research program. The cancellation of alternate experiments by DOE in 1991 confirms the importance to NASA initiating its space fusion program. Even if restarted later, fusion is too important to NASA's future to have an off-again, on-again fusion research program.

12.17 Currently, NASA's space fusion reactor future and its developmental pace is determined by the requirements, the goals, funding levels, and economics of the terrestrial application for commercial electrical power. The rate of the terrestrial fusion research program's progress, i.e., its funding level, is established by the world-wide commercial costs of energy which are not necessarily compatible with NASA's interests. Fusion COE estimates today cannot demonstrate a viable economic trade since crude oil costs are relatively low. The current price, as is well known, is a potentially volatile number, based on international conditions, not on space needs and requirements. Thus, even if net

12.0 Conclusions

power were demonstrated from the fusion reactors today, the commercial fusion plants would not be used simply because they are not presently economically competitive with today's relatively cheap price of energy. On the other hand, space fusion, if available, would be used for immediate application, effecting cost savings, producing new space science results, enabling new missions, enhancing others, and increasing safety. *The cost trades for space fusion and terrestrial fusion are different.* Those are strong motivating factors why NASA should consider space fusion now.

12.18 The fusion fuel of choice for space is aneutronic, but that, in the pure sense, does not appear to be possible at this time. For space energy conversion processes, the generation of only charged particles from the reactor is preferred. Neutrons, gamma rays, or X-rays serve only as impediments to missions or cause the vehicle to transport protective devices serving in effect as "ballast." Therefore, deuterium-helium-3 because of its inherently greater safety and reasonably good specific power output for performing space missions is the fuel of preference. D-^{3}He, while it is beneficial in the design process, does not completely alleviate the neutron problem which remains as a significant design consideration. Also, that reaction will be more difficult to achieve based upon theoretical considerations of physics, but benefits are expected in applying the technology to the development of a practical reactor. That is a consequence of simplifications of the engineering involved and the improved operational considerations which, therefore, make it the preferred fuel cycle. The engineering aspects, as well as the operational and safety considerations, are considered to be simplified sufficiently that a D-^{3}He reactor is more likely to be made a practical development and acceptable reactor first, although demonstration of the lower ignition temperature fuel D-T reactors will be achieved first. The space program can take full advantage of the higher energies of D-^{3}He for gains in the propulsion system specific impulse which are needed for the high performance demanding stellar missions.

12.19 Demonstration by tests of an engineering fusion reactor for application to a flight vehicle system design which yields high specific power (>1 kW/kg) is of utmost importance to space and the ability to efficiently convert the plasma energy into efficient, variable high specific impulse performance ranging from 5×10^3 to 10^6 seconds.

12.20 The confinement scheme preferred is magnetic confinement as being the one most likely to ultimately best satisfy the space system requirements. With recognition and considerations to the status of existing designs and their theoretical potential, the reactor having the greatest chance of meeting the space specfic power requirement is the FRC (Field Reversed Configuration) because of its high β. Declassification of the ICF could change this provided D-^{3}He can be burned. Plasma stability must be demonstrated at power regimes of interest to space as a major objective before we can consider this approach as viable. Other key parameters like steady state operation, power level, life, temperature balance, are important and require analysis/testing. Any of the other approaches having a high β, like compact toroids, dipole, or tandem mirrors are viable options to this approach.

12.21 It is incumbent upon NASA to have <u>alternate approaches</u> because the FRC's developmental status has not been advanced toward the demonstration of reactor physics at this time, but instead the developmental program has focused on plasma experiments. Hence, it is too early to know whether or not this concept can definitely be developed into a satisfactory space reactor and backup approaches are mandatory. Scaling laws and plasma stability when the design is scaled to reactor (net power) configurations are concerns.

12.22 <u>Other concepts</u> are available for consideration. These include a large variety of options (such as Migma and RACE), although there has been practically no examination of them in the context of space applications. The NASA Lewis EFBT was terminated before it had been extensively explored for net power and has not been exposed to the intensive review of the main DOE approaches.

12.23 Fusion systems by their nature tend to be large in size. Accurate, dependable modeling of plasma confinement is complex and not dependable, particularly with regard to understanding transport. <u>Full scale experiments</u> are key to the accomplishment of answers concerning concept feasibility. While these full scale test devices are obviously more expensive than small experiments, they are expected to provide an overall less costly approach to space fusion energy development when one takes the cost and safety benefits into consideration.

12.24 The most difficult mission to achieve of those considered herein is the <u>stellar mission class</u>. It is a viable mission, provided that high specific power reactors can be developed at a 10 kW/kg level or better. To

12.0 Conclusions

perform that mission class, the use of D-^{3}He is mandatory. The great demands placed by the stellar mission on its space propulsion system include:

- 12.23.1 specific power of 10 kW/kg or greater
- 12.23.2 multiple in-space restart capability
- 12.23.3 high jet power, on the order of 20 GW or higher if higher specific power systems are developed
- 12.23.4 high thrust, 5×10^4 N at 400,000 seconds average specific impulse
- 12.23.5 high rf transmitted electrical power capability of 10-20 MW output
- 12.23.6 long, steady state firing durations on the order of 50 years.

12.25 All of the <u>high energy planetary missions</u> examined had an enormous benefit in either the achievement of significant reductions in flight time or by carrying more massive payloads to enhance science mission objectives.

12.26 <u>Direct conversion</u> of the plasma to electrical power has been researched, but considerably more work is needed to obtain high efficiency, passively cooled, light weight, highly reliable, long duration designs for space.

12.27 The question of the <u>availability of space fusion fuel</u>, particularly lunar helium-3, must be fully addressed. An option is to breed ^{3}He using lithium-6 which should be studied for implementation feasibility. Another is to burn semi-catalyzed D-D and remove it from the plasma. The production of tritium in the quantities required for space missions must also be given extensive consideration as an option. Tritium production of the magnitude required for space flight operations could have as adverse impact on Earth's environment. If proven, the tritium manufacturing facility could be shifted to the moon where those concerns would be eliminated.

12.28 Because of the size of fusion vehicles, <u>on orbit assembly</u> will be necessary. This suggests that an orbiting space station could play a

12.0 Conclusions

key role in the operational mission support function for both assembly and staging for continued operations.

12.29 The cost of the reactors and propulsion systems makes <u>space basing</u> and <u>reuse</u> essential for fusion's space practicality. The inherent nature of fusion systems indicates from first considerations the possibility of being able to meet those needs.

13.0 SPACE PROGRAM OPTIONS FOR DEVELOPMENT OF A FLIGHT FUSION SYSTEM

Program options for pursuit of space fusion are presented in this section.

The program activities listed below range from a minimal interest-indicating level to one designed to make space fusion a reality. These options focus upon a generic program for the use of fusion energy, that is, it is not intended to be mission application specific. Although fusion has been stated to be both an enabling and enhancing technology, it's potential provides such a quantum leap for NASA's space transportation and power capability that it must be considered primarily as an enabling technology. The basis for the enhancing designation is that chemical and fission can perform the Manned Mars Missions although the high operational costs with using those sources do not permit a practical, affordable continued space presence program as discussed in Section 10.

Listed are four funding levels for program options. The recommended strategy is presented in Section 14.0. Recommendations for future high energy mission tasks are provided in Sections 14 and 15.

SPACE FUSION RESEARCH PROGRAM OPTIONS
OPTION 1: ANALYTICAL AND SYSTEMS STUDY TASKS $1-3M/YEAR
OPTION 2: SPACE FUSION EXPERIMENTS SUPPORT PROGRAM $10-15M/YEAR
OPTION 3: DEDICATED NASA SPACE FUSION PROGRAM $50-100M/YEAR
OPTION 4: EXPEDITED DEVELOPMENT OF PROTOTYPE SPACE FUSION FLIGHT SYSTEM $300-500M/YEAR

As shown by this study the requirements for space propulsion and power fusion requirements diverge sufficiently from the terrestrial program such that a separate space program is essential to serve NASA's space objectives and interests. It is important for NASA to control its energy destiny by providing funding for a space fusion program. The wisdom of this recommendation is

13.0 Space Program options for Development of a Flight Fusion System

nowhere better illustrated than by the DOE decision in 1991 to cancel alternate experiments. The terrestrial fusion program has advanced sufficiently that space can seize upon the technology now, particularly as new DOE endeavors can be expected to focus more upon the production of net electrical power experiments. It is expedient for NASA to implement a space fusion program to accomplish its missions and program goals, to meet its objectives, its research needs, its program plans, and to determine the rate of progress for fusion applications for space.

Below are examples of some of the key differences that must be taken into consideration in the development of research options. Space flight systems require radiation cooled, light weight designs capable of high performance propulsion. That capability is absolutely crucial to NASA. The space program fuel of preference is deuterium and helium-3, the physics of which is more difficult to achieve than the mainline terrestrial program's fuel cycle, D-T. The terrestrial application can more readily accommodate the D-T cycle. NASA has a requirement for a propulsion system capability and preferably a reactor that will serve a dual function of electrical power generation and propulsion. NASA's space operational environment of vacuum and zero gravity provide design options not available to the terrestrial application. The life time and maintenance needs are different. The differences in applications require different reactor designs using different approaches in the fundamental reactor configuration. Furthermore, what is not widely recognized outside of the fusion community is that the resolution of the physics details in one design will not necessarily resolve problems in another.

13.1 OPTION 1: ANALYTICAL AND SYSTEMS STUDY TASKS

A minimally funded program of $1-3 M annually would permit the conduct of studies on issues which affect space fusion. This level of effort is a minimal amount to acquire an improved management perspective of fusion energy. It does not address the key issues but indicates a minimal interest in space fusion on the part of NASA. Nine topical tasks and task objectives are listed below in a prioritized order. All of the tasks cannot be performed for the indicated annual funding level of $1-3M.

13.0 Space Program options for Development of a Flight Fusion System

OPTION 1: ANALYTICAL AND SYSTEMS STUDY TASKS ($1-3 M PER ANNUM)
• Field Reversed Configuration Space Fusion Reactor Design • Alternative Space Fusion Reactor Designs • Dual Mode Power Conversion • Magnetic Confinement Flight System Analysis • Stationary Propulsion Power • Inertial Confinement Space Fusion System • Alternate Sources of Fusion Fuels • Aeronautical Propulsion

13.1.1 FIELD REVERSED CONFIGURATION SPACE FUSION REACTOR DESIGN

Using mission performance requirements for high energy mission classes, conduct a 2-phase approach for a FRC reactor design for space propulsion and electrical power. The first part: (1) establishes an appropriate set of detailed requirements, (2) performs a design to meet those requirements, and (3) from the design establishes the fusion vehicle's performance capability. It features a FRC designed for the space application. Heat balances are a part of this activity as is first wall life. Also, one of the key items to address is the means by which to achieve specific impulse and thrust variations at the optimal efficiency. In the second phase of the activity, key experiments are conducted to confirm the phase 1 analyses. One objective is to model the propellant mixing scheme for propulsion and subsequently to verify the model by testing to demonstrate uniform mixing of the diluent into the plasma exhaust. A corollary task is to test the means for converting plasma energy to thrust at the magnitude needed to accomplish missions of the type described herein.

13.1.2 ALTERNATIVE SPACE FUSION REACTOR DESIGNS

Perform a second focused task similar to 13.1.1 for another reactor design concept. Candidates are the tandem mirror, other compact toroids, Migma, RACE and the EFBT. The solution for the MCF approach should be based upon the results of 13.1.4.

13.1.3 DUAL MODE POWER CONVERSION

Using the baseline mission requirements data from Section 2.0, conduct a preliminary design analysis on techniques for designing a dual propulsion and electrical power fusion reactor system, identifying performance, efficiencies, and mass trades. Testing should be accomplished later to demonstrate concepts for feasibility and then, subsequently, for realizing improvements in efficiency.

13.1.4 MAGNETIC CONFINEMENT FLIGHT SYSTEM ANALYSIS

Conduct a detailed system study of a flight MCF fusion energy vehicle system designed to produce propulsion and electrical power. Determine the range of specific powers which may be achievable for MCF propulsion systems. Study the feasibility of:

    (1) very low, <20 MW output,

    (2) low 20-60 MW,

    (3) moderate 200-300 MW, and

    (4) high, 1-20 GW jet power ranges for the FRC.

Perform the same study for a second option. This task is similar to Task 13.1.1 but is generic, whereas 13.1.1 is specific to the FRC.

13.1.5 REACTOR IN-SPACE RESTART AND ENERGY STORAGE

This study will evaluate the techniques for restart of fusion reactors in space. It should include techniques, restart duty cycles, power levels, mission system requirements, and energy requirements. Investigations include the means by which a fusion vehicle's energy storage requirements can be met, including the values for the restart plus mission electrical power requirements during reactor shutdowns.

13.1.6 STATIONARY PROPULSION POWER

A system study on the use of fusion electrical power to drive laser powered spacecraft either as energy for an ablation driven rocket or energy to power ion engines is performed under this task. The missions for application of this propulsion system include Earth–Moon, Moon–Mars, and Mars–Martian moons. Establishment of parametric energy levels for flight times and payload masses comprise parts of this task.

13.1.7 INERTIAL CONFINEMENT SPACE FUSION SYSTEM

The focus for this project is the conduct of conceptual evaluations of approaches to make ICF a viable source of fusion energy for space. It presumes that a satisfactory target gain for space will be demonstrated by the DOE program. The VISTA study considered a gain of 1500 for DT; for D-^{3}He the value is higher. The major emphasis comprises approaches to decrease the driver size. Fueling is another concern. The use of advanced fuels should also be evaluated.

The following constitute a minimum effort research activity for ICF:

Research Project	Manpower costs	Hardware and software costs
Excimer Laser & Optics Development	$1.20M	$2.0M
Advanced D-D–Fueled Pellet Designs	$0.30M	$1.0M
Induction Power System Development	$0.45M	$0.3M
Radiator Development	$0.45M	$0.5M
Plasma Conductivity and Drag Determination	$0.75M	$1.1M
Totals	$3.15M	$4.9M

13.1.8 ALTERNATE SOURCES OF FUSION FUELS

Where frequent missions are planned, the optimal approach is to refuel spacecraft at its destination as opposed to carrying both outbound and inbound fuels simultaneously. This is a small task to consider new non-terrestrial and lunar sources of deuterium and helium-3, the means for their exploration, the chances of finding the fuels, recovery techniques, reserves, and cost figures of merit.

13.1.9 AERONAUTICAL PROPULSION

A very preliminary analytical effort should be conducted to determine whether fusion could serve as a means for aeronautical propulsion. This includes the determination of requirements for achieving this capability. The effort should be directed at a mission of providing propulsion to Earth orbit either from the ground or from a flying launch platform. The objective is to seek an alternative means to reach Earth orbit in a safer, more economical manner than with

13.0 Space Program options for Development of a Flight Fusion System

chemical systems. Higher payload mass fractions for passengers and/or cargo are the study goals. Projections on thrust and performance of fusion reactors of the category required for this mission class are to be included.

OPTION 1 SUMMARY: ANALYTICAL AND SYSTEMS STUDY TASKS	
PROS	CONS
• PROVIDES FURTHER IN-DEPTH ANALYSIS OF MERITS	• FAILS TO PROVIDE NECESSARY DATA TO VALIDATE STUDY ASSUMPTIONS
• MINIMAL LEVEL TO INDICATE INTEREST	• FUNDING INADEQUATE TO ACCOMPLISH MEANINGFUL RESULTS
	• CONCLUSIONS MAY BE INVALID

13.2 OPTION 2: SPACE FUSION EXPERIMENTS SUPPORT PROGRAM

A low level test program, funded at $10-15M annually, is the minimal designed to produce experimental test results which may effectively assist NASA in obtaining fusion energy for space. The funding would be used to support DOE's current programs that are of interest to space but which are in danger of being terminated due to budget limitations. At this level, one experiment could be funded at a reasonably rapid rate to be of value to NASA and still permit some analytical work to be performed.[1]

This is a leveraged, cooperative activity using DOE facilities to accomplish these objectives:

(a) To expedite testing of reactor concept(s) already in the DOE program. It would supplement those current experiments having application to space but which have been assigned a low priority in the terrestrial program. These experiments could be expedited with additional funding. Testing includes the FRC and other key compact space-related confinement experiments. As this report is being completed the Los Alamos FRX-c is being dismantled and the Spectra Technology LSX experiment is being terminated.

---

[1] Since this recommendation was initially prepared, DOE's alternative experiments are being terminated. Some have been rapidly dismantled. This option may now be more costly-it would have to be examined.

13.0 Space Program options for Development of a Flight Fusion System

    (b) To provide a space focus by expanding tests and objectives using on-going confinement experimental programs. The conduct of fusion experiments which have space applications will be performed. This includes power and propulsion conversion techniques.

    (c) To test new plasma confinement concepts which are attractive to space, but which are not being funded otherwise.

    (d) To expedite D-^{3}He fusion reactor research.

The testing should ultimately lead to understanding of confinement physics for space fusion reactors. This is considered to be the best use of resources since, as stated in the text of this report, successful test demonstrations are mandatory in order to ultimately make available high energy sources for new high energy space missions.

OPTION 2:
SPACE FUSION ALTERNATE EXPERIMENTS SUPPORT PROGRAM
($10-15 M PER ANNUM)

- FIELD REVERSED CONFINEMENT TESTING FOR SPACE FUSION EXPERIMENT SUPPORT
- COMPACT TORUS TESTING FOR SPACE FUSION EXPERIMENTS
- HIGH RISK–HIGH GAIN CONCEPTS

13.2.1 FIELD REVERSED CONFINEMENT TESTING FOR SPACE FUSION EXPERIMENT

The preferred current plasma confinement design approach is the FRC for reasons stated in the text. Testing support should address space related aspects of the FRC and should expedite the overall test schedule. Neutral beam injection and increased scaling with burning of D-^{3}He (80 keV) are the objectives. Demonstration of the capability for long burn durations is an important function of this task. The preferred approach is – test at anticipated net power conditions first – understand theory later, even following first proof if necessary. Two experiments active at the time of report preparation are being terminated, probably eliminating this option at this funding level.

13.0 Space Program options for Development of a Flight Fusion System

13.2.2 COMPACT TORUS TESTING FOR SPACE FUSION EXPERIMENT

This is a potentially useful reactor for space. Test support would be at a minimal amount and possibly yield significant dividends.

13.2.3 HIGH RISK–HIGH GAIN CONCEPTS

Some testing should be accomplished which is best categorized as potential reactor experiments. The concept feasibility should be the goal of this general program. The funding could be at a level between $500K to $1,000K to evaluate critical aspects of new concepts. Specific detailed follow-on test and analytical projects can proceed later where warranted.

OPTION 2 SUMMARY SUPPLEMENTAL FUNDING FOR ACTIVE DOE ALTERNATIVE EXPERIMENT PROGRAMS	
PROS	CONS
• Minimal level to indicate produce needed test data	• Limited to current, on-going experiments • May not yield necessary data the quickest
• Can be quickly implemented -FRC testing	• Deactivated experiments cannot be tested • New concepts not tested
• Efficient use of funds by leveraging with on-going work	• Does not provide NASA in-house capability. • Not a quick approach. • May ultimately cost more.

13.0 Space Program options for Development of a Flight Fusion System

OPTION 2a SUMMARY FUND (LOW LEVEL) DOE FOR NEW CONFINEMENT OPTIONS	
PROS	CONS
• NEW CONCEPTS TESTED • QUICK TO IMPLEMENT AND MAKES USE OF EXISTING FUSION EXPERTISE	• GREATER TECHNICAL RISK • FRC TESTING NOT SUPPORTED • DOES NOT PROVIDE NASA IN-HOUSE CAPABILITY. • MAY ULTIMATELY COST MORE.

13.3 OPTION 3: DEDICATED NASA SPACE FUSION RESEARCH PROGRAM

A moderate funding level of $50M-100M will fund a minimum level for a serious research program. A NASA in-house fusion capability is one major objective of this option. This is a low level program to develop an in-house fusion energy expertise to better understand the issues that space fusion needs to consider before making a commitment. Thus, it is intended to commit the Agency to serious research investigations of the space fusion energy technology issue, but it does not pursue a heavily committed flight program activity. It requires, for example, demonstration of critical milestones before making heavy flight related commitments.

OPTION 3: DEDICATED NASA SPACE FUSION RESEARCH PROGRAM ($50-100M PER ANNUM)

• PROVIDE AN IN-HOUSE DESIGN EXPERTISE AND TEST CAPABILITY
    - NASA FACILITIES
    - OPTION: NATIONAL LABORATORY STAFFED WITH NASA PERSONNEL.
• GOAL: DEMONSTRATE A FULL SCALE FLIGHT FUSION ENERGY SYSTEM INCLUDING ALTERNATIVE CONFINEMENT DESIGNS AND TEST CAPABILITY

The expertise for space fusion flight equipment ultimately should reside within NASA as the operational organization to provide critical support for meeting mission success and system safety goals. This approach for acquiring an

13.0 Space Program options for Development of a Flight Fusion System

advanced space propulsion and power technological capability is no different than that used for any of NASA's other high technology hardware, such as chemical propulsion systems, fuel cells, solid rocket motors, life support systems, and so on. To develop an in-house test and design capability would not be a unique experience since it had already been implemented at the Lewis Research Center (Appendix A). The additional funding would either support NASA facilities or, as an option, a National Laboratory with NASA personnel. Demonstration of a full scale flight system is the goal including alternative space concepts, designs, and a test capability. The technical approaches recommended earlier are still applicable, but they would be accomplished by NASA staff.

OPTION 3 SUMMARY: DEDICATED NASA SPACE FUSION PROGRAM	
PROS	CONS
• ESTABLISH PROGRAM PRIORITIES TO SUIT NASA NEEDS	
• PROVIDES NASA IN-HOUSE CAPABILITY.	• LESS QUICK TO IMPLEMENT
	• LEARNING CURVE
• FLEXIBILITY TO CONDUCT ALTERNATE EXPERIMENTS IN A QUICK RESPONSE MODE	• HIGHER NEAR-TERM FUNDING REQUIRED
	• FUNDING LEVEL SUPPORTS ONE EXPERIMENT
• MINIMAL LEVEL TO PRODUCE NEEDED TEST DATA	• DOES NOT EXPEDITE SOLUTIONS
	• MAY BE LESS COST EFFECTIVE IF FUSION FLIGHT SYSTEMS CAN BE PROVEN MORE NEAR TERM
• DOES NOT OVERLY COMMIT NASA TO AN UNPROVEN CAPABILITY	• LESS QUICK TO IMPLEMENT
• EFFICIENT USE OF FUNDS	• DOES NOT EXPEDITE THE CAPABILITY
	• IF SPACE FUSION PROVEN, MORE EXPENSIVE IN LONG TERM

13.0 Space Program options for Development of a Flight Fusion System

13.4 OPTION 4: EXPEDITED DEVELOPMENT OF A PROTOTYPE SPACE FUSION FLIGHT SYSTEM

A high level of funding in excess of $300-500M per annum is necessary in order to expeditiously develop a prototype flight system.

OPTION 4:
EXPEDITED DEVELOPMENT OF PROTOTYPE SPACE FUSION FLIGHT SYSTEM
($300-500M PER ANNUM)

- GOAL: DEMONSTRATE A FULL SCALE FLIGHT FUSION ENERGY SYSTEM INCLUDING ALTERNATIVE CONFINEMENT DESIGNS AND TEST CAPABILITY
- PROVIDE AN IN-HOUSE DESIGN EXPERTISE AND TEST CAPABILITY
    - NASA FACILITIES
    - OPTION: NATIONAL LABORATORY STAFFED WITH NASA PERSONNEL.
- EXPEDITE THE SPACE FUSION CAPABILITY

This higher level of funding would expedite the Option 3 approach and would provide serious funding for an alternate confinement experiment. This option assumes that there is a high degree of meeting with the successful application of fusion energy in a flight vehicle. After a concept has been demonstrated to have a reasonable chance of providing a source of fusion energy for space, a flight-like, full scale system would be designed, built, tested, and flown. This approach funds the parallel development of key system related issues to expedite the implementation.

13.0 Space Program options for Development of a Flight Fusion System

OPTION 4 SUMMARY EXPEDITED DEVELOPMENT OF PROTOTYPE SPACE FUSION FLIGHT SYSTEM	
PROS	CONS
• ESTABLISH PROGRAM PRIORITIES TO SUIT NASA NEEDS	• COMMITS NASA TO AN UNPROVEN TECHNOLOGY-MAY NOT WORK
• EXPEDITES MISSION ENABLING CAPABILITY WITH TREMENDOUS PAY BACK OF INVESTMENT IF SUCCESSFUL	
• PROVIDES NASA IN-HOUSE CAPABILITY	
• VERY COST EFFECTIVE IF FUSION FLIGHT SYSTEMS CAN BE PROVEN MORE NEAR TERM	• HIGHER NEAR-TERM FUNDING REQUIRED
• FLEXIBILITY TO CONDUCT ALTERNATE EXPERIMENTS IN A QUICK RESPONSE MODE	
• ADEQUATE LEVEL TO PRODUCE NEEDED FUSION AND FLIGHT SYSTEM DATA	

13.5 SUMMARY

To commence addressing fusion in a manner that will have a practical benefit for space, a cooperative test program at a minimal level of $10 to 15M annually is needed. An in-house capability and program designed to conduct independent experiments is estimated at no less than $50M-100M. The options are summarized in Table 13-1.

13.0 Space Program options for Development of a Flight Fusion System

TABLE 13-1. Summary of fusion annual funding options.

Program Options	Program products
Option 1- $1 to 3 M	Studies only. Does not produce needed test data.
Option 2- $10-15 M	Focuses on the generation of data from the most space relevant experiment.
Option 3- $50-100 M	Is the minimal level to develop an alternate concept for demonstration of net power.
Option 4- $300 to 500 M	Pursues an extensive program for conduct of confinement options, development of propulsion, development of electrical power, and the conduct of system analyses and key experiments for a flight vehicle system. It develops a prototype flight system on an expedited basis.

14.0 RECOMMENDED SPACE FUSION STRATEGY AND PROGRAM PLAN

Four program options were presented in Section 13.0. The recommended strategy and program plan based upon this study's results are presented below as the preferred approach. The program plan addresses critical topics for space fusion application in a prioritized order. Tasks are suggested to validate assumptions which relate to the demonstration of fusion and flight system feasibility. Options are presented in tasks as appropriate. A flight strategy is included as well. The overall schedule, which is presented in Section 14.3.2.1, shows the importance of initiating the program at the earliest possible time.

This strategy will differ from many strategies that have been forwarded for propulsion in that it is more forward looking and that an in-depth analysis was made of advanced requirements and of the fusion technology in particular. For example, as one recent strategy, the Aerospace Industries Association prepared a "National Rocket Propulsion Strategic Plan," dated February 15, 1990. (Anom90) Under the "Advanced Propulsion Concepts" chapter the document recommends as "Program N0. 7 – Fusion, 1991-2000, $5 million" as the suggested effort for this crucial technology. That level of funding and priority is assumed to be based upon the text earlier where it is stated that

> **"Far-Term Systems**
>
> "...All of the far-term programs in Advanced Propulsion Concepts are funded at a level that allows small research programs. These small programs will determine the feasibility of these propulsion systems. Many of the technologies required for these advanced propulsion ideas are not available for near-term demonstrations. A continuing assessment of these propulsion systems is needed; many advancements are possible in the basic understanding of energy conversion techniques (using fusion or antimatter, etc.). When a system concept has demonstrated a practical capability on an acceptable theory, a proof of concept demonstration or experiment should be conducted using the theoretical design."
> (Anom90, p147)

With that strategy we will get nowhere for the reasons stated throughout the text of this report. It reflects unfortunately a lack of understanding of the real needs of propulsion in NASA, the benefits of high specific power coupled with high specific impulse and of fusion technology in general. The $5 million, just to stay abreast of the technology status, is considered an expenditure of funds best made on more productive matters since the status is well known, nor is it possible at that level to conduct the research experiments needed to develop fusion energy for space.

14.0 Recommended Space Fusion Strategy

Based upon considerable thought given to the projected mission needs and technical issues, this section presents a comprehensive program and logic for the efforts that are considered important to the future of the space program.

14.1 BASIS FOR THE STRATEGY

The objective of this report has been to consider high energy space missions, their energy requirements, the appropriate system requirements, and the energy means to meet those missions. This report shows fusion as the energy source having the greatest chance of fulfilling those requirements. But it is the least developed of the possible sources. Fusion energy appears attainable, but key fusion experiments and vehicle system analyses remain. Since there are fundamental differences in agency goals and mission requirements between NASA and DOE, the initiation of a space fusion program by NASA is considered necessary.[1] This program is needed to develop a high specific power propulsion capability, one that meets NASA's mission requirements and program priorities.

What should be the approach? Four funding options were presented in Section 13.0. The purpose of this section is to forward a coherent total program strategy using the information accumulated in the study as presented in the text. The strategy includes a description of what should be done, who should do it, and how it could be accomplished. It, therefore, forwards the preferred program approach as derived from this activity.

Fusion experiments were examined in Sections 7 and 8 to determine whether the current terrestrial program may be pursuing or has pursued concepts which are attractive from a space propulsion and power application standpoint. The conclusion is that the DOE program has been sufficiently broad-based to reveal approaches that NASA can now explore in depth to achieve a high energy space fusion capability. Further, it is concluded that the DOE program is spread too thin to pursue space related matters. Lastly, it is the function of NASA to develop its propulsion and space power systems. Thus, the space program can leverage the work accomplished to date, taking advantage of the fact that fusion technology has come a long way.

If NASA is to make use of this potentially highly rewarding technology, the time has arrived for the initiation of a space fusion program. While none of the reactor concepts being developed fits the requirements for spacecraft propulsion and power systems – nor for that matter were they expected to fulfill those needs as a result of requirements and mission

---

[1] As an illustration, note that current (FY91) budgetary restrictions are eliminating support for the alternate experiments considered to be key to NASA (Fpa90).

14.0 Recommended Space Fusion Strategy

differences – there are attractive alternative reactor concepts which NASA should explore. The most attractive concepts are the least developed. The ones preferred from this study are those having high β.

A thesis has, therefore, been developed in this study around which the recommended strategy is developed:

THESIS

Fusion energy for the space application is mission enabling and safety enhancing. If key technical assumptions can be validated, the use of fusion will become feasible and become the prime element in the space transportation and power infrastructure.

CAVEAT

A fusion energy research program for space has not been implemented because no space mission requirement has existed for high energy missions. No space mission requirement exists because fusion has not been researched and developed for space. The space program is awaiting the development of fusion for commercial application to electrical power generation before considering fusion energy for space applications. But a ground based fusion energy utility electrical power plant technology will not be suitable for meeting the space program's requirements. This situation has become a vicious circle causing a significant loss in NASA's ability to conduct more ambitious and safer space missions.

14.2 ASSUMPTIONS

In the development of a strategy, we must first consider the most critical technical issues. In the case of this study, that implies the most technically impacting assumptions, structured to a hierarchical order of technological difficulty. These critical assumptions are prioritized below. Thus,

---

Space fusion energy would be of great benefit to NASA space science and exploration objectives, provided:

14.0 Recommended Space Fusion Strategy

Fusion Program Assumptions

1. Controlled, stable plasma burning in space fusion reactors is feasible.

2. Stable plasma burning, considering assumptions 4 through 8, and the space operational use of low neutron producing reactions like D-^{3}He are feasible and preferably fuels exhibiting no neutron reaction products.

3. Helium-3 will be available in the amount required and at a cost effective price for space missions.

4. Specific power propulsion and flight reactor systems, producing a minimum of ~1 kW/kg, are viable, preferably 10 kW/kg.

    4.1 A reactor start and multiple remote restart capability in space can be designed within the necessary specific powers ranges.

    4.2. Cooling and shielding mass requirements are acceptable.

5. Space reactors can be designed to produce 20 MW to 300 MW power output for solar system science and manned exploration missions. For interstellar missions, the reactor jet power requirements needed are on the order of 30 GW for specific power systems of 10 kW/kg.

6. Fusion propulsion is practical.

    6.1 High thrust up to 50,000 Newtons.

    6.2 Throttling efficiently performed.

    6.3. Specific impulses efficiently varied from several thousand seconds to 10^6 seconds.

7. A fusion reactor and propulsion system can be designed to burn reliably for: 4 months for a Manned Mars Mission, 10 months to 5 years for solar science missions, and 50 years minimum for stellar missions. Reuse of space based solar system fusion propulsion system reactors can be achieved, a necessity for reducing fusion vehicle flight operational costs.

8. Efficient direct conversion of charged particle energy to electrical power in space on the order of up to 20-30 MW is feasible.

14.3 FUSION PROGRAM TO ADDRESS KEY ASSUMPTIONS AND ULTIMATELY TO DEVELOP A FLIGHT CAPABILITY

The progression of the major program steps necessary to address the assumptions and to make an operational flight fusion system is presented in Fig. 14.1:

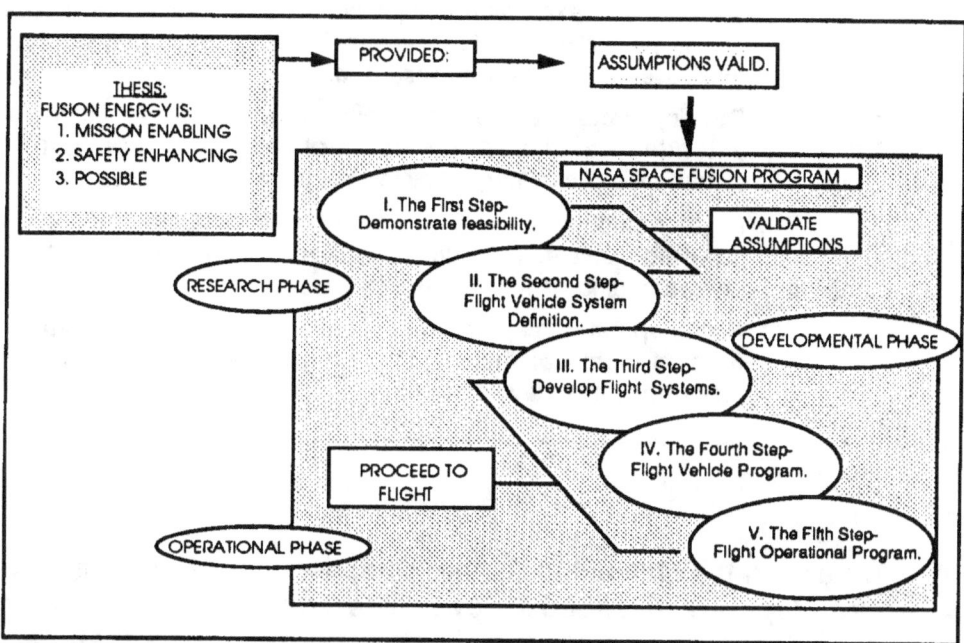

Fig. 14.1. Progression of major program steps.

If the advantages of fusion energy are to be realized, a program needs to be initiated that addresses the assumptions. Twelve tasks were identified to validate the assumptions.

These 12 program element tasks have been divided into 5 steps which range from research, to development, to flight operations. The major objectives of each of these steps are shown in Table 14-1.

14.0 Recommended Space Fusion Strategy

TABLE 14-1. Objectives: Major Space Fusion Program Steps:

I. Demonstrate feasibility: Commit to a NASA Space Fusion Program (SFP) and prove principles via the conduct of key fusion energy conversion demonstration experiments.

II. Flight vehicle systems definition: Determine and refine specific system requirements. Show that the fusion flight system aspects are feasible. Include investigations of the benefits to NASA's science and exploration objectives. Evaluate fusion's specific power capability. Perform fusion flight system analyses, system related experiments, and testing of concepts.

III. Develop prototype flight systems: Conduct test verifications and extended analyses up to flight system configurations. Commence preliminary vehicle design options studies.

IV. Flight vehicle program: Design, build, and fly a flight fusion powered vehicle. Conduct technology maintenance to enhance understanding of fusion systems.

V. Flight operational program: Conduct science and exploration mission flights of fusion powered vehicles. Pursue advanced fusion systems development.

Details on the element tasks that comprise the major goals for each of the five steps are provided below. The content of the program may be considered to be accomplished in one of either, a research, developmental, or operational phase. Twelve major program tasks are identified in Table 14-2 to validate the assumptions:

14.0 Recommended Space Fusion Strategy

TABLE 14-2. Summary of major program tasks to validate assumptions.

MAJOR PROGRAM TASKS	1. RESEARCH 2. DEVELOPMENT/ QUALIFICATION 3. OPERATIONS				
	PROGRAM STEPS				
	I	II	III	IV	V
1. INITIATE PROGRAM	✔				
2. DEMONSTRATE NET POWER FROM FUSION ENERGY FOR SPACE USE	✔				
3. PREPARE BROAD BASED FUSION DEVELOPMENT PLAN	✔				
4. ASSURE FUEL AVAILABILITY	✔	✔	✔	✔	✔
5. DEVELOP MISSION REQUIREMENTS		✔			
6. CONDUCT SPECIFIC POWER ANALYSIS	✔°	✔			
7. DEMONSTRATE FLIGHT SYSTEM FEASIBILITY		✔	✔	✔	
8. EVALUATE SPACE START OPTIONS		✔			
9. PERFORM CONFINEMENT OPTION PROGRAM	✔°	✔	✔		
10. CONDUCT PLASMA ANALYSIS	✔	✔	✔	✔	✔
11. PERFORM PRELIMINARY DETAILED FLIGHT VEHICLE DESIGN OPTIONS			✔	✔	✔
12. ASSURE LIFE AND RELIABILITY CAPABILITY.			✔	✔	✔

° – Option

14.0 Recommended Space Fusion Strategy

Space Fusion Program: Research Phase

14.3.1 PROGRAM DEFINITION: TASK DESCRIPTIONS

14.3.1.1 *THE FIRST STEP* – DEMONSTRATE FUSION FEASIBILITY: COMMIT TO A SPACE FUSION PROGRAM. SHOW OF PROOF OF PRINCIPLE

Objective: Demonstrate that fusion for a space application reactor is feasible.

The suggested overall flow for the initiation of a space fusion program is shown in Fig. 14.2.

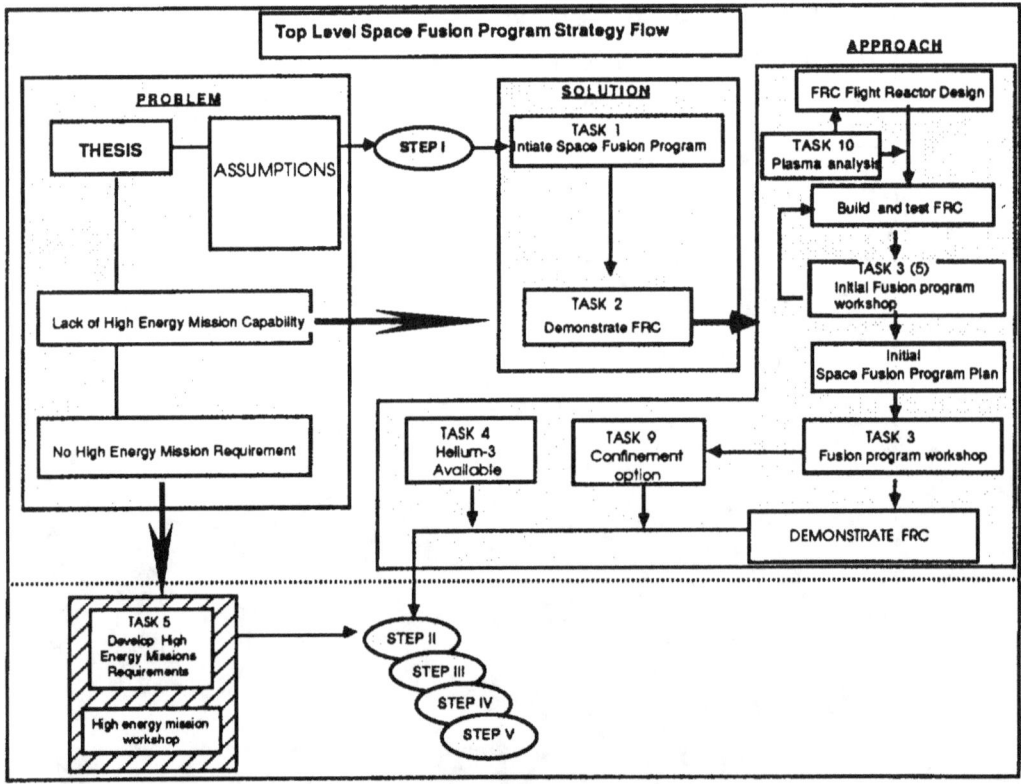

Fig. 14.2. Step I – PROGRAM INITIATION STRATEGY.

This step contains four major tasks:

14-8

14.0 Recommended Space Fusion Strategy

1. initiate a space fusion program
2. initiate a FRC experiment program
3. develop an initial comprehensive program plan.
4. initiate a fuel availability program.

Proposed Space Fusion Program Tasks for Step I

14.3.1.1.1. TASK 1. INITIATE AN ACTIVE "NASA SPACE FUSION ENERGY CONVERSION PROGRAM" TO ADDRESS THE KEY SPACE FUSION TECHNOLOGY ISSUES

This recommendation addresses the thesis. It is designed to break the "vicious circle conflict" between mission requirements and mission capability. A need exists to actively pursue a high energy space mission capability. If the capability existed, then NASA would use it. NASA has the charter for the conduct of space and aeronautical research, including the development of the energy conversion means necessary for the conduct of its broad category of space and aeronautical missions.

Space fusion energy is within NASA's domain and best interest. Higher energy density and more powerful energy sources are critical to NASA's mission. Without high levels of energy, space missions – and hence NASA's future – becomes energy limited and highly mission constrained; and ultimately the lack of a high specific power, high specific impulse capability would stagnate NASA's space programs and prevent it from accomplishing bold new high energy missions. NASA, a research organization which operates on the leading edge of technology, accordingly should possess the hands-on technical competence in those technical disciplines which it utilizes. The capability is necessary to minimize program risks. Otherwise, management decisions cannot be made from an independent and knowledgeable judgment basis, a factor particularly critical whenever research and technology become more advanced and the energy levels increase.

A Space Fusion Program would not be new. NASA had previously conducted a very successful fusion program which was space focused and which made important contributions to fusion technology (Appendix A). The pragmatic approach is to make the maximum use of available resources. Consequently, as one option, the prudent solution, and one which offers optimal technical leverage, is to use existing fusion personnel and facilities at a national laboratory where there would be

14.0 Recommended Space Fusion Strategy

already in existence a dedicated facility and staff to conduct a NASA space fusion program but which would avoid conflicting with the objectives of the terrestrial fusion program. The two separate programs, NASA's and DOE's, have the advantage that they could complement one another in some aspects of the technology. The relationship suggested here would parallel that between NASA and the Jet Propulsion-California Institute of Technology for planetary spacecraft design, mission operations, and other functions. LLNL, for example, is operated under contract to DOE via the University of California. The other labs are similarly operated. Another option is to build a separate NASA facility and staff it with civil servants. That, too, has advantages, but it would require a longer time to initiate the program. Facility costs may be greater although this would have to be examined. Another option is to use a national laboratory to perform the FRC work while building up a NASA civil servant capability on a confinement option. Another is to contract the work completely. A university operated reactor is still another option. University and industry will provide a valuable component to the program regardless of the option taken.

14.3.1.1.2. TASK 2. INITIATE A "FRC SPACE FUSION EXPERIMENTS AND REACTOR TEST PROGRAM"

To address assumptions numbers 1 and 2, demonstration of fusion energy for space is essential. This program's task objective is to test the FRC burning D-^{3}He. The goal is to develop a minimum neutron producing reaction, and preferably one consisting solely of energetic charged particles. The absence of neutrons at this point can only be considered a goal. In the ultimate sense it cannot be achieved with current technology projections, but it can be reduced to approximately 1 to 2 per cent. The reactor having the most favorable characteristics for achieving that goal and for NASA's application is considered to be the FRC. Other compact toroids are options. The dipole is another. FRC progress can be expedited by funding a program to increase the magnetic field strength and plasma temperature to provide 40 keV ion temperatures for burning D-^{3}He as discussed below.

The preferred program approach is to initiate the design of a minimal series of large step, high risk FRC experiments aimed at quickly demonstrating a space fusion reactor capable of burning D-^{3}He. The plasma is believed to be capable of being heated to ignition using neutral beam injection and of being maintained stable by the beam flux. Plasma stability and ignition verification experiments using high energy neutral beam injection to increase the energy level to the required values should, therefore, be conducted as a key part of the technical strategy.

This empirical approach, by-passing the depth of understanding desired by a science program, is appropriate for an engineering developmental

14.0 Recommended Space Fusion Strategy

program and has, in fact, been a path successfully taken to implement prior inventions. This must be accepted as an expedited but high risk approach. The magnitude of the gain to space programs justifies the risk level and warrants the recommendation. In considering the "at risk" funding level later in this section it should be emphasized that those cost estimates are no more than educated estimated judgments to demonstrate plasma stability in an FRC. To better define the risk, more definitive cost estimating must be performed concurrent with the results from more refined analysis and experiments, an iteration process.

The objective, then, is to demonstrate for flight conditions the burning of D-^{3}He as quickly as possible without gaining a scientific characterization of the plasma. Thus, the first goal is to advance the FRC reactor to a D-^{3}He burn configuration.

While the approach is to advance the FRC to burn parameters without understanding of the plasma science, that understanding is considered important to the program in the long term. Hence, there is a necessity to conduct analyses of the reactor's plasma to characterize the confinement sensitivities to variations in operational control parameters (Task 10).

Table 7-3 characterizes the basic reactor parameters for a 500 MW reactor. If D-^{3}He burning is not successful, the D-T operational regimes can be tested. D-T, although not the preferred fuel, could still serve a very useful function in the accomplishment of space science and exploration although considerable advantages are lost. If that situation occurs, efforts should continue to further pursue either the FRC D-^{3}He fuel cycle or alternative advanced plasma confinement concepts in this or other reactors.

Using a high risk FRC approach we would proceed without delaying for the workshop discussed below. As a different confinement approach using an optional design (Task 9), the SFP could conduct experiments[2] on non-Maxwellian systems for example. The goal is to avoid a critical single failure point with regard to conducting research on one fusion confinement scheme. It is also to advance technology by examining less defined concepts that may hold high promise for space applications. The funding availability will obviously determine whether or not options are possible. It would be preferred to conduct a parallel multiple experiment program on the order of 4-7, funded at a cost of $10M per program. This avoids falling into the trap of relying on a single approach. Suggested

---

[2] There is a distinction made between "reactor" and "experiment." "Reactor" in this report refers to energy conversion equipment producing net power. "Experiment" refers to investigations of machines to verify concepts, analyses, and theory. Hence, "experiments" are not necessarily of an operational regime where net power is produced.

approaches for confinement options will be a major product of the workshop.

14.3.1.1.3. TASK 10. CONDUCT PLASMA ANALYSES

The plasma analysis task supports understandings of the reactor, thrust conversions, and electrical power conversion systems. A level of effort must be expended to understand the experimental results attained. This knowledge is essential for the successful use of the reactor, engine, and electrical power generation systems over the anticipated flight operational regime.

14.3.1.1.4. TASK 3. PREPARE A PRELIMINARY PROGRAM PLAN FOR THE DEVELOPMENT OF SPACE FUSION ENERGY

This task initiates a workshop of fusion scientists and engineers having an interest in the application of fusion energy to space, organized for the purpose bringing forth the best set of ideas, requirements, and issues to address in a Space Fusion Energy Program. The objective, then, is to provide at the workshop conclusion, the elements of an Initial Space Fusion Program Plan which has been defined using the best thinking available. A follow-up workshop after the completion of the initial FRC testing and after the mission requirements have been more thoroughly defined in Task 5 would be appropriate.

A comprehensive space mission workshop is suggested in Step II, Task 5. An alternative is to hold a joint working group between the flight system users and the fusion technologists at the time of the initial planning. An initial set of mission requirements would thus be established by conducting an advanced mission workshop devoted to establishing science and exploration missions.

14.3.1.1.5. TASK 4. ASSURE AVAILABILITY OF FUEL, PARTICULARLY HELIUM-3

The long lead time anticipated to provide ^{3}He indicates that an early study to establish the sources for a dependable supply would be appropriate. The production technology and economics are to be a part of the study. Fuel supply options are to be evaluated.

A reliable source and storage means of ^{3}He, assumption 3, is vital to the space fusion technology application. This task must also consider the means to provide and store large quantities of deuterium. Storage in the solid state, for example could be considered. The establishment of a ^{3}He production facility is a long lead time item to support the development, qualification, and flight program activities. Study efforts, options, and planning are performed in Step I to prepare for the Step II ^{3}He production operations. Also, the technical feasibility of the fuel

14.0 Recommended Space Fusion Strategy

processing technology for deuterium and ^{3}He is to be studied, and preferably a pilot model demonstrated. This study identified terrestrial and extraterrestrial sources for ^{3}He. For the space application it eventually may become sufficiently important to manufacture the element. This may be necessary in the event that a limited terrestrial source can be made available prior to a lunar supply. A clear definition of the fuel demand, based upon mission objectives planning, and the means of its supply should be addressed. The plans for acquiring the anticipated quantity of ^{3}He should have been completed for implementation in Step II.

The minimal amount envisioned for testing and initial flights (based on Table 2-7a) is (in kilograms) provided in Table 14-3:

TABLE 14-3. Preliminary estimate of ^{3}He requirements (kg).

Testing (based on 2x Manned Mars mission firing time)	12
Jupiter	7
Manned Mars	6
Pluto	11
Asteroid hopping (6 visited)	10
Total	46

The above assume a 1 kW/kg system operating at 70% efficiency. One kilogram of ^{3}He will produce 19 MW-years of fusion power. There is a sufficient ^{3}He supply to commence a test program, but a larger supply is required to conduct a space program. ^{3}He may also be conserved by the use of hydrogen for selected tests once the D-^{3}He reaction has been demonstrated. That is a function of test objectives and a subject for examination after the program has progressed.

Options:

14.3.1.1.6. TASK 9. CONFINEMENT OPTIONS

This task could be defined as one which serves as a precursor to any program activity. It would consist of detailed design studies of the preferred approaches that would be a product of the workshop suggested above. If the top 10 suggestions were studied in depth at a level of $5M per approach, the program would be initiated at a cost of $50M, spread over 2 years. The disadvantage is that the program would be delayed for 3 years—which would be very costly as discussed in

Section 10.0. The other problem is that the study may show concepts to be invalid when the converse may apply in reality.

14.3.1.1.7. TASK 6. MCF VEHICLE ANALYSIS OPTION

A preprogram vehicle analysis study could be conducted to better quantify the capability of a fusion powered vehicle. The concern resulting from a preliminary study of this nature is that the level of understanding of a flight reactor is inadequately developed to permit a high level of confidence in the results.

14.3.1.2. *THE SECOND STEP* – DEMONSTRATE FLIGHT VEHICLE SYSTEM FEASIBILITY AND PERFORM AN INITIAL FLIGHT SYSTEM DESIGN

Objective: Show space fusion flight system feasibility, determine requirements, produce test verifications, and conduct extended analyses up through flight system prototypes. Perform fusion system analyses and related experiments and testing.

Once fusion net power has been demonstrated in Step I, a better understanding of fusion's performance capability will be possible. At this point it would be appropriate to define high energy mission requirements. The FRC is considered a prime contender and is used as an illustrative example.

As the first step, this study's strategy has assigned the top priority to the conduct of net power experiments on the reactor configuration(s) considered most likely to succeed. Without reactor test data, there always will be the question of viability of vehicle flight system study results. From the results of the test program in Step I and the mission requirements, more accurate, and therefore credible, preliminary vehicle system designs can be accomplished. Thus, in Step II the program demonstrates system feasibility to meet space mission requirements.

The Step II overall flow for the determination of the feasibility of space fusion systems technology is shown in Fig. 14.3.

14.0 Recommended Space Fusion Strategy

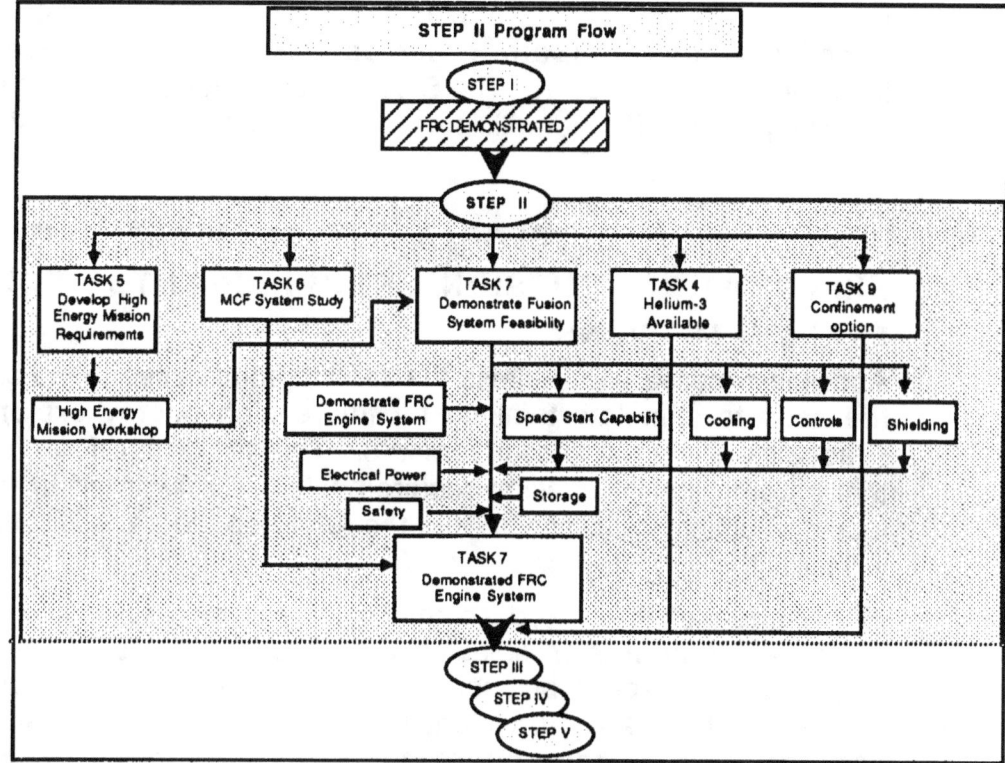

Fig. 14.3. Step II program flow.

This step contains 7 major tasks:

1. establish a fuel supply (Task 4 continuation from Step I)
2. develop mission requirements (Task 5)
3. establish specific power (Task 6, Step I option)
4. show system feasibility (Task 7)
5. demonstrate space start capability (Task 8)
6. conduct alternate experiments (Task 9, Step I option)
7. perform plasma analysis (Task 10, continuation from Step I).

Proposed Space Fusion Program Tasks for Step II

14.3.1.2.1. TASK 4 (CONTINUED). ASSURE AVAILABILITY OF FUEL, PARTICULARLY HELIUM-3

In Step I of the program the fuel production options for deuterium and ^{3}He will have been studied. Preferably the validity of the production

14-15

14.0 Recommended Space Fusion Strategy

concept is demonstrated, and the plans for acquiring the anticipated quantity of helium-3 are expected to have been developed for implementation in Step II. The Step II task develops the fuel production capability in time for the anticipated need, most likely by the start of Step III.

14.3.1.2.2. TASK 5. DEVELOP A SET OF SPACE SCIENCE AND EXPLORATION MISSION REQUIREMENTS FOR HIGH ENERGY MISSIONS

A task is necessary to address the second point raised in the caveat, i.e., the need to stimulate thinking for future high energy exploration and space science mission requirements. This task, therefore, is to insure that proper science mission requirements are forwarded to the system task and ultimately to assure that the future space transportation system will meet the user needs.

A workshop of interested scientists should be assembled that more precisely define mission objectives. These objectives are necessary to fully reflect the space science and exploration research requirements. Scientists who are not traditionally associated with the space program activities could broaden the program to most effectively use the new capability. The purpose is to include science objectives which to date may not have been considered due to the lack of a high energy demonstration.

This task is thereby initiated as one critical element to terminate the circuitous situation between missions and the lack of a mission energy capability as presented in the study thesis. This subject is discussed in further depth in Section 15, the report's recommendations. This workshop complements the Space Fusion Technology Research Workshop by further delineating system requirements. A better understanding of the requirements for the power levels, durations, payload masses, etc. are examples of workshop products. The implementation of workshop originated missions could be time phased in an ascending order of difficulty to match anticipated fusion progress. At this time, the 250 MW technology is considered to be more likely to meet the 1 kW/kg performance level than the 80 MW size. With technology developmental progress, smaller high performance reactors may become feasible. From the results of the workshop a space science mission requirements document is prepared as an input to the system feasibility activity discussed below.

14.3.1.2.3. TASK 6. DEFINE AND ESTABLISH SPECIFIC POWER CHARACTERISTICS MORE PRECISELY

Specific power, the third assumption, and specifically the assumption of the feasibility of attaining 1 kW/kg to 10 kW/kg, is crucial to the attainment of the mission objectives discussed in this report. The very preliminary systems analysis work accomplished is encouraging. Because of the lack of detailed investigations for space, specific power design work should be undertaken to better quantify this parameter and to establish approaches and research endeavors for mass reductions. That design activity should be performed to provide a preliminary assessment as to the feasibility of meeting the necessary level of specific power for attaining advanced science mission objectives. This entails the design of reactors of the power output level required for the conduct of space missions.

A preliminary system design study could be accomplished as the first step in the strategy rather than to initiate testing as suggested. From the studies and the very preliminary experimental test data, it was concluded that another study, regardless of the funding level, will lack the necessary accuracy of system parameters to produce credible results. Instead, the emphasis has been placed upon spending funds on testing so that more dependable vehicle system analyses can be subsequently performed.

14.3.1.2.4. TASK 7. DEMONSTRATE SYSTEM FEASIBILITY

CONDUCT ENGINE, PROPULSION, AND FLIGHT VEHICLE SYSTEM RELATED ANALYSES, DESIGN, AND TESTS. DEMONSTRATE: SPACE START, POWER LEVEL, THRUST, SPECIFIC IMPULSE, DIRECT ELECTRICAL POWER CONVERSION, COOLING, AND SHIELDING PERFORMANCE

This is a very extensive project to address the critical system aspects of the fusion space flight vehicle, assumptions 4, 5, 6, and 8. This is also a space vehicle research integration activity, in other words, a program which researches the science of the integration of all space vehicle systems critical for the successful, safe operation of a fusion powered spacecraft. The very unique issues that are associated with fusion energy conversion for space systems applications are investigated here. The goal in this task will be the ultimate demonstration of the reactor's ability to perform in a space flight system. This includes the propulsion, life projections, performance, safety, vehicle system controls including AI, reliability, and the operational aspects such as supplying electrical power to the flight system for remote restarting of reactors in the space environment while meeting the specific power requirements and to produce controllable thrust from an FRC or other fusion reactor. Therefore, the specific power potential (assumption 4) as well as the capability to deliver the variable high thrust and specific impulse performance capabilities (assumption 6) are part of the task. Task 7 uses

14.0 Recommended Space Fusion Strategy

the space start/remote restart results from Task 8 as an input. There is an option of using the Step I reactor in Step II or to design a new reactor, depending upon the results achieved in Step I.

Reactor power output levels and efficiency are to be demonstrated. Of interest here, too, is the ability to throttle the specific impulse down to approximately 5,000 seconds using uniform, efficient mixing of the diluents while maintaining efficient fusion burning (assumptions 5 and 6). Optimal diluents from the standpoint of system considerations are a part of this effort. The diluent of preference from the reactor's operational characteristics may vary from the system's optimal mass, and that question is an important one to investigate in this project. The means for control of the reactor and fusion system is also a part of this major task. The efficient conversion of plasma energy directly to electricity (assumption 8) will have profound system implications but may be a difficult engineering feat to achieve, particularly at the high level of efficiency desired, at the low level of rejected heat desired, and for meeting the long life reliability requirements.

Other key points raised in this study are also to be addressed in Step II. These include determination of the requirements and the means to control the propulsion system. The analysis conducted in the study shows great safety advantages, but assumptions are necessary due to the early state of the technology at this time. At the time of the initiation of Step II space fusion reactor technology should be better understood and fusion spacecraft safety should be analyzed. The economics reported herein are predicated upon the feasibility of fusion engine system space storage and reuse capabilities. A better definition of the reactor's operational characteristics accomplished at this point in the program should likewise assist in the development of that important assumption.

14.3.1.2.5 TASK 8. DEMONSTRATE A SPACE START/REMOTE RESTART CAPABILITY

As part of the task to better evaluate and ensure that the required specific power is achievable, this project provides for the conduct of those analyses necessary to define the requirements for a space reactor start capability and to establish concepts that will best meet those requirements, assumption 4.1. This is a critical subject for space programs, but it is one where no work has been performed. Following the analysis, optional techniques and design approaches for implementation are to be evaluated by experiments and testing. From those test evaluations, the technology for fusion flight system starts in space will be better defined empirically, thereby enhancing our understanding of the vehicle mass. The testing should indicate developmental areas for reducing the start-up mass. The experiments should be instrumental in establishing which start-up approaches will provide achievable capabilities. The means to store the start-up reactor energy while still maintaining the necessary specific power level is a very significant matter to be included in this task.

14.3.1.2.6 TASK 9. CONDUCT ALTERNATE EXPERIMENTS AND UPGRADE PERFORMANCE

This task uses the results from the space fusion technology workshop to incorporate alternate confinement approaches into the experimental test program. The content and planning for it are products of the workshop and the progress of the FRC testing. The idea is to provide an initiative to stimulate thinking for alternate approaches.

This report emphasizes magnetic confinement fusion. Some consideration should also be given to inertial confinement. That is considered, too, to be a product of the fusion workshop. Suggestions provided in Section 13 are applicable here.

Part of this task's effort should include investigations of enhancements like spin polarization and very high risk, high gain approaches having potential for significant improvements in specific power and other key space reactor performance parameters such as reduced space start power.

It is expected that as the FRC or other reactor test program and test data are obtained, there will be opportunities for the initial design to be improved. Hence, this task includes performance improvements, upgrades, and verifications of the redesign.

An experiment option could also be considered an appropriate task to implement under Task 2. An alternate non-Maxwellian approach has already been suggested there. Clearly an optional confinement approach is desirable to avoid a critical research program single failure point. The initiation of alternate, improved approaches are suggested early in the program due to the lead times involved with fusion research and due to the long operational mission firing durations.

Following the completion of this step, there is a need for technology maintenance. A level of effort task, one which is oriented towards specific flight applications, is included. Also, there is an effort identified to research large gigawatt propulsion systems and systems of the small several megawatt size.

14.3.1.2.7 TASK 10 (CONTINUED). SPACE PLASMA PHYSICS REACTOR ANALYSIS

The plasma analysis task supports reactor, thrust conversions, and electrical power conversion systems.

14.0 Recommended Space Fusion Strategy

At the conclusion of Step II the research phase of the program is complete.

Space Fusion Program: Developmental Phase

14.3.1.3 *THE THIRD STEP* – DEVELOP FLIGHT SPACECRAFT SYSTEMS – SPACECRAFT PROPULSION SYSTEM DESIGN AND DEVELOPMENT OF PROTOTYPE HARDWARE

Objectives: With the demonstration of net power from fusion (Step I) and the development and understanding of the requirements for a flight vehicle and the demonstrated feasibility of meeting those system requirements (Step II), the program can advance toward the design and development of the critical flight weight systems. The propulsion system must be demonstrated first since it is clearly the long lead critical item. Using the preprototype hardware from Step II, Step III of the program extends through the flight developmental phase, leading up to flight hardware qualification, in the fourth step. The hardware developed for this step is prototype flight hardware. A typical Phase A aerospace vehicle design option study will be accomplished in this step to provide configuration options for optimal performance determinations.

Step III contains five program tasks. It continues 3 tasks and initiates 2 new ones:

1. maintain a fuel supply (Task 4 continued)
2. show system feasibility (Task 7 continued)
3. technology maintenance (Task 9 continued)
4. conduct Phase A flight system study (Task 11)
5. conduct life and reliability demonstrations (Task 12).

Proposed Space Fusion Program Tasks for Step III

14.3.1.3.1 TASK 3 (CONTINUED). FUEL AVAILABILITY

A larger quantity of helium-3 can be anticipated for the conduct of this phase of the program. A supply should be in production.

14.3.1.3.2 TASK 7 (CONTINUED). FUSION SYSTEM

The SFP program advances to the design, manufacture and test of prototype flight hardware based upon the accomplishments in Steps I and II.

14.3.1.3.2.1 7.6. PROTOTYPE PROPULSION SYSTEM

The system test configuration should match the projected flight designs as closely as possible, and the operational test duty cycles and environments should reflect the anticipated flight use.

14.3.1.3.2.2 7.7. PROTOTYPE ELECTRICAL POWER SYSTEM

This task produces an advanced power system to investigate critical parameters for the successful, safe use of the space electrical power conversion system. It includes design, manufacturing, and test of the critical equipment necessary for the efficient production of electrical power by fusion energy in space including: heat rejection systems, efficiency, power level, controls, environmental effects on operational performance, system level analyses, and safety evaluations. Dual operational modes, i.e., propulsion and power combined, should be demonstrated.

14.3.1.3.3 TASK 9 (CONTINUED). TECHNOLOGY MAINTENANCE

Since this study is focused on flight programs and the implementation of fusion energy to enable science and exploration flight programs and for improving ground and flight safety, research endeavors beyond the initiation of the capability are not provided in this document. There is no further discussion concerning maintenance of the existing technology although such an activity is crucial to NASA's capability and therefore its operational success with using fusion energy. A task is provided to indicate the need in Step II and beyond. As one project within the task it could, for example, continue the confinement option task. Advancements in the current technology are also included as part of this task.

14.3.1.3.4 TASK 11. DESIGN A FLIGHT VEHICLE SYSTEM. PHASE A FLIGHT VEHICLE STUDY

This task, a typical Phase A flight system study to establish space flight configuration options, is accomplished at the end of of Step III. The Phase A study will provide a better definition of the system level requirements which the fusion system will be required to meet, vehicle design/performance options, and ultimately the configuration(s) to be built in Phases C-D.

14.3.1.3.5 TASK 12. ESTABLISH LONG RELIABLE LIFE

An initial phase of Task 12 will provide a plan for the optimal means to achieve and to demonstrate high reliability. One verification technique for consideration is to test a fusion reactor which has been designed to perform at lower energy levels for the solar system missions and to operate it during the conduct of those lower energy missions but at higher power output level which is more representative of the high power demanding missions. That would at least appear to stress test the design concept. The viability of that approach along with other options needs to be studied. The study can be at least in part accomplished in the plasma analysis task. Once the proof of concept has been obtained for predicting and verifying the overstress effects, the reactor design could be upgraded to achieve an interstellar capability with a high confidence factor of meeting mission success without requiring real time life testing. As an option, operational experience in off-nominal conditions may become key for the attainment of high reliability. Another option is hardware redundancy. The life and reliability task is continued in all subsequent steps of the program to indicate the need to continue an activity which acts in a building block fashion to produce reliability data.

The demanding high reliability mission goal, assumption number 7, can be best accomplished by a full understanding of: the space reactor's physics, its sensitivities to plasma instability, performance effects of variations and deviations for all critical reactor operational parameters, effects of process variables, and system operational failure modes. Involved here are analysis and testing. This program commences early, recognizing the length of firing durations involved with high energy, low thrust missions, and the newness of the technology. The program should progress from the simpler, less demanding to the more severe test parameters to establish a data base. Proceeding into a real time reliability life test verification program would be prohibitively long for missions like the Oort Cloud and Alpha Centauri missions. The alternative is to use the understandings of plasma physics and/or to conduct system and component overstress testing. The principal approach recommended by this study has been to advance the reactor to ignition without attaining a comprehensive understanding of plasma scaling laws. If that approach is successful, Task 12 would provide an in-depth understanding of the demonstrated FRC reactor design's operational characteristics. That understanding will serve as one key means to achieve engine and power system reliability.

14.0 Recommended Space Fusion Strategy

> **Space Fusion Program: Flight Operations Phase**

14.3.1.4 *THE FOURTH STEP* – FLIGHT VEHICLE PROGRAM ERA

Step IV contains two major programs, one to address the manufacture of flight vehicles, the other to continue and to advance space fusion technology. The technology support is needed to maintain a technical capability and to understand life limiting processes. The advancements are necessary to support the more difficult science missions and to support applications.

14.3.1.4.1 FLIGHT PROGRAM

Objective: Design, manufacture, and qualify a high quality space fusion flight vehicle.

A flight vehicle is the goal of Step IV using information from the space fusion program's studies, analyses, experiments, and testing and the preliminary vehicle design options produced by Step III. Included here are the typical aerospace flight vehicle program activities for Phases B through D: flight system design, manufacture, qualification testing, integration, vehicle assembly, flight checkout, and flight test. From the results of the mission analysis of this study, two reactor sizes are of interest, a 250 MW and a 80 MW (jet power output). The 250-300 MW size is considered appropriate for the Manned Mars Mission while the 50-80 MW size captures the science missions. The 250 MW reactor, applied first to an unmanned science mission, qualifies the manned flight. That flight is considered the flight test demonstration qualification program, one that "man-rates" the system. That approach is discussed further in Step V.

The list below summarizes the activities needed to complete this phase of Step IV.

1. A continuation of Task II, here a flight fusion propulsion system is built and qualified. One option is to conduct a portion of the qualification test program in space in LEO. It should be initially flight tested in the operational environment for checkout purposes if the qualification testing is not partially accomplished in space.

2. Phase B. This is the standard Phase B design selection and flight vehicle design process used on NASA's aerospace programs.

3. Phase C. Flight hardware is built and qualified for space flight. Two flight vehicles are initially built to support the suggested mission scenario: a manned vehicle, adaptable to an unmanned flight and a smaller science vehicle. Economic trades, after more definitive characterizations have been obtained, will validate the approach or indicate alternate approaches. At the present time the above is presented as a program strawman approach.

4. Phase D. The flight vehicle is assembled, made ready for launch, placed into orbit, and flight tested in space. Unlike the small chemical propulsion systems, orbital flight checkout and testing are particularly important to fusion systems. That is attributed to the size of the equipment involved which at this time is believed to make good, representative space simulators very expensive. There is also the safety aspects where the neutron emissions will activate materials used in the vehicle. In space the hazard can be controlled through isolation and the activated material simply discarded by a disposal mission at the conclusion of the test program. The vehicle is sent into a safe orbit around the sun.

5. Task 4. The program to produce ^{3}He and deuterium to support the test requirements and to prepare for the flight operational program is continued.

14.3.1.4.2 TECHNOLOGY

14.3.1.4.2.1 TASK 12. LIFE DURATION AND RELIABILITY

The data base and and long duration testing are important to future missions, and therefore both are continued. This program will identify product improvements and verify the reuse capability.

14.3.1.4.2.2 TASKS 7 & 9. TECHNOLOGY MAINTENANCE

This program continues to pursue technology understandings and provides a means for the development of advanced technological improvements and their incorporation into flight systems.

14.3.1.5 *THE FIFTH STEP* – FUSION FLIGHT MISSION ERA

The program content at this point contains a three tier activity:

(A) flight operations,

(B) manufacturing/production to expand the fleet size and capability, and

(C) technology enhancements.

14.3.1.5.1 (A). SPACE FUSION FLIGHT OPERATIONS

The major goal and interest in fusion energy here is ultimately its flight application for the safe conduct of missions that can otherwise not be accomplished. It is particularly suited for enhancing the safety of the Manned Mars Mission as no other energy source and for ultimately making the possible the settlement of Mars. The first space application of fusion energy is conservatively suggested as an unmanned mission for the conduct of science, such as, for example, a sample return mission from Jupiter. A suggested flight program mission scenario is provided in Fig. 14.4.

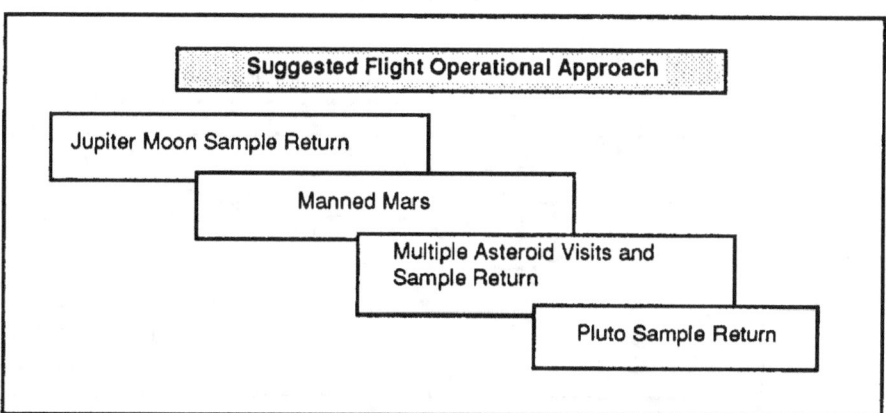

Fig. 14.4. Initial fusion missions.

A reusable reactor designed to power a fusion rocket engine for an unmanned mission at the Manned Mars Mission power level will provide key programmatic advantages by providing operational experience, space environmental exposures, and longer flight mission durations. Reuse is considered as the appropriate course to take for flight qualification. The flight duration of a Jupiter mission (~1.7 years) will more than adequately qualify the system for a manned flight to Mars (~0.5 years). That approach offers the most expeditious method to allow the earliest possible use of fusion energy for manned space flight while obtaining science data. Next, following the Manned Mars Mission, multiple asteroid science mission visits are anticipated as a good application for fusion; it was included as the third mission in this step. It is an intermediate duration mission (~4 years) to build-up flight operational experience over longer duration missions. A science sample return

14.0 Recommended Space Fusion Strategy

mission to Pluto could be subsequently undertaken. That mission would otherwise require a significantly longer flight time for non-fusion powered flight systems to accomplish and is out of reach of current propulsion technology. It is the last of the planets to be explored.

Those missions are typical, i.e., for planning purposes, and to observe the potential for the application of high energy to space flight. The science mission workshop and subsequent reviews of science priorities among the science community would obviously be the best source to determine the most appropriate missions to be accomplished. The suggested operational approach for bringing the fusion engine system into the NASA space transportation infrastructure is illustrated in Fig. 14.5.

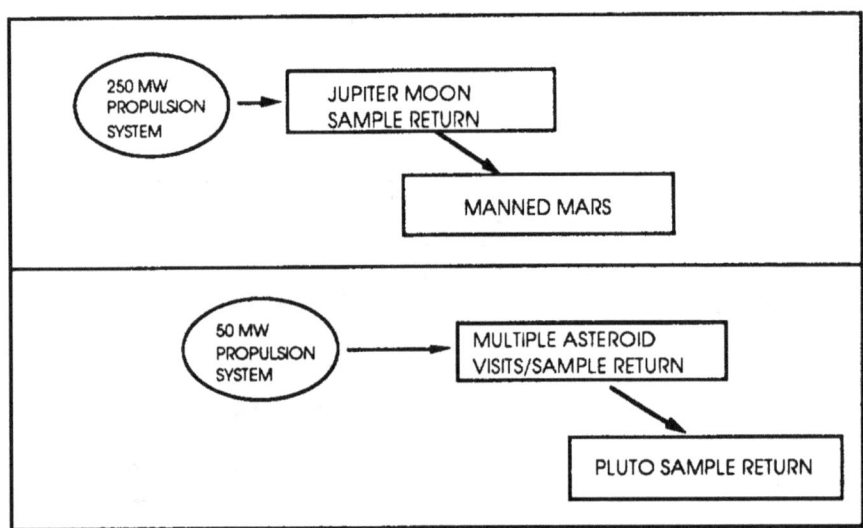

Fig. 14.5. Initial fusion propulsion space infrastructure.

14.3.1.5.2 (B). MANUFACTURING

In the operational phase the means for the production of fusion fuel is continued. Also, the manufacture of the 50-80 MW vehicle is completed during this era. The fleet size is increased to support the mission requirements.

14.3.1.5.3 (C). SPACE FUSION TECHNOLOGY

A need exists to continue the technology under a separate program to conduct research and development, to maintain technology, to build up the space fusion fleet, and to design new advanced, reduced mass systems. This is a distinct program activity, but one in the fusion operational era, which is identified separately to maintain a clear separation between flight operations and technology development. These two programs, operations and technology, should be budgeted separately and managed separately. This task develops the large gigawatt and small size reactors, for example. The technology need is identified rather than an extensive R & D program defined at this early stage. Production of the fuel, Task 4, continues.

14.3.2 PROGRAM DEFINITION AND SCHEDULES

From a user's perspective, the main interest is in understanding when fusion can be made available. The objective herein has been to point out the immediate need for fusion energy for space and to point out the differences in the two applications, ground and flight, which consequently warrant the initiation of a separate space fusion energy conversion program. It appeared worthwhile to make educated guesses to provide an indication of the availability of fusion energy in response to schedule questions from NASA staff. An educated guess is the limit to which one should interpret the following schedules. They serve useful purposes, however. They show the course of events that need to be undertaken. They point to the length of time that will be required, thereby reinforcing the point that NASA needs to initiate a fusion program now. Just imagine the impact to NASA if we could today conduct our space flight operational planning and its related budget planning to reflect the ability to provide for the Manned Mars Mission's energy requirements based upon only one Shuttle launch! That single launch would include the propulsive energy to transmit the flight vehicle, the landing vehicle and the crew to Mars and to return them safely to Earth, more safely than with the use of chemical or fission propulsion. Actually, when the in-situ production of propellants becomes fully operational, and a reusable Martian lander-ascent vehicle is developed, the trans Earth–Mars transport spacecraft can be used to more fully devote its performance to delivering habitability and life quality masses to the Martian settlers rather than transporting propulsion system and propellant mass.

The basic tasks that must be accomplished to obtain that high energy capability and the anticipated set of events that have to occur for flight demonstration are provided below–identified by program step. Hence, from the events in those steps, we can derive an indication of the schedule and the budget level to support the development of fusion energy for space.

14.3.2.1 OVERALL PROGRAM SCHEDULE

The overall integrated developmental, qualification, and early flight program for a fusion system is presented in a 40-year plan in Fig. 14.6.

14.0 Recommended Space Fusion Strategy

To simultaneously accomplish the other planetary science missions that have been identified in this report, a space fusion fleet is suggested. The fleet would be comprised of additional flight fusion propulsion systems and vehicles of the class mentioned – 250 MW and 80 MW. In reviewing the flight times for science mission objectives, the average solar system mission time is ~5 years. A fleet of three fusion powered vehicles in the <80 MW range provides an excellent program capability for capture of solar system science. For manned missions two of the 250 MW sized vehicles would permit four 0.5 year manned trips between Earth and Mars annually. This small fleet size is, of course, based upon a reuse capability, assumption 7.

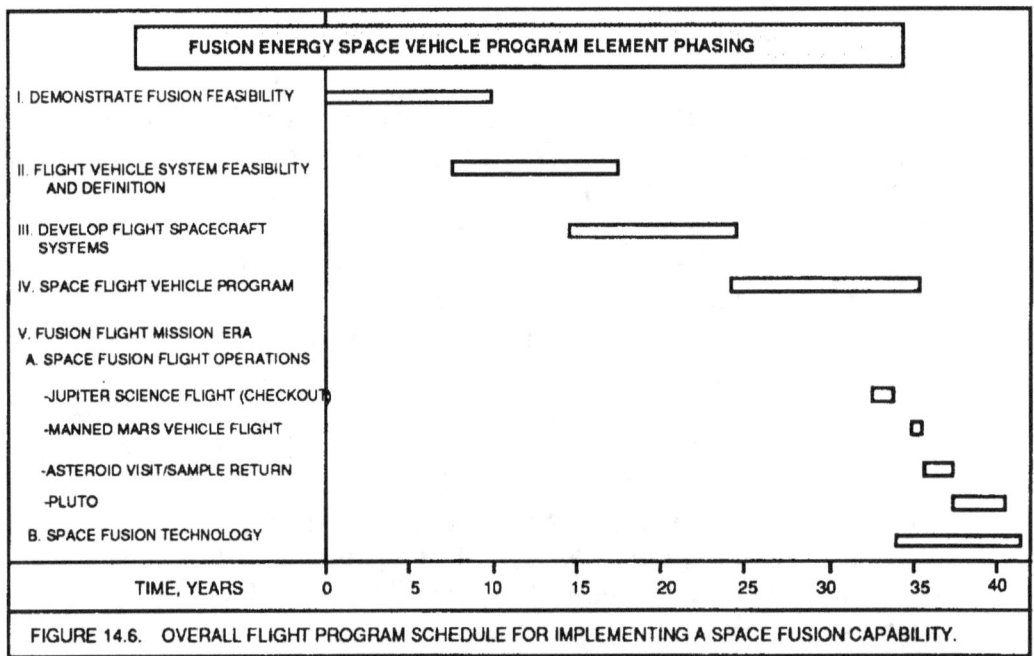

FIGURE 14.6. OVERALL FLIGHT PROGRAM SCHEDULE FOR IMPLEMENTING A SPACE FUSION CAPABILITY.

The schedule above, which shows each of the program steps, is in one key sense, optimistic. It assumes that mother nature will cooperate and not produce any new unknowns with plasma instabilities, Step I, and that the above design parameters for the reactor and the flight vehicle systems are achievable, Step II. The flight vehicle design is anticipated to be straightforward, but since this is a new energy system, a longer pre-design and developmental period is provided. The entire success or failure of this program will depend upon:

1. NASA's commitment towards the development of fusion energy,

14–28

14.0 Recommended Space Fusion Strategy

    2. acts of nature, particularly unknowns, and our ability to solve them,

    3. the overall management approach for the program.

Although the 30-35 year time to a space flight operational status may appear long, approximately the length of NASA's life at the time of this report, it is not extraordinarily so considering the magnitude of the tasks. The Apollo Program required 10 years to land man on the moon using relatively simple chemical propulsion technology, a type of energy conversion system, that had been in development since ~1910. The success therein can be attributed a very strong commitment by the nation and NASA, plus a strong internal propulsion and vehicle technical management team that had a significant research, hands-on background in their areas of expertise at MSFC and JSC. The Shuttle program had its origins in the mid-1960's with the first flight occurring approximately 15 years later. The first flight in the fusion program is shown to occur in 30-35 years, based on the key assumption that the plasma confinement and specific power/specific impulse research progresses as planned.

Even with the first demonstration of breakeven by the tokamak, there is reason to expect that the current approach of the space program relying on a DOE research program that develops commercial electrical power production will take a much longer period than that shown in this report. For space flight propulsion use, it will be an infinitely long period since the tokamak lacks the high specific power needs of space propulsion. An alternative program must be explored for space. In an era of a declining DOE fusion program budget, and with DOE lacking a space mission charter, NASA cannot expect DOE to fund NASA's space propulsion systems. So, while the 30-35 year time frame is long, it is better than the option of waiting. The job will be difficult; and even with nature's cooperation, it will be time consuming, again the rationale to commence with development now.

The optimized program outlined herein stresses the importance of fusion development by limiting funding at the beginning to the task of highest priority, namely, the demonstration of net power fusion energy followed by the development of the essential ancillary flight systems. The following shows a suggested schedule for each of the program Steps and the content of each Step. To compress the total schedule there is overlap in the scheduling of Steps. The practicality of the overlaps is a function of the research progress and the NASA funding commitment. The 12 tasks in the prior section are more fully described below.

14.0 Recommended Space Fusion Strategy

14.3.2.2 STEP I — FUSION FEASIBILITY DEMONSTRATION

Demonstration of the feasibility of fusion energy is the single most important task. The strategy presented is to design the preferred experimental plasma confinement approach, namely, those having high β such as the FRC, to D-^{3}He burn parameters. The rationale for suggesting this approach is based in part upon the difficulty experienced in understanding plasma transport. Hence, until burning is proven on a space applicable reactor design, there will always be a question regarding the use of fusion in space.

The second most important is the question of specific power. That question cannot be answered until fusion has been demonstrated and a number of key system related issues have been resolved. However, once we have demonstrated fusion, sufficient parametric design information should be available where a first approximation can be obtained on the viability of achieving the specific powers being considered in this report, particularly after the system work has been accomplished in Step II.

The phasing of tasks to demonstrate fusion feasibility is identified in Fig. 14.7.

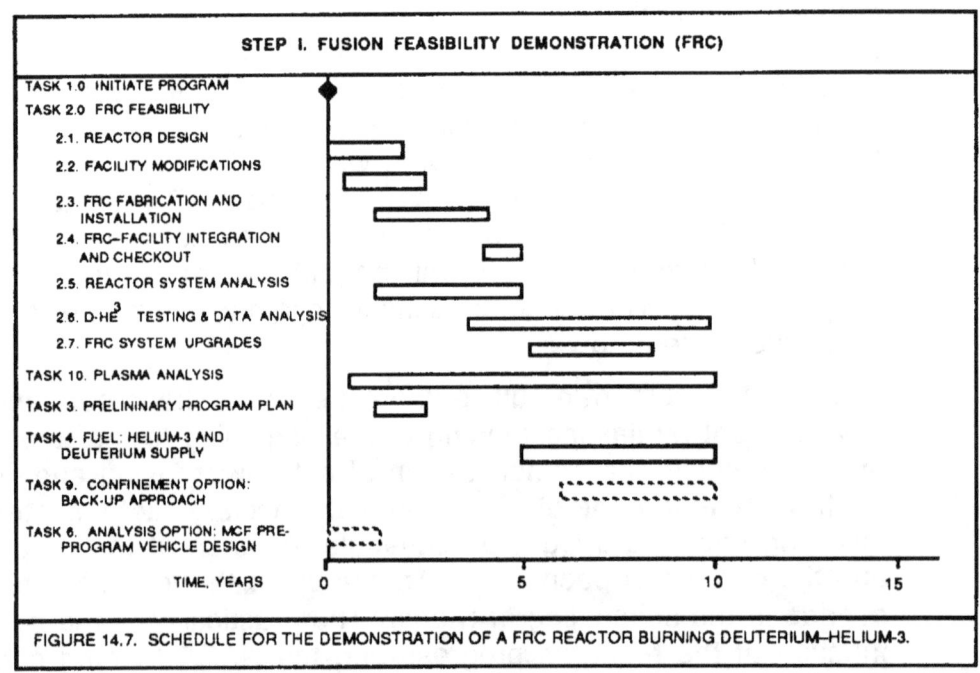

FIGURE 14.7. SCHEDULE FOR THE DEMONSTRATION OF A FRC REACTOR BURNING DEUTERIUM–HELIUM-3.

14.0 Recommended Space Fusion Strategy

14.3.2.2.1 TASK 2. FRC FEASIBILITY

This section uses the FRC as illustrative of the process required. While the discussion focuses on the FRC the timing and steps are illustrative of the work ahead if another approach should be deemed more appropriate. There is no doubt that the preferred fuel of choice is D-^{3}He, so this study suggests that the plasma confinement design proceed directly to one that burns D-^{3}He with recognition that the selected fuel is more difficult to ignite than the mainline program's D-T.

14.3.2.2.1.1 TASK 2.1. REACTOR DESIGN

It is anticipated that the D-^{3}He advantages warrant this approach and that the engineering features will allow for a more rapid implementation of a flight reactor and more efficient propulsion system for space. That opinion is reflected by many of the researchers in this field. The schedule assumes that the burning of D-^{3}He can be accomplished by current technology magnets ~5 T field strength coil without relying upon technological breakthroughs in magnet technology. The plasma is to be ignited by neutral beam injection. After ignition the plasma temperature is further increased by the reactor's internally produced energy to provide the desired level of net power. It is assumed that neutral beam injection will provide an adequate stability margin. Thus, as a goal in Step I, when the product of fuel confinement time and fuel density ($n\tau$ product) is sufficiently large (for D-^{3}He, $n\tau \geq 2\times10^{15}$ cm^{-3} sec where $T_i = 40$ keV, for example), the charged fusion product heating can balance plasma energy losses from conduction, convection, and radiation as bremsstrahlung and synchrotron radiation. When this condition occurs, the plasma is said to be ignited; the burn can proceed without further input of energy from external auxiliary heating systems, and net power can be achieved by designing a reactor of sufficient size to produce excess power above that critical level.

Once net power has been demonstrated, the program should pursue steady state burning for the new design. The current FRC's only operate in the pulse mode. Steady state burning research is also necessary to improve upon fuel burning efficiency.

The time allowed for FRC design is one year. That is tight, but with a sufficiently high level of funding it is considered possible. The FRC is a more simple reactor. Additional time to support the FRC build completion and integration is allowed.

14.3.2.2.1.2 TASK 2.2. FACILITY MODIFICATIONS

A new FRC design, one producing net power will require a safe location of the facility to accommodate the expected neutron flux. It is assumed that the more economical and expedient approach, at least for the near term will be to use an existing facility, appropriately modified for the FRC.

The installation design and planning are commenced early to expedite the facility availability. A one year redesign period after a data base has been generated is allowed to design the modifications, to construct the hardware, and to install the mods into an existing facility.

14.3.2.2.1.3 TASK 2.3. FRC FABRICATION AND INSTALLATION

This allows three years to construct the FRC and to install it into the test facility. An additional 2 years for upgrades is included.

14.3.2.2.1.4 TASK 2.4. FRC-TEST FACILITY INTEGRATION AND CHECKOUT

Three years are allowed to integrate the FRC with the test facility and to debug the operation.

14.3.2.2.1.5 TASK 2.5. REACTOR SYSTEM ANALYSIS

An initial analysis and design project supports the integration of the reactor into the facility. Analysis is continued during testing under Tasks 2.6 and 10.

14.3.2.2.1.6 TASK 2.6. D-^{3}HE TESTING AND DATA ANALYSIS

The testing would initially proceed with hydrogen as the test fluids for checkout of the FRC design parameters and the integration of the FRC with the facility. Next it would advance to D-^{3}He. A series of experimental steps are envisioned that increase the beam power and magnetic field strength to net power from D-^{3}He. It will also be important to establish early whether the reduced neutron flux expected from D-^{3}He will be sufficiently low to permit reactor life durations of the length noted for missions and their reuse. That flux will play a key role in determining the nature of the course that the program will take subsequently.

Based upon the assumption that no major surprises are experienced, this task would continue testing to obtain a data base for design information leading to understanding of life degradation factors.

14.3.2.2.1.7 TASK 2.7. FRC SYSTEM UPGRADES

An allowance is made for the fact that this is a new technology research program and that lessons learned will soon fall out. This includes the redesign and possible need for maintenance. The manufacture and installation of the modifications are accomplished in Task 2.

14.3.2.2.2 TASK 10. PLASMA ANALYSIS

While the approach is to proceed to design the reactor to net power without understanding the plasma characteristics in a minimal number of sequentially phased experiments, there is an important requirement to fully establish plasma dynamic characteristics ultimately as testing progresses, and therefore an effort is identified for plasma analysis. The plasma characterization will be important for extrapolation of designs and for obtaining high levels of system safety and reliability. The plasma analysis task is commenced early to permit specialists to participate in the design phase and in the facility and diagnostics work.

14.3.2.2.3 TASK 3. PRELIMINARY PROGRAM PLAN

The development of an Initial Space Fusion Program Plan is suggested. It can be developed from a workshop of fusion specialists to assist in the establishment of a FRC program. That early workshop would also serve for preliminary planning for a more comprehensive workshop that would examine alternatives and would produce refinements of the initial plan. The early workshop could be a joint meeting with users, also in preparation for a more comprehensive high energy mission workshop to be held after this program has advanced to demonstrate net power.

14.3.2.2.4 TASK 4. FUEL, ^{3}HE AND DEUTERIUM, SUPPLY ANALYSIS

The planning for the fuel (^{3}He and deuterium) production task is commenced in this step since it is a critical, long lead technology item.

14.3.2.2.5 TASK 9. CONFINEMENT OPTION: BACK-UP APPROACH

Depending upon a new start funding level, this may be initiated in either Step I or II. If there is sufficient funding to initiate it in Step I, then the workshop planning would be accordingly adjusted.

14.3.2.2.6 TASK 6. ANALYSIS OPTION: MCF PRE-PROGRAM VEHICLE DESIGN STUDY

One of the difficulties with providing better performance numbers for fusion system specific power capabilities is due to the lack of good MCF data and analytical programs. Because there is tremendous latitude with projections for future performance capabilities, it is difficult to attach confidence values to the projections. Hence, the approach taken here is to demonstrate fusion as rapidly as possible in a high risk technical approach.

An option is to make performance assumptions and perform a detailed study prior to committing to the program. It should be noted, however, that the mass sensitivities and variability unknowns are sufficiently great that considerable time would be spent and study costs expended with the result being uncertainty. It does not answer the question, "What does it take to make fusion work for space?"

14.3.2.3 STEP II — FLIGHT VEHICLE SYSTEM DEFINITION

Next to the importance of the demonstration of fusion energy conversion in Step I, the second most important effort is to demonstrate the flight fusion system feasibility aspects, that is, to prove that a flight propulsion-vehicle system of the size of interest to the space program can be constructed which has the desired specific power of 1 to 10 kW/kg. Six system related tasks to demonstrate that performance capability are conducted in Step II, Fig. 14.8. Thus, in Step II the necessary analyses, trade studies, research, and testing tasks are performed to demonstrate the feasibility of fusion system technology.

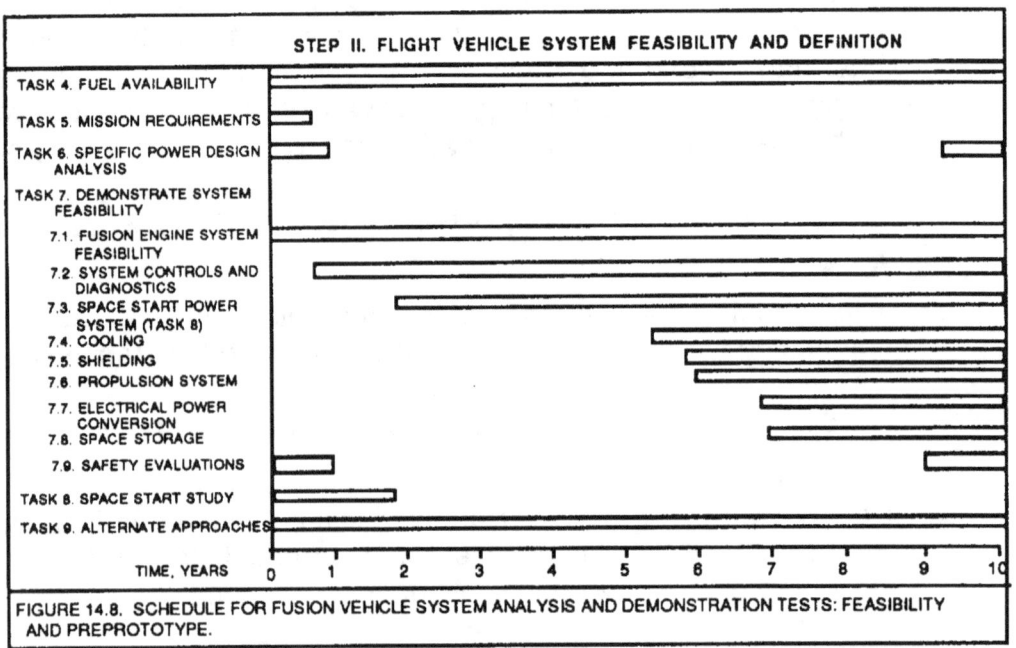

FIGURE 14.8. SCHEDULE FOR FUSION VEHICLE SYSTEM ANALYSIS AND DEMONSTRATION TESTS: FEASIBILITY AND PREPROTOTYPE.

The key subjects to be investigated for converting fusion energy into propulsive power, Task 7, included:

1. conversion of the plasma to thrust directly while maintaining a high level of reactor plasma stability,
2. variation and measurement of the engine's specific impulse,
3. the capability to throttle the engine's thrust level such that the trajectory optimizations can be performed,
4. determination of plasma stability characteristics, and
5. understanding the fundamental parameters leading to instability.

Means to control the instabilities and the design's sensitivity to perturbations of the plasma stability inducing mechanisms shall be determined as part of the task. Techniques for cooling the engine without sacrificing reactor performance and system mass, while maintaining high reliability and long life are determined and tested in this task. It will better define the reactor's and the ancillary supporting equipment mass. The flight control system necessary for the safe and optimal performance of a space flight reactor shall be identified and testing commenced to meet new reactor control requirements for space applications and to validate existing instrumentation designs for flight use. Upon completion of the aforementioned tasks, a detailed design of the reactor and propulsion and power system shall be conducted to accurately establish a preliminary specific power design and to identify areas for research for producing improvements.

In Step II, then, the following specific tasks are indicated as essential to verify system feasibility:

14.3.2.3.1 TASK 4. FUEL AVAILABILITY

The purpose in this task is to initiate the necessary action to make a source of ^{3}He and deuterium available for use as required to support the test and flight programs. This schedule continues from the Step I planning and demonstration of process. A production process is selected and work commenced to produce fuel. A decision is made for lunar and/or terrestrial production based upon the results of the Step I task. By the end of Step II or early in Step III the fuel plants should be in production.

14.3.2.3.2 TASK 5. DEVELOP HIGH ENERGY MISSION REQUIREMENTS

One year is allowed for the conduct of a high energy space exploration and science mission workshop including the time necessary for the preparation of a document of the mission requirements from the proceedings.

14.3.2.3.3 TASK 6 SPECIFIC POWER

The capability of fusion powered flight vehicles to deliver specific powers ranging between 1 kW/kg to 10 kW/kg needs to be analyzed and demonstrated. Upon completion of defining, designing, and demonstrating the key system mass components, a preliminary analysis of a MCF system shall be made to establish the specific power potential and to identify the subjects that need further research. At the completion of Step II an updated analysis is made to more accurately quantify the capability.

14.0 Recommended Space Fusion Strategy

14.3.2.3.4 TASK 7. DEMONSTRATE FUSION SYSTEM FEASIBILITY

Task 7.1. Fusion engine system feasibility

At the top of the technology development list is fusion propulsion, namely, the means for efficiently converting plasma energy into vehicle kinetic motion, controlled in a manner to allow trajectory optimization while maintaining the reactor at the desired high level of performance. This task accomplishes the analysis, design, and testing to demonstrate the propulsion capability up through a prototype hardware level. This task demonstrates the capability to meet the fusion performance requirements. Means shall be identified for protecting the magnet and critical vehicle components from the neutron flux, and from the effects of heating, radiation, electrostatic charges, EMI, etc. This activity includes testing of an engine system including flow control devices, reactor power, thrust conversion, thrust and specific impulse control, thermal management, etc. in an anticipated operational flight environment.

Fusion rocket engine system activities are shown in Fig. 14.9. Four tasks are presented:

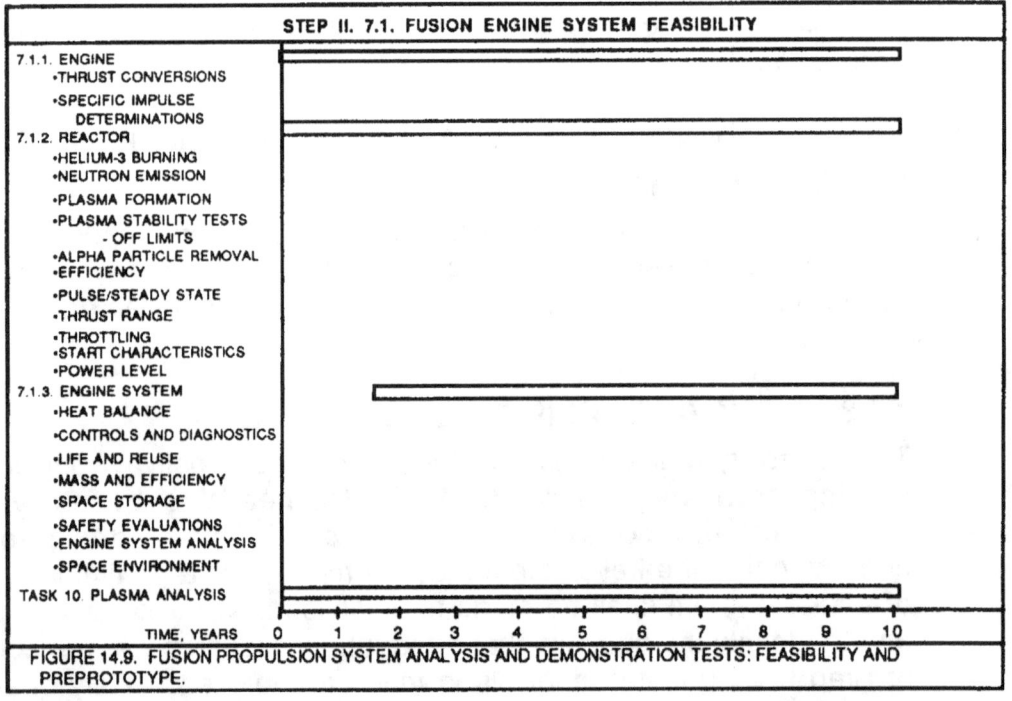

FIGURE 14.9. FUSION PROPULSION SYSTEM ANALYSIS AND DEMONSTRATION TESTS: FEASIBILITY AND PREPROTOTYPE.

Task 7.1.1. Engine

Engine-related work is performed in which the program demonstrates that the reactor's plasma energy can be diverted to thrust directly and that the means for varying the thrust can be achieved. Another objective is to obtain propulsion performance data.

Task 7.1.2. Reactor

Reactor-related activity continues the work which was commenced in Step I in an updated design that will be more flight-like. A list of test program objectives is provided in Fig. 14.9. One of the major program assumptions, the reactor power output (Task 7) capability of the FRC to deliver a range of jet power from approximately 50 MW to 10 GW for a space reactor design, is to be analyzed. The jet power level for the Manned Mars Mission, identified as approximately 250-300 MW from the results of this study, is to receive the program's initial emphasis.

Task 7.1.3. Engine system

The engine system comprises the reactor (7.1.2), thrust conversion means (7.1.1), and other ancillary systems necessary to provide a propulsion device that will produce controllable thrust. The system requirements to be either demonstrated for feasibility by test or by analysis from extrapolations of test data are:

1. Thrust level of 1 N to 50 kN,
2. Efficient throttling of thrust,
3. Burn durations ranging from 4 months up to 7 years and extrapolated to 50 years per mission,
4. Reusable engines and vehicles,
5. Specific impulse of 5,000 to 1,000,000 seconds ultimately.

14.3.2.3.5 TASK 10. PLASMA ANALYSIS IS CONTINUED IN ORDER TO CHARACTERIZE THE PLASMA UNDER OPERATIONAL MODES FOR SPACE FLIGHT PROPULSION, POWER, AND SYSTEM LIFE.

Task 7.2. System controls and diagnostics

The automated means to control the reactor and the propulsion systems shall be accomplished in this task. Included are thrust, specific impulse, thermal control, redline limits, etc.

14.0 Recommended Space Fusion Strategy

Task 7.3. Space engine start/remote restart capability (Task 8)

Following the Task 8 investigations of options, space start/restart system trade studies are to be performed. Designs of components are performed, hardware built, and tests conducted on the options studied to prove concepts and assumptions. To complete the energy storage and the system investigation, prototype systems are to be designed, built, and demonstrated in the fusion propulsion system.

Task 7.4. Cooling

The means for maintaining a cooled helium-3 tank and other cryogenic fluids as well as the means to cool the flight vehicle shall be identified. The cooling capability for propulsion and power is included. Design trades are to be performed, and new concepts required to conduct the reactor cooling are tested to verify performance.

Task 7.5. Shielding

Shielding is provided to protect the magnet, engine, and vehicle systems from the effects of neutrons. Concepts are evaluated and design trades performed to minimize the vehicle mass and then tested to verify performance assumptions.

Task 7.6. Propulsion system

The key components are assembled into a configuration resembling that anticipated as a flight configuration. Testing is accomplished in a simulated space environment. This includes the engine system, propellant and fuel feed and storage, thermal shielding, start system, and controls.

Task 7.7. Electrical power conversion

The electrical power task provides for trade studies and power system analyses. In this task testing is performed to demonstrate a high level of efficiency with a flight-weight, direct space power converter. The design is to be integrated with a FRC propulsion design. The capability of the reactor to produce an efficient, direct conversion of fusion energy to electrical power at an output up to 20 MW shall be shown.

Task 7.8. Space storage

The capability of the flight system to remain in a space operational environment shall be evaluated and tested.

Task 7.9. Safety evaluations

A study evaluation of the ground and flight safety aspects of the operational use of a space vehicle shall be analyzed. Safety related tests, where identified, shall be conducted.

14.3.2.3.6 TASK 8. SPACE START/REMOTE RESTART STUDY

This capability is of sufficient concern that a special task was identified to commence early a focused evaluation of engine space start system requirements, power level, options, technology status, and trade studies in preparation for the experiments and demonstrations that are to be conducted under the system tasks in Task 7.

14.3.2.3.7 TASK 9. ALTERNATE APPROACHES TO THE FRC

Alternate approaches are to be undertaken based upon the workshop proceedings and, later, based upon the progress of the FRC.

14.0 Recommended Space Fusion Strategy

14.3.2.4 STEP III – DEVELOPMENT, FLIGHT SPACECRAFT SYSTEM

Five tasks are to be performed in Step III, Fig. 14.10.

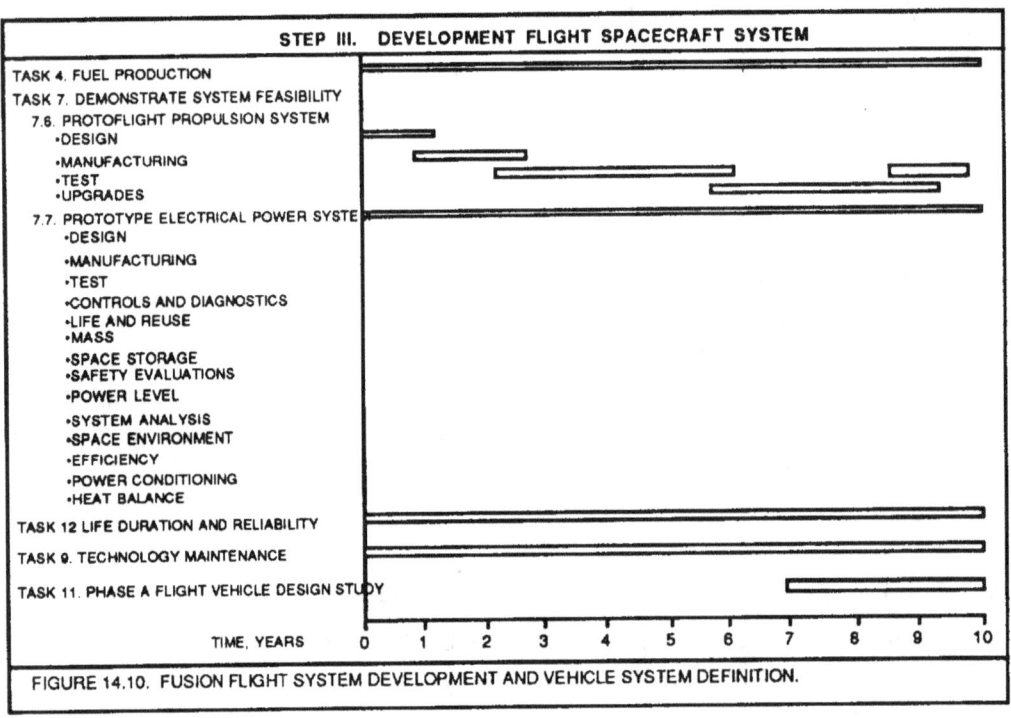

FIGURE 14.10. FUSION FLIGHT SYSTEM DEVELOPMENT AND VEHICLE SYSTEM DEFINITION.

14.3.2.4.1 TASK 4. FUEL AVAILABILITY

Fusion fuel is produced in this step.

14.3.2.4.2 TASK 7. DEMONSTRATE FUSION SYSTEM FEASIBILITY

Task 7.6. Protoflight Propulsion System

Propulsion development traditionally leads the development of all other hardware in space vehicle programs. In the case of the Shuttle, the contract for the main engine preceded the award of the vehicle by several years. Competitive engine contracts had been awarded even preceding that date. The same situation is particularly applicable in this case. The preprototype hardware from the previous steps is designed into flight prototype configurations.

Task 7.7. Prototype Electrical Power System

A prototype electrical power system is developed similarly as the propulsion system was developed in Step II.

14.3.2.4.3 TASK 12. LIFE DURATION AND RELIABILITY

Life duration and reliability analysis, including plasma characterizations and understandings, and testing are accomplished.

14.3.2.4.4 TASK 9. TECHNOLOGY MAINTENANCE

The confinement options technology initiated earlier is continued under this task. Also, additional development programs are performed here to characterize the technologies being developed for flight use.

14.3.2.4.5 TASK 11. PHASE A FLIGHT VEHICLE DESIGN STUDY

Phase A flight vehicle design studies are initiated in this task. These studies will explore flight vehicle design options conducted by aerospace contractors. Long lead critical hardware procurements are commenced, in this case the propulsion system.

14.3.2.5 STEP IV – FLIGHT SPACECRAFT SYSTEM

This Step comprises the standard space activity for a Phase B through D flight program. A final design is selected and the hardware is qualified for flight, built, and space flight tested as shown by Fig. 14.11.

Two sizes of propulsion systems are produced for flight, 250 MW and 80 MW, the former for manned spaceflight, the latter for science missions. The 80 MW system is delivered at the time of the Step V flight program.

The technology program contains two tasks – one to demonstrate life and reliability, the other to perform technology maintenance functions. It is important to continue the reliability analysis and testing to develop a data base for the long duration missions as well as to assist the flight programs.

14.0 Recommended Space Fusion Strategy

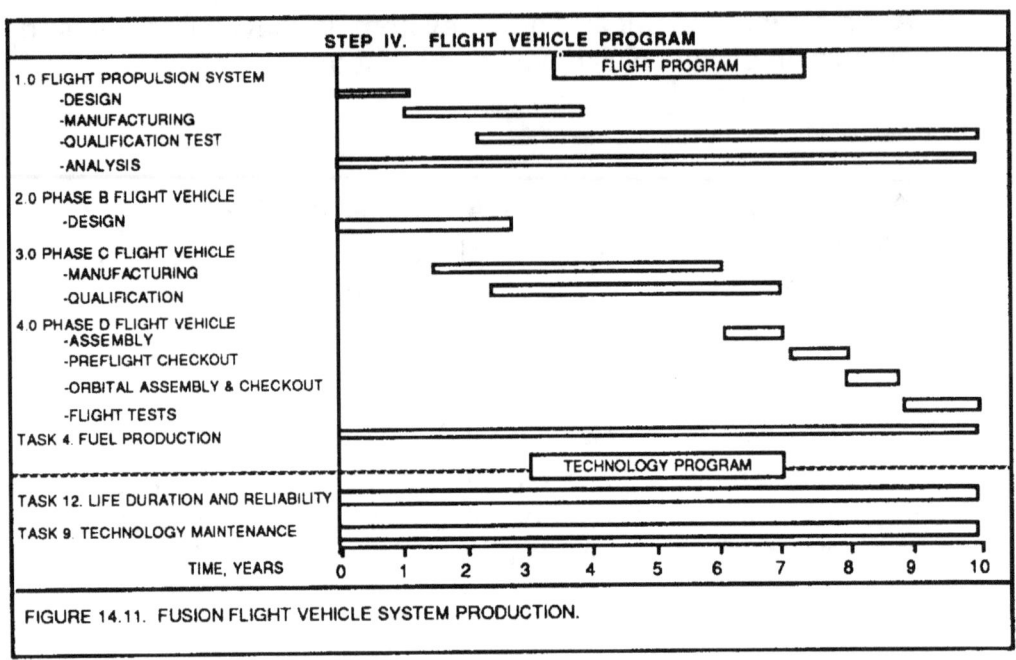

FIGURE 14.11. FUSION FLIGHT VEHICLE SYSTEM PRODUCTION.

14.3.2.6 STEP V – FUSION FLIGHT MISSION ERA

By Step, V fusion propulsion and power have been proven and are available for use in the space transportation infrastructure. There are now two major program elements, space operations – one to fly the current technology spacecraft, and program development – to advance the technology to accomplish other more technically demanding missions. Those missions identified by this study indicate fusion reactors on both sides of the power scale, small/compact which are necessary to power vehicles for launch to Earth orbit, and the other, reactors sufficiently great in jet power output to perform stellar missions.

14.3.2.6.1 SPACE FUSION FLIGHT OPERATIONS

As a mission scenario, typical flight programs are presented in Fig. 14.12. The operational phase will ultimately be determined by events that are too difficult to project at this time, most likely depending upon many factors both scientific and technological as well as political. In the meantime the scenario forwarded offers a view toward how such a system might be brought on line and the timing thereof.

14-42

14.0 Recommended Space Fusion Strategy

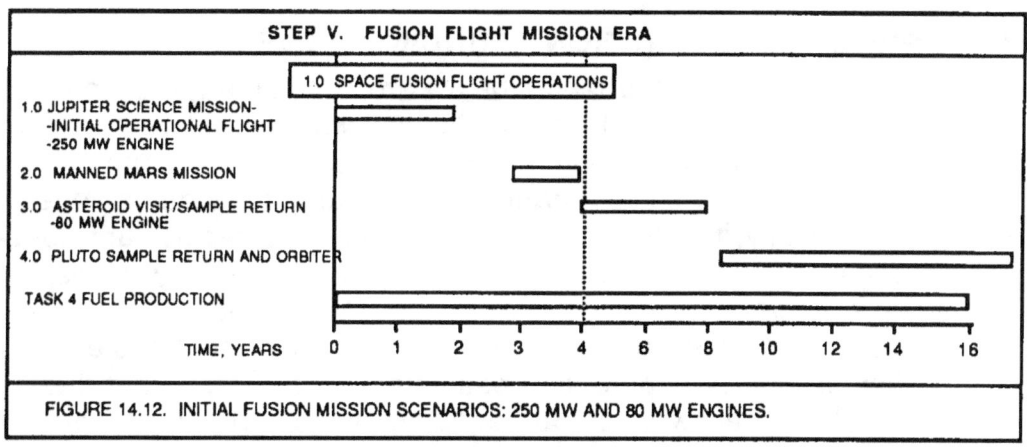

FIGURE 14.12. INITIAL FUSION MISSION SCENARIOS: 250 MW AND 80 MW ENGINES.

Under the flight scenarios shown, the Jupiter mission qualifies the system for manned flight to Mars. Production of the science vehicle flight fusion system follows the higher powered manned system. The manned research program activity assists the unmanned program by developing the prototype science mission vehicle class. Note that the initial vehicle production is expected to extend beyond the initial operational flight for a manned vehicle to accommodate the unmanned flight system's production.

14.3.2.6.2 SPACE FUSION TECHNOLOGY

Program activity is indicated in Fig. 14.13. This work advances the technology to the more difficult fusion applications, identified as Task 9. The life duration and reliability testing efforts, Task 12, are continued.

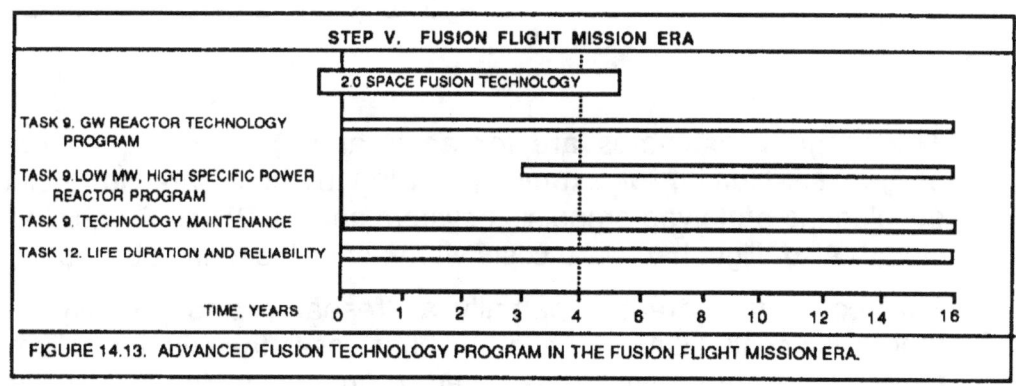

FIGURE 14.13. ADVANCED FUSION TECHNOLOGY PROGRAM IN THE FUSION FLIGHT MISSION ERA.

14-43

14.0 Recommended Space Fusion Strategy

14.4 PROGRAM DEFINITION: FUNDING LEVEL

The costs listed in the tables below are considered to be no more than very rough estimates. That was the best which could be presented since many of these tasks require further research. Also, no FRC reactor design studies have made that would provide a cost guide. Consequently, the costs associated with a FRC reactor are not defined, nor are the various ancillary systems, where in many cases, research may also be needed. The last year (1990) funding for the two FRC experiments was approximately $5M. The budget tables follow the schedule charts above. The estimated values represent 1990 dollars without any allowances for escalation. There is no reserve, and overhead estimates are not included.

The program budget summary per step is provided in Table 14-4.

Budget Summary:

TABLE 14-4. Program **R & D** budget.

Step	1st year of step	2nd year of step	3rd year of step	4th year of step	5th year of step	6th year of step	7th year of step	8th year of step	9th year of step	10th year of step	TOTALS
I	$12.5	$169.1	$231.3	$166.3	$79.1	$91.0	$116.0	$61.0	$96.0	$86.0	$1,108.
II	$29.3	$39.2	$105.8	$97.0	$97.9	$94.5	$98.3	$94.2	$94.2	$89.4	$840
III	$61.5	$108.5	$187.8	$164.4	$115.4	$83.4	$143.4	$133.4	$133.4	$123.7	$1,255
IV	$50.0	$50.0	$50.0	$50.0	$50.0	$50.0	$50.0	$50.0	$50.0	$50.0	$500
V (1)	$0.0	$0.0	$0.0	$0.0	$0.0	$0.0	$0.0	$0.0	$0.0	$0.0	$0.0
V (2)	$250.0	$350.0	$450.0	$250.0	$210.0	$210.0	$210.0	$210.0	$210.0	$210.0	$2,560
											$6,263

The program estimates are for an R and D budget only, not flight programs. Hence, for example, in Step III costs are not included to develop the prototype propulsion system nor the Phase A vehicle system study. Also, flight hardware costs were not included in subsequent steps.

Fuel costs were treated partially as research and partially as flight production costs. One of the large unknowns is the cost for helium-3 and deuterium in the quantities required to support testing and flight. The cost for producing helium-3 is currently under study. The University of Wisconsin made preliminary estimates which indicated that it could be retrieved from the lunar surface at a cost of approximately $1,000 per kilogram.

14.0 Recommended Space Fusion Strategy

The funding levels in the summary may appear low to some, high to others, representing expenses that cannot be afforded at this time. The question is not can we afford the costs. The question instead is, can we afford not to undertake the research and still maintain a viable space program for the future of the United States.

14.5 PROGRAM MANAGEMENT

To establish a management approach that will most effectively initiate a space fusion program we consider five primary factors. Refer to Table 14-5. The first factor reflects who will be the "ultimate user" of the technology. The capability of the using organization having the technology expertise is considered an essential element for the safe, successful flight operational phase. The second factor applies to the organization's experience with aerospace systems integration technology. The importance of the fusion experience is self explanatory. The subjective cost factor includes the complexity of management systems as well as the current availability of equipment and facilities. Implementation includes the readiness of an organization to initiate the critical experiments. A subjective evaluation is provided in Table 14-5.

14.0 Recommended Space Fusion Strategy

TABLE 14-5. Evaluation of options to manage and meet technical capability for expediting a space fusion program.

Program management prime responsibility	Ultimate user	Aerospace experience	Fusion experience	Cost	Implementation timeliness
NASA	+	+	−	+	−
External to NASA					
1. DOE	−	−	+	−	+
2. National Laboratory	−	−	+	+	+
3. DOD	−	+	−	−	−
4. Industry					
a. Fusion	−	−	+	−	+
b. Aerospace	−	+	−	−	−
5. University	−	−	+	−	−
Combined team: NASA management/··· technical					
1. DOE	−	+	+	−	+
2. National Laboratory	+	+	+	+	+
3. DOD	−	+	−	−	−
4. Industry					
a. Fusion	−	+	+	−	+
b. Aerospace	−	+	−	−	−
5. University	−	+	+	−	−

Because the great importance of this capability indicates an urgency to initiate a program, timeliness is considered a priority factor. To initiate the program quickly, the preferred approach is to use the National Laboratory personnel as part of the program personnel under NASA management similarly as contracted with the Jet Propulsion Laboratory. A team effort where NASA, industry, and universities work together in a well defined program, will be required to meet success as achieved by the Apollo Program.

14.6 PROGRAM SUMMARY

A recommended strategy for a space fusion research program has been presented in this section to address the key issues raised. It is an orderly program, structured to advance to more demanding requirements as progress is attained while providing a demonstrated technical foundation prior to committing to the subsequent prioritized task. It relies upon and emphasizes full scale testing as the most rapid approach and therefore, potentially the most cost effective assuming that an active advancing space program continues.

The recommended approach is to proceed directly to a D-^{3}He FRC reactor design which in the best judgment is capable of producing net power. This is a high risk–high gain approach. The program flow for the FRC is envisioned in Fig. 14.14 as illustrative of the effort required.

14.0 Recommended Space Fusion Strategy

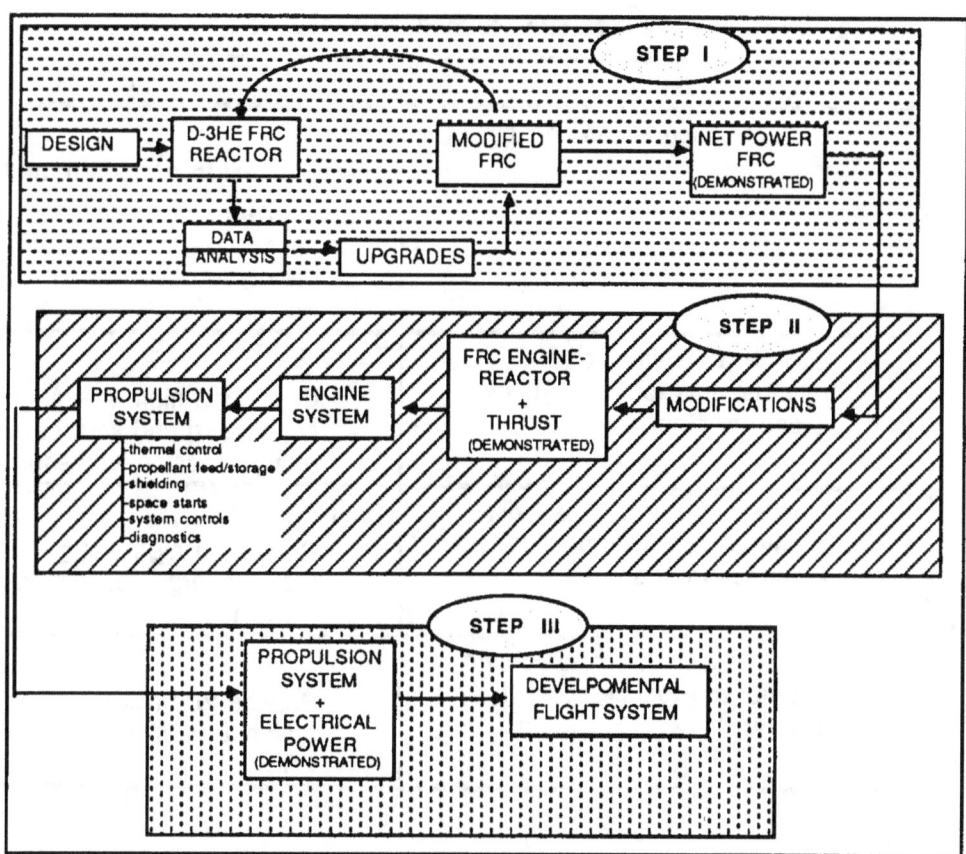

Fig. 14.14. Fusion propulsion capability: FRC reactor configuration flow.

Because the next energy step required for space missions will not be quick nor simple, NASA must plan its research and development well in advance. This is actually even a late date for a fusion program initiation from the viewpoint that fusion applications exist for space now. Science missions to the outer planets would be conducted to great advantage from a quick, high data yield capability. The Manned Mars Mission, where fusion's potentially high specific power capability would be enhanced by safety and mission performance advantages, could make use of fusion if the technology were available. Its presence would aid greatly in the planning and budgeting process by simplifying operational requirements, i.e., less mass into LEO.

To assist in addressing this critical matter, a strategy has been provided. It addresses key managerial and technical issues as well as the assumptions made in projecting the availability of fusion for space use. Fusion energy development is considered to become a critical space program necessity. The managerial issues and activities needed to resolve those issues are presented in summary format in Table 14-6.

14.0 Recommended Space Fusion Strategy

TABLE 14-6. Managerial issues concerning space fusion.

Issue	Program Task
1. Lack of space fusion capability.	Task 1: (1) Commit to a program and (2) Implement contents of this program for a feasibility demonstration and prototype development (Table 14-3).
2. Lack of a definition of requirements for high energy space missions.	Task: 5: Conduct a high energy space exploration and science high energy mission workshop.
3. Lack of a fusion program or a review of high energy propulsion for space missions.	Task 3: Conduct a high energy space fusion energy applications workshop.
4. FRC would be a single concept attempt to produce space fusion energy.	Task 9: Provide for back-up and alternate or advanced fusion approaches to the FRC.

The statement that fusion is an enabling energy technology for future space missions for safe, economical, high mission gain for solar system and outer space programs where science and exploration missions reaching out to the stars can be performed is predicated upon 8 assumptions. A program designed to address each of the assumptions is summarized in Table 14-7.

The cost information provided represents only estimates. The numbers would be more accurately stated to be a 1990 level of effort, that is, it reflects work to perform a concentrated program on one confinement concept. Cost escalations are not included. Overhead is not included. Additional full scale experiments are not included.

14.0 Recommended Space Fusion Strategy

TABLE 14-7. Activities to address technical aspects for space fusion energy developments through a prototype design.

Assumptions	Relevant Program Tasks

1. Space fusion feasible.	Task 2:
	  Test the FRC at burn parameters using D-^{3}He.
	Task 9:
	  Consider alternate approaches to the FRC.

2. Helium-3 can be burned in a space reactor and exhibit stable burning properties.	Task 2:
	  Test the FRC at burn parameters using D-^{3}He.
	Task 10:
	  Perform plasma analysis.

3. Fuel: Helium-3 and deuterium are available.	Task 4: Analyses of supplies and demands on ^{3}He.

4. Specific power = 1 kW/kg to 10 kW/kg.
  4.1. Space restart is possible.
  4.2. Cooling and shielding are possible.
	Task 6:
	  MCF system study.
	Task 7:
	  System analysis, design and test demonstrations.
	Task 8:
	  Restart concepts, analyses, and demonstrations.
	Task 11:
	  Vehicle system design.

5. Reactor power output = 20 MW to 300 MW.	Task 6:
	  MCF system study.
	Task 7:
	  System analysis, design, and test demonstrations.

6. Thrust feasible
  6.1. F= 1 N to 50 kN and throttable.
  6.2. Efficient throttling of thrust can be achieved.
  6.3. Isp=5 k to 1000 k seconds.
	Task 7:
	  System analysis, design, and test demonstrations.
	Task 10:
	  Perform plasma analysis.

7. Burn durations = 4 months up to 50 years; reusable engines.	Task 12:
	  Reliable long life analysis, stress testing, and demonstration
	Task 10:
	  Perform plasma analysis

8. Efficient, direct conversion electrical power output to 20 MW can be produced.	Task 7: System analysis, design, and test demonstrations

15.0 RECOMMENDATIONS FOR A HIGH ENERGY SPACE MISSION PROGRAM USING FUSION ENERGY

A program strategy for a space fusion high energy system has been determined and presented in Section 14 as a recommended space research initiative. That strategy constitutes the study's primary recommendation concerning high energy conversion. Section 15 completes the study's recommendations with a general set of recommendations beyond those concerning the development of a space fusion reactor. This section focuses upon a more general set of recommendations which are oriented toward the implementation of a fusion capability and the mission aspects of high energy space missions.

15.1 GENERAL

1. NASA should adopt a world leadership role in the initiation of space fusion research. The recommended goal is to develop fusion energy for serving space science and for accomplishing the manned exploration of space.

2. An aggressive, forward looking new start program is recommended at a level commensurate with the potential benefits and at a level sufficient to achieve meaningful results. The amount of funding for high specific power and high specific impulse propulsion systems, ≥ 1 kW/kg and 5×10^3 to 10^6 seconds, should be substantial, a level that will expedite attainment of an enhanced payload delivery capabilities at the earliest possible time. On an agency-wide basis, NASA spends on the order of $500M for chemical propulsion in order to use and to maintain current chemical propulsion technology. A level of 10%, i.e., $50M annually, of that expenditure as an investment for future dividends is considered a minimal level. One other funding figure of merit is to consider the economic impact on flight operations. Based upon our current launch capability for a 1 kW/kg system, a 1.0% investment from the launch savings of only **one** Manned Mars Mission, when amortized over a 10-year period, is over $80M per annum. The savings for additional manned flights to Mars and for planetary programs is billions more. The potential for the space mission enabling gains and cost savings warrants at least a modest 1.0% per year investment of the operational costs for one Martian trip. Using that criteria, NASA would be committing, as a minimum, approximately $80 to $100M annually for developing a high energy – high specific power capability. $150M per annum is recommended by the strategy in Section 14.

15.0 Recommendations

3. NASA should initiate a High Energy Space Mission Class capability. The program presented in Section 15.2 is recommended.

4. The concept of technical leverage should be used to the fullest extent to maximize the timeliness of experimental results and to minimize costs. Internal NASA expertise has proven to be absolutely essential to past space successes. It should, therefore, be provided for space fusion energy. The existing National Laboratory staff will provide an essential leverage element to NASA for direct benefit. The NASA-National Laboratory-University-small industrial research laboratory team approach would offer a directed approach to efficiently achieve results at a minimal cost level with expectations for the greatest yield – aided by early participation of the aerospace industry. After progress has been made on the key fusion demonstrations, aerospace industry should be encouraged to fully participate with fusion space vehicle system designs.

15.2 NATURAL SCIENCES AND EXPLORATION MISSIONS

1. Initiate a high energy mission space science program to encompass all the natural sciences which may derive benefit from the availability of large energy sources in space. Initiate the program by sponsoring a dedicated high energy space mission workshop to stimulate science objectives with planning for this mission class and to develop mission concepts for the advancement of the natural sciences and space exploration using energy levels not previously considered. A wide range of scientific disciplines should be represented, including not only planetary scientists but solar scientists, astrophysicists, astronomers, life scientists, plasma scientists, physicists, chemists, geologists, biologists, and others deriving benefit from high energy space missions, taking non space scientists' interests into consideration as well.

2. Provide a new, advanced vision for NASA by initiating programs to accomplish Oort Cloud and Alpha Centauri rendezvous stellar missions. Include search for extraterrestrial solar system planetary systems and life as major NASA goals. Because of their demanding technical requirements, these missions, particularly the stellar mission, will establish the limit for energy required, both propulsive and power, as well as the fusion vehicle system requirements. A visit to the nearest star will be a long term program commitment, and its careful, up-front expenditures and planning will have an enormous benefit on the overall mission success and on NASA's ability to conduct the mission in the most economical, safe manner. Because programs of this nature involve long term investments, the need to optimize the advanced planning is absolutely essential. A funding level which is phased to be

commensurate with space fusion engineering progress should be the guide. To commence the program, funding is not initially anticipated to be great – primarily a mission analysis, planning, and requirements activity.

3. Conduct studies to obtain a better definition of the outbound science data requirements and the sample return science mission objectives and thereby provide a better definition of their mass and energy-time mission requirements for this high energy payload class. The assumption used in this study for the planetary missions was a 20 MT outbound payload and 10 MT returned payload. For the Oort Cloud and stellar missions, a 10 MT payload appears to be a minimal mass. These payloads need to be studied in greater depth to establish system requirements.

4. Conduct technical requirements definition studies of "Remote In-situ Laboratories" operating as scientific outposts.

5. While additional studies are not needed to show the value of fusion energy, some limited mission evaluations should be conducted as part of existing mission analysis activities. The purpose is to establish the power and mission definition parameters for a wider range of high energy missions such as solar, comet rendezvous/sample return, inner planets, materials processing, life support, multiple payloads to multiple planets, and others not examined in this study. The study objective is a broadened definition of high energy fusion system requirements.

15.3 SPACE FUSION REACTOR (SFR) SYSTEM DEMONSTRATION

This section assumes the incorporation of the Section 14 recommendations and elaborates upon those recommendations from system consideration perspectives.

1. <u>The most important, fundamental objective of this recommended SFR demonstration is to provide test results showing fusion confinement in a confinement concept having space application.</u> The *first priority* of the space fusion energy program must be given to experiments which demonstrate space fusion's viability. Thorough critical design studies conducted first could be useful in evaluating which approaches are most likely to meet with success. Further, they would point the way to critical issues to address. Expenditure of resources on fusion experiments and testing, including the related theoretical work for the space application of fusion energy, must be emphasized. Minimize paper studies for now that do not directly contribute toward fusion since the mission benefits and many innovative uses will follow once fusion is demonstrated just as they have with chemical propulsion. Thus, we

15.0 Recommendations

would be focusing upon the most valuable application of resources to achieve definitive test experimental results.

2. The FRC reactor exhibits the best β and thrust conversion characteristics to potentially serve as a space fusion reactor, and it should be given priority for immediate experimental test support. Solutions to instabilities need to be developed. A study should be accomplished to evaluate the feasibility of the FRC operating at a 20 to 60 MW jet power level for planetary missions, 250 MW for manned Mars, and at the 5 GW and 50 GW levels for stellar applications. Weight reduction concepts and analyses are also important tasks that are necessary to investigate means to improve upon the specific power characteristics.

3. MCF ash removal studies of various approaches should be conducted to determine techniques for recovery of unburned fuels. High efficiency reactor systems are important for achieving efficient, high specific power flight systems.

4. Plasma confinement-stability studies of various approaches should be conducted for the more attractive reactor configurations for space related reactor considerations as options to the FRC.

5. Research should be initiated on superconducting magnets having a space fusion application. Efficiency and mass reductions are the goals.

6. The impact of neutrons on reactor materials for the D-^{3}He reaction and on the adjacent vehicle structure and materials needs to be better defined. Work related toward their reduction should be pursued, neutron elimination being the target.

7. Analytical efforts and confinement experiment test programs should be performed to investigate alternative advanced reactor technology as a hedge against any uncertainty with the FRC. Recommendations from the Space Fusion Energy Workshop are anticipated to be incorporated into this activity. At least one backup reactor confinement concept should be included in a space fusion research developmental program in order to avoid a program critical single failure point.

8. The fusion program should have some provisions to test and to evaluate new, innovative concepts, for it is by no means clear that the final thought has been generated.

9. Review of ICF, if it becomes declassified, as a viable concept should be performed.

10. Thermal control at the reactor and vehicle level must be analyzed and researched.

15.0 Recommendations

15.4 INTEGRATED SPACE FUSION POWER SYSTEM INITIATIVE

1. Conceptual studies and experiments should be conducted that consider the development of a "Solid State Propulsion System" – one with no moving parts as a goal – to attain high system reliability, one having large MTBF values of tens of years.

2. Direct electrical conversion design and testing activities for space applications, including the collection of electrons should be initiated. The high voltage concern for operations in a space environment where fusion reactors will be operated should be investigated and appropriate experiments performed for feasibility demonstrations.

3. Based upon requirements and design criteria for space propulsion and power, analyses should be conducted and testing performed to develop dual mode propulsion and electrical power reactor designs, including separate and combined propulsion and power mode operation. A study of the level of understanding of the current joint propulsion and power technology, the generation of the theoretical design options and concepts, and subsequently, the conduct of the appropriate research and experiments to demonstrate the concept(s) should be accomplished. As part of this activity, the means by which to vary the propulsion system's thrust level should receive considerable analytical attention.

4. Analyses to characterize the vehicle's electrical power and data transmission requirements for a stellar mission should be conducted. Conceptual design studies of the electrical power system should be performed to establish areas in need of key experiments. One parameter of interest is the power system's high voltages required for operation in a vacuum environment. (Refer to number 2 above.) The other is the heat balance required and the means for achievement.

5. Long term liquid helium-3 thermal analyses and test demonstrations relevant to the size(s), storage time, and physical environment for space fusion powered vehicle(s) are needed. Heat balance must also be addressed.

6. Conceptual designs and feasibility studies for low mass drivers for use in ICF concepts should be developed.

7. Magnetic plasma thrusters and other techniques for vehicle trim functions should be pursued for conceptual design and feasibility approaches. System diagnostics need to be investigated – the objective being improved mass fractions and a reliable vehicle. Also, consideration of the possibility for recovery of reactor waste heat is important for improvements in efficiency in the generation of supplemental electrical power as, for example, in providing attitude control using ion thrusters or alternatively, control moment gyros, as well as for the spacecraft electrical power system in general.

15.0 Recommendations

8. Other possible space based fuel sources, recovery and in-situ manufacturing techniques, and economics for deuterium, hydrogen, and helium-3, particularly on other planets and moons, should be investigated.

15.5 RECOMMENDATIONS CONCERNING OTHER TOPICS ORIGINATING FROM THE STUDY

1. Feasibility and trade studies for ablation and laser energy transmitted ion propelled propulsion systems should be conducted. Power levels for mission scenarios should be defined.

2. A study of thermal power production capabilities for Mars should be conducted and consideration given to alternative energy sources.

3. Life support research to enable utilization of local planetary resources for providing planetary habitability should be performed. One subject of particular interest is the power level requirements.

4. A study on the use of fusion energy for aeronautical propulsion should be conducted.

16.0 REFERENCES

Abd85   Abdou, M., Mattas, R., Bartlit, J., Berry, L., Berwald, D., et al "TPA Fusion Technology Status Report," University of California, Los Angeles Report, UCLA-ENG-85-38 (1985)

Abt76   Abt, H. A., & Levy, S. G., *Astrophysical Journal Supplement*, 30, p. 273 (1976)

Anom   JPL Analytical Ephemerides, section 3.12, PLCON subroutine

Anom58   "National Aeronautics and Space Act of 1958," GPO: 1958 O - 29011 (416) Washington, D.C. (July 29, 1958)

Anom83   "Planetary Exploration Through the Year 2000, A Core Program," Solar System Exploration Committee of the NASA Advisory Council, NASA, Washington, D. C. (1983)

Anom86   "Advanced Mission and Information Studies," Vol. III, technical proposal, Triton Sample Return Mission - sample problem, submitted to NASA Headquarters, Code EL, Science Applications International Corporation (SAIC) (May 30, 1986)

Anom87   "Manned Mars Mission and Program Analysis," Mid-term Progress Review, SRS Technologies to NASA/MSFC, p. 84 (December 1987)

Anom87   "Starpower, The U.S. and the International Quest for Fusion Energy" Office of Technology Assessment, Congress of the United States (October 1987).

Anom88   "Fusion Propulsion Study" Phase 1 Interim Report Presentation, McDonnell Douglas, Air Force Contract No. FO4611-87-C-0092 (May 17, 1988)

Anom88   "NASA Lunar Helium-3 Fusion Power Workshop," sponsored by NASA Lewis Research Center (April 25-26, 1988)

Anom89   "1989 Long-range Program Plan," NASA Headquarters, Washington (November 1989)

Anom89   "Report of the 90-Day Study on Human Exploration of the Moon and Mars," NASA Headquarters, Washington (November 1989)

Anom89   "US National Space Policy, White House," Washington (November 2, 1989)

Anom90   "Summary and Principal Recommendations of the Advisory Committee on the Future of the U. S. Space Program," NASA Headquarters, Washington (December 10, 1990)

Anom90   "National Rocket Propulsion Strategic Plan," Aerospace Industries Association, Washington, (February 15, 1990)

16.0 References

Bak87   Baker, C. C., Abdou, M. A., Callen, J. D., Dean, S. O., et al. "Technical Planning Activity Final Report," Argonne National Laboratory Report, ANL/FPP-87-1 (1987)

Bal88   Balsiger, H., Fechtig, H., and Geiss, J. "A Close Look at Halley's Comet," *Scientific American,* $\underline{259}$, 3 (September 1988)

Bar77   Barr, W. L., Moir, R. W., Kinney, J. D., "Experimentral and Computational Results on Direct Energy Conversion for Mirror Fusion Reactors," *Nuclear Fusion,* $\underline{17}$, 1015 (1977)

Bar83   Barr, W. L., and Moir, R. W., "Test Results on Plasma Direct Converters," *Nucl. Technology/Fusion,* $\underline{3}$, 98 (1983)

Bil89   Bilder, R. B., Cameron, E. N., Kulcinski, G. L., and Schmidt, H.H., "Legal Regimes for the Mining of Helium-3 from the Moon, WCSAR-TR-AR-8901-1, University of Wisconsin (February 27, 1989)

Blo78   Bloomquist, C., et al, "On-orbit Spacecraft Reliability," Planning Research Corporation, PRC R-1863, p. 404 (September 30, 1978)

Bor87   Borowski, S. K., "A Comparison of Fusion/Antiproton Propulsion Systems for Interplanetary Travel," AIAA/SAE/ASME/ASEE 23rd Joint Propulsion Conference, PRA-SA-APRI (June 26, 1987)

Bre87   Brereton, S. J. and Kazimi, M. S., "Safety and Economic Comparison of Fusion Fuel Cycles," MIT, PFC/RR-87-7, DOE/ID-01579-1 (August 1987)

Bro82   Bromberg, J. L., *Fusion,* MIT Press, Cambridge, MA (1982)

Bus91   Bussard, R.W., "Some Physics Considerations of Magnetic Inertial Confinement: A New Concept for Spherical Converging-flow Fusion," *Fusion Technology,* $\underline{19}$, pp. 273-293 (March 1991)

Cal86   Callen, J. D., Santarius, J. F., Baldwin, D. E., Hazeltine, R. D., Linford, R. K., et al, "TPA Plasma Science Final Report," University of Wisconsin Report (1986)

Can68   Canetti, G. S., "Final Report, Definition of Experimental Tests for a Manned Mars Excursion Module," SD 67-755-1, North American Rockwell (January 1968)

Cha88   Chapline, G. F., Dickson, P. W., and Schnitzler, B. G., "Fission Fragment Rockets– A Potential Breakthrough," UCRL-99474, 1988 International Reactor Physics Conference (September 18-22, 1988)

Cha88   Chapline, G., "Fission Fragment Rocket Concept," *Nuclear Instruments and Methods in Physics Research,* A271, 207-208 North-Holland, Amsterdam (1988)

16.0 References

Cha89   Chapman, R., Miley, G., Kernbichler, W., and Heindler, M., "Space Fusion Propulsion with a Field Reversed Configuration," *Fusion Technology*, 15, 1154-1159 (1989)

Che74   Chen, F. F., *Introduction to Plasma Physics*, 1st Edition, Plenum Press, New York (1974)

Dav85   Davidson, R. C., et al, "Magnetic Fusion Advisory Committee Report on High Power Density Fusion Systems," submitted to DOE Office of Energy Research (May 1985)

Dea85   Dean, S., Cohn, D., Crocker, J., Flanagan, C., Gordon, J., et al, "TPA Fusion System Group Status Report," Fusion Power Associates Report (1985)

Div83   Divine, N., and Garrett, H. B., "Charged Particle Distributions in Jupiter's Magnetosphere," Journal of Geophysical Research, V88, A9, pp 6889-6903, (Sept. 1, 1983)

Don88   Donahue, T. chair., et al, "Space Science in the Twenty-First Century, Imperatives for the Decades 1995 to 2015, Overview," National Research Council, Space Science Board, Commission on Physical Sciences, Mathematics, and Resources, Task Group Report, Washington: National Academy Press (1988)

         "Astronomy and Astrophysics"

         "Fundamental Physics and Chemistry"

         "Planetary and Lunar Exploration"

         "Solar and Space Physics"

Emm90   Emmert, G. A., et al., "Apollo-L2, An Advanced Fuel Tokamak Reactor Utilizing Direct Conversion," Proceedings, IEEE Thirteenth Symposium on Fusion Engineering, p. 1043 (IEEE, NJ, 1990)

Eng62   Englert, G. W., "Towards Thermonuclear Rocket Propulsion," *New Scientist*, 16:16, 307 (1962)

Eng66   Englert, G., "High Energy Ion Beams Used to Accelerate Hydrogen Propellant Along Magnetic Tube of Flux," NASA TND-3656 (1966)

Eps71   Epstein, S. and Taylor, Jr., H. P., "O^{18}/O^{16}, Si^{30}/Si^{28}, D/H, and C^{13}/C^{12} Ratios in Lunar Samples," *Proceedings of the Second Lunar Conference*, V. 2, pp. 1421-1441, M. I. T. Press (1971)

Fle86   Fletcher, J. C., "NASA's Vision and Goals Statement," Memorandum to All NASA Employees (December 9, 1986)

For83   Forward, R. L., "Alternate Propulsion Concepts," Air Force Rocket Propulsion Laboratory, AFRPL TR-83-067 (December 19883)

Fpa90   "Executive Newsletter," Fusion Power Associates, Gaithersburg, MD (December 1990)

16.0 References

Fri88   Friedlander, A. and McAdams, J., Science Applications International Corporation, memo to NASA Headquarters, Norman Schulze, October 13, 1988

Fri88   Friedlander, A., Cole, K., "Power Requirements for Lunar Base Scenarios," Lunar Bases & Space Activities in the 21st Century," paper no. LBS-88-211 (April 5-7, 1988)

Fri89   Friedlander, A., McAdams, J., Schulze, N., "Performance of Advanced Missions Using Fusion Propulsion," AAS/GSFC International Symposium on Orbital Mechanics and Mission Design, NASA Goddard Space Flight Center, Greenbelt, MD. (April 24-27, 1989)

Gar88   Garrison, P. W. and Stocky, J. F., Future Space craft Propulsion," *Journal of Propulsion*, AIAA, 4, 6, 520 (Nov.-Dec .1988)

Ger89   Gerwin, R. A., et al, "Characterization of Plasma Flow Through Magnetic nozzles," AL-TR-89-092, Space Division, Air Force Systems Command (February 1990)

Gla60   Glasstone, S. and Loveberg, R. H., *Controlled Thermonuclear Reactions*, An Introduction to Theory and Experiment, Van Nostrand, Princeton, N.J. (1960)

Hal89   Haloulakas, V. E. and Bourque, R. F., "Fusion propulsion Study," McDonnell Douglas Space System Co. and General Atomics, Air Force Technology Center, Space Systems Command, AL-TR-89-005 (July 1989)

Ham88   Hammer, J. H., "Experimental Demonstration of Acceleration and Focusing of Magnetically Confined Plasma Rings," Lawrence Livermore National Laboratory, UCRL-98988 (June 22, 1988)

Har77   Harrington, R. S., *Astronomical Journal*, 82, p. 753 (1977)

Har82   Harrington, R. S., "The Frequency of Planetary Systems in the Galaxy," *Extraterrestrials, Where Are They?*, ed. M. H. Hart & B. Zuckerman, New York: Pergamon Press, p. 142 (1982)

Har88   Hartman, C. W., "Acceleration of Compact Plasma Rings for Fusion Applications," Lawrence Livermore National Laboratory, UCRL-98504 (August 26, 1988)

Has86   Hasegawa, A., "Magnetically Insulated Fusion: A New Approach to Controlled Thermonuclear Fusion," *Physical Review Letters*, 56, 2 (January 13, 1986)

Hol88   Holdren, J., et al, "Exploring the Competitive Potential of Magnetic Fusion: The Interaction of Economics with Safety and Environmental Characteristics," *Fusion Technology*, 13 (January 1988)

Jon86   Jones, S. E., et al, *Physical Review Letters*, 56, p. 588 (1986)

16.0 References

Kam87 Kammash, T. and Gilbraith, D. L., "A Fusion Reactor for Space Applications," *Fusion Technology,* 12, pp. 11-21 (July 1987)

Kam88 Kammash, T. K., and Gilbraith, D., "Mars Missions with the MICF Fusion Propulsion System," University of Michigan, AIAA -88 - 2962 (July 11-13, 1988)

Ker88 Kernbichler, W., Miley, G. H., and Heindler, M, "D-^{3}He Fuel Cycles for Neutron Lean Reactors," DOE/ER/52127/39, Eighth Topical Meeting on the Technology of Fusion Energy, American Nuclear Society (October 9-13, 1988)

Kha90 Khater, H. Y., et al., "Activation and Safety Analyses for the D-^{3}He Fueled Tokamak Reactor Apollo," Proceedings, IEEE Thirteenth Symposium on Fusion Engineering, p. 728 (IEEE, NJ, 1990)

Kol88 Koloc, P. M., "The PLASMAK™ Configuration and Ball Lightning," International Symposium on Ball Lightning, Tokyo (July 1988) and "PLASMAK™ Star Power for Energy Intensive Space Applications," Phaser Corp, College Park, MD., Draft (September 1, 1988)

Kul87 Kulcinski, G. L., et al, "SOAR: Space Orbiting Advanced Fusion Power Reactor," University of Wisconsin, UWFDM-722 (September 1987)

Kul87 Kulsrud, R. M., "Polarized Advanced Fuel Reactions," *Muon-Catalyzed Fusion and Fusion with Polarized Nuclei,* ed. Brunelli, B., and Leotta, G. G., Plenum Press, New York, p 167 (1987)

Kul89 Kulcinski, G. L. et al., "Apollo–An Advanced Fuel Fusion Power Reactor for the 21st Century," *Fusion Technology,* 15, 1233 (1989)

Log86 Logan, B. S., "Economical D-^{3}He Fusion Using Direct Conversion of Microwave Synchrotron Radiation," Fusion Power Associates Annual Meeting, Washington (April 24, 1986)

Log88 Logan, B. G., "Initiative for the 21st Century: Advanced Space Power and Propulsion Based on Lasers," Lawrence Livermore Laboratory, UCRL 98520 (April 1988)

Log88 Logan, G., "Optimum Strategy for Magnetic Fusion Powering Interplanetary Spacecraft," Lawrence Livermore Laboratory, UCRL - 98520 (April 25-26, 1988)

Mag85 Maglich, B. C., "Proton Chain Fusion of Lithium Nuclei as 'Aneutronic' Energy Source," United Sciences, Inc., Princeton, N. J., STA 14:13 (January 1, 1985)

Man87 Mankins, J.; Olivieri, J.; and Hepenstal, A.; "Preliminary Survey of 21st Century Civil Mission Applications of Space Nuclear Power," JPL-D-3547 (March 1987)

16.0 References

Mas59   Maslen, S. H., "Fusion for Space Propulsion," IRE Trans. Military Electronics, Mil-3, 52 (1959)

McA88   McAdams, J., Science Applications International Corporation, memo to NASA Headquarters, Norman Schulze (November 16, 1988)

McN82   McNally, Jr., J. R., "Physics of Fusion Fuel Cycles," *Nuclear Technology/Fusion*, 2, 9-28 (Jan. 1982)

Men89   Mendell, W. ed., "Lunar Bases and Space Activities of the 21st Century," (Lunar and Planetary Institute, Houston) (1989)

Mil76   Miley, G. H., *Fusion Energy Conversion,* American Nuclear Society (1976)

Mil78   Miley, G., H. and Gilligan, J. G., "Preliminary Design of a Self Sustained Advanced-fuel Field Reversed Mirror Reactor - SAFFIRE," Trans Am. Nuc. Soc., 30, 47 (1978)

Mil79   Miley, G., H. and Gilligan, J. G., "SAFFIRE - A D-^{3}He Pilot Unit for Advanced-fuel Development," EPRI Review Meeting, San Diego, CA (1979)

Mil87   Miley, G. H., et al, "Advanced Fusion Power: A Preliminary Assessment," Committee on Advanced Fusion Power, Air Force Studies Board, Commission on Engineering and Technical Systems, National Research Council, Washington (1987)

Mil88   Miley, G. H., "^{3}He Sources for D-^{3}He Fusion power," *Nuclear Instruments and Methods in Physics Research*, A271, 197-202, North-Holland Amsterdam (1988)

Moe72   Moeckel, W. E., "Comparison of Advanced Propulsion Concepts for Deep Space Exploration," *Journal of Spacecraft*, AIAA, 9, 863 (1972)

Mor73   Moir, R. and Barr, W., "Venetian - Blind Direct Energy Conversion for Fusion Reactors," Lawrence Livermore Laboratory, *Nuclear Fusion*, 13 (1973)

Obe82   Oberg, J., "Terraforming," *Extraterrestrials, Where Are They?*, ed. M. H. Hart & B. Zuckerman, New York: Pergamon Press, p. 62 (1982)

Ort87   Orth, C. D., Hogan, W. J., Klein, G., Hoffman, N., Murray, K., Chang Diaz, F., "Interplanetary Propulsion Using Inertial Fusion," The 4th Symposium on Space Nuclear Power Systems, Lawrence Livermore National Laboratory, UCRL- 95275 Rev 1 (January 1987)

Ort87   Orth, C., et al, "The VISTA Spacecraft–Advantages of ICF for Interplanetary Fusion Propulsion Applications," Lawrence Livermore National Laboratory, UCRL-96676 (October 2, 1987);

Pai86   Paine, T. O., et al, "Pioneering The Space Frontier, The Report of the National Commission on Space," Bantam Books (May 1986)

Peg87	Pegoraro, F., "Depolarization of Spin Polarized Plasmas By Collective Modes," Scuola Normale Superiore, Pisa, Italy; *Muon-Catalyzed Fusion and Fusion with Polarized Nuclei,* ed. Brunelli, B., and Leotta, G. G., Plenum Press, New York, p 180 (1987)
Pen85	Peng, Y.-K. M., "Spherical Torus, Compact Fusion at Low Field," Oak Ridge National Laboratory Report ORNL/FEDC-84/7 (1985)
Per88	Perkins, L. J., Miley, G. H., Logan, B.G., "Novel Fusion Energy Conversion Methods," *Nuclear Instruments and Methods in Physics Research A271,* Elsevier Science Publishers B. V., pp. 188-196 (1988)
Pos69	Post, R. F., "Fuel Cycles, Loss Reduction and Energy Recovery," Proceedings of the British Nuclear Energy Society Conference on Nuclear Fusion Reactors, Culham (Sept. 17-19, 1969)
Raf87	Rafelski, J. and Jones, S., "Cold Nuclear Fusion," *Scientific American,* 257, 1 (July 1987)
Rid87	Ride, S. K., "NASA, Leadership and America's Future in Space, A Report to the Administrator" (August 1987)
Rie88	Riehl, J., Mason, L., Gilland, J., Sovey, J., Bloomfield, H., "Power and Propulsion Parameters for Nuclear Electric Vehicles," NASA Lewis Research Center, Draft, version 1, release 1 (July 1988)
Rob81	Robert, F. and Epstein, S., "The concentration and isotopic composition of hydrogen, carbon, and nitrogen in carbonaceous meteorites," rev. September 15, 1981, *Geochimica et Cosmochimica Acta,* 46, pp. 81-95, Pergamon Press (1982)
Rot59	Roth, J. R., "A Preliminary Study of Thermonuclear Rocket Propulsion," Paper 59-944, *American Rocket Society* (1959)
Rot86	Roth, J. R., *Introduction to Fusion Energy,* Ibis Publishing, Charlottesville, VA (1986)
Rot89	Roth, J. R., "Space Applications of Fusion Energy," *Fusion Technology,* 15 (May 1989)
San53	Sänger, E., "The Theory of the Photon Rocket," Ing. Arch. 21, 213 [in German] (1953)
San88	Santarius, J. F., et al, "Critical Issues for SOAR: The Space Orbiting Advanced Fusion Power Reactor," University of Wisconsin, UWFDM-753 (August 1988)
San89	Santarius, J. F., "Lunar ^{3}He, Fusion Propulsion, and Space Development," UWFDM-764, University of Wisconsin (1989)
San89	Santarius, J. F., "Magnetic Fusion Energy and Space Development," 24th IECEC, 5, 2525, IEEE, NY (1989)

16.0 References

Sar88   Sargent, M. G., "A Comparison of Magnetic Confinement Fusion (MCF) and Inertial Confinement Fusion (ICF) for Spacecraft Propulsion," Jet Propulsion Laboratory, JPL D-5878 (October 1988)

Sch91   Schulze, N. R. and Roth, J. R., "The NASA-Lewis Program on Fusion Energy for Space Power and Propulsion 1958-1978," *Fusion Technology*, 19, 1, pp11-28, (January 1991)

Sha89   Shaw, G. L., Shin, M., and Dalitz, R. H., "Growing Strange Drops of Matter," *Nature*, 337, No. 6206, pp. 436-439 (February 2, 1989)

Stu64   Stuhlinger, E., *Ion Propulsion for Spaceflight*, McGraw Hill N.Y. (1964)

Tel85   Teller, E., *Plasma Physics and Controlled Nuclear Fusion Research*, Vol I and II International Atomic Energy Agency, Vienna (1985)

Tel91   Teller, E., Glass, A. J., Fowler, T. K., Hasegawa, A., and Santarius, J., "Space Propulsion by Fusion in a Magnetic Dipole," *Fusion Technology* and presented at the First International A. D. Sakharov, Conference on Physics, (Moscow, USSR), May 27-31, 1991, LLNL report UCRL-JC-106807, (April 12, 1991)

Tus88   Tuszewski, M., "Status of the Field-Reversed Configuration as an Alternate Confinement Concept," Los Alamos National Laboratory, LA-UR-88-2821 (October 1988)

Tus91   Tuszewski, M., Barnes, D. C., Chrien, R. E., Cobb, J. W., Rej, D. J., Siemon, R. E., Taggart D. P., and Wright, B. L., *Physical Review Letters*, A.P.S., 66, 6 (February 11, 1991)

Wil82   Wilkening, L. L., *Comets*, The University of Arizona Press, pp 57, 637, 659 (1982)

Wit86   Wittenberg, L., Santarius, J., Kulcinski, G., *Fusion Technology*, 10, 167 (1986)

Yea85   Yeates, C. M., et. al, Galileo: "Exploration of Jupiter's System," NASA SP-479, Washington (1985).

Zuc82   Zuckerman, B., "The Frequency of Planetary Systems in the Galaxy," *Extraterrestrials, Where Are They?*, ed. M. H. Hart & B. Zuckerman, New York: Pergamon Press (1982)

Appendix A

THE NASA-LEWIS PROGRAM ON FUSION ENERGY FOR SPACE POWER AND PROPULSION, 1958-1978

Norman R. Schulze
National Aeronautics and Space Administration Headquarters
Washington, D.C. 20546

J. Reece Roth
University of Tennessee
Department of Electrical and Computer Engineering
Knoxville, Tennessee 37996-2100

Received August 24, 1989
Accepted for Publication April 27, 1990

A retrospective summary and bibliography of the National Aeronautics and Space Administration research program on fusion energy for space power and propulsion systems conducted at the Lewis Research Center are presented. This effort extended over a 20-yr period ending in 1978, involved several hundred person-years of effort, and included theory, experiment, technology development, and mission analysis. This program was initiated in 1958 and was carried out within the Electromagnetic Propulsion Division. Within this division, mission analysis and basic research on high-temperature plasma physics were carried out in the Advanced Concepts Branch. Three pioneering high-field superconducting magnetic confinement facilities were developed with the support of the Magnetics and Cryophysics Branch. The results of this program serve as a basis for subsequent discussions of the space applications of fusion energy, contribute to the understanding of high-temperature plasmas and how to produce them, and advance the state of the art of superconducting magnet technology used in fusion research.

NOTE: This article was originally published in *FUSION TECHNOLOGY*, Journal of the American Nuclear Society, 19, 1, 11-28 (January 1991) and has been approved for publication in this report (January 25, 1991). The article is copyright protected 1991 by the American Nuclear Society, La Grange Park, Illinois.

Appendix A

INTRODUCTION

The National Aeronautics and Space Administration (NASA) fusion program was initiated at the Lewis Research Center in 1958 to take advantage of the high specific energy content of fusion energy for application to space power and propulsion systems and to attain in-house technical capability to conduct space-related fusion research and development. The NASA program was set up as the result of studies that began in 1956 in the National Advisory Committee for Aeronautics (NACA), NASA's predecessor.[1] A carefully planned and increasingly successful program of fusion-related plasma physics research and technological development was carried out in the areas of basic plasma physics, high-temperature plasma confinement, cryogenic and superconducting magnet development, and analysis of ambitious manned and unmanned interplanetary missions using fusion space power and propulsion systems. During the mid 1970s, when the space program budget decreased after successful accomplishment of the Apollo program and came under heavy pressure from development of the Space Shuttle, NASA's long-range program of research and development on the space applications of fusion energy was one of the resulting program casualties. This program was terminated in 1978.

SCOPE AND ACCOMPLISHMENTS OF PROGRAM

The former NASA fusion program was conducted at the Lewis Research Center, in Cleveland, Ohio, with a budget estimated to be approximately one million dollars per year. This relatively small budget achieved a number of first-of-a-kind accomplishments, new discoveries of physical processes in plasmas, patentable inventions, and advances in the technological state of the art. Some, but not all, of these accomplishments have been incorporated into the mainline fusion program for electric utility applications. Although it was a relatively small program, known now to only a very few of the current NASA staff, U. S. Department of Energy (DOE) researchers, or other fusion scientists, significant contributions to the field of fusion energy resulted and are documented in the NASA and archival journal publications listed in this paper. This is the only known bibliography relating to the NASA-Lewis fusion program. A summary of several aspects of the fusion program before 1969 can be found in Ref. 1.

The goal of the NASA-Lewis program was to define the overall approach and to identify and pursue the critical physics and technology required to develop fusion energy for application to space power and propulsion. A balanced program was pursued – one involving theory, high-temperature plasma physics experiments, large-scale superconducting magnet technology development, a plasma containment experiment having reactor relevance, and advanced mission analysis. The plasma research represented a significant

Appendix A

effort that investigated basic physical processes relevant to magnetoelectric plasma heating and containment concepts with potential applications to space power and propulsion. The part of the program dealing with magnet technology was initiated in 1958 to efficiently produce larger and stronger magnetic fields. For plasma heating, NASA-Lewis at first adopted steady-state ion cyclotron resonance heating (ICRH), in contrast to other fusion programs, which used pulsed radio-frequency (rf) power. Ion cyclotron resonance heating and steady-state operation appeared to be more relevant to space propulsion applications.

The organizational structure of the fusion program at NASA-Lewis included two branches in the Electromagnetic Propulsion Division: the Advanced Concepts Branch, with Eli Reshotko, Warren D. Rayle, and George R. Seikel as branch chiefs; and the Magnetics and Cryophysics Branch, with James Laurence and later Gerald Brown as branch chiefs. The basic research in high-temperature plasma physics was accomplished in the Advanced Concepts Branch, and the Magnetics and Cryophysics Branch developed the coils for the high-field superconducting magnet facilities.

Individuals who were prominent in the program as managers or in a technical supervisory capacity include Abe Silverstein, John Evvard, Wolfgang E. Moeckel, Eli Reshotko, Warren D. Rayle, George R. Seikel, J. Reece Roth, Gerald W. Englert, and John J. Reinmann. Universities under contract to NASA-Lewis in connection with this program included the University of Texas-Austin, Brigham Young University (BYU), Texas Tech University, the University of Illinois, Massachusetts Institute of Technology, and Rensselaer Polytechnic Institute.

Within the Electromagnetic Propulsion Division at NASA-Lewis, initial magnetoelectric confinement experiments were conducted on the Pilot Rig, the mirror machine predecessor to the more sophisticated Electric Field Bumpy Torus (EFBT). Both experiments were conceived and carried out by J. R. Roth, a member of the Advanced Concepts Branch. There were about a half dozen professionals involved with these high-temperature plasma physics experiments on the EFBT. There were also another half dozen in John J. Reinmann's High-Temperature Plasma Section in the same branch. In order to analyze the space applications of fusion energy, mission studies were done by the Mission Analysis Branch in the Advanced Systems Division, as well as by the Electromagnetic Propulsion Division. The mission analysis technique used to compare advanced propulsion methods was developed by Wolfgang E. Moeckel, chief of the Electromagnetic Propulsion Division.

Some of the fusion-related accomplishments and program areas covered by the NASA-Lewis effort during this 20-yr period include the following:

1. basic research on the EFBT magnetoelectric fusion confinement concept, including identification of its radial transport mechanism and confinement time scaling[2-31]

Appendix A

2. operation of the Pilot Rig mirror machine, the first superconducting magnet facility to be used (in 1964) in high-temperature plasma physics or fusion research[32-61]
3. operation of the superconducting bumpy torus magnet facility in 1972, the first such facility used in fusion research to generate a toroidal magnetic field[62-66]
4. steady state production of neutrons from deuterium-deuterium (DD) reactions, starting in 1967 (Ref. 29)
5. studies of the direct conversion of plasma enthalpy to thrust for a direct fusion rocket by propellant addition and magnetic nozzles[67-73]
6. power and propulsion system studies, including D-3He power balance, neutron shielding, and refrigeration requirements[74-79]
7. development of large-volume, high-field superconducting and cryogenic magnet technology
8. advancing the state of the art in cryogenic and liquid helium handling technology[62-66,80-82]
9. ICRH of plasmas at high power and in the steady state[83-105]
10. steady-state, dense, magnetoelectric plasma generation by crossed electric and magnetic fields in the burnout geometry[106-117]
11. advancing the state of the art in plasma diagnostics, including fluctuation-induced transport, heavy ion beam probes, charge-exchange (cx) neutral particle analysis, optical spectroscopy, and Thompson scattering[118-127]
12. development of ferrofluids and studies of ferrofluidics technology[1,78,79]
13. mission analysis and system studies of fusion propulsion systems for interplanetary missions.[79,128-139]

Many of these accomplishments were the first of their kind, resulted in patents, and/or have been incorporated into the world fusion program.

An evaluation[140] of the program was conducted in 1977-1978 by G. H. Miley, a faculty member of the University of Illinois. His report points out some of the accomplishments noted above, such as the pioneering use of superconducting magnets for fusion experiments: "The NASA devices have provided more operating experience with superconducting coils than any other fusion experiment in the U.S."

Dr. Miley's report also addressed the NASA experimental program and status:

> Clearly the strong points of the NASA experiments, as already suggested, are the steady state operation and good heating (high T_i). It is also noteworthy that while the NASA experiments remain below the leading magnetic devices in $n\tau$ and T_i, the slopes of the curves are as

steep as (or steeper) than for the other devices. In other words, the rate of progress had been comparable to that at other fusion laboratories . . .

The NASA experiments are as much as ten orders of magnitude below the gain and average power that would eventually be required for a practical reactor. While considerably more funding has gone into mainline DOE experiments, they still remain as much as 4 to 5 orders of magnitude low in these respects . . . the only other bumpy torus being ORNL's [Oak Ridge National Laboratory] Elmo Bumpy Torus or EBT [cf. the NASA EFBT (Electric Field Bumpy Torus)]. Focusing on the latter two devices ... , we observe that NASA device performance has generally been comparable to that for the EBT but falls below that for 2X-II [Lawrence Livermore Laboratory experiment]. The latter is not surprising, however, since it employs a volume and heating power which are about five times that of SUMMA.

MAGNETIC CONFINEMENT RESEARCH

The first major superconducting magnet facility at NASA-Lewis was a magnetoelectric confinement experiment using a modified Penning discharge in a superconducting magnetic mirror known as the Pilot Rig,[32-61] which is shown in Fig. A-1.

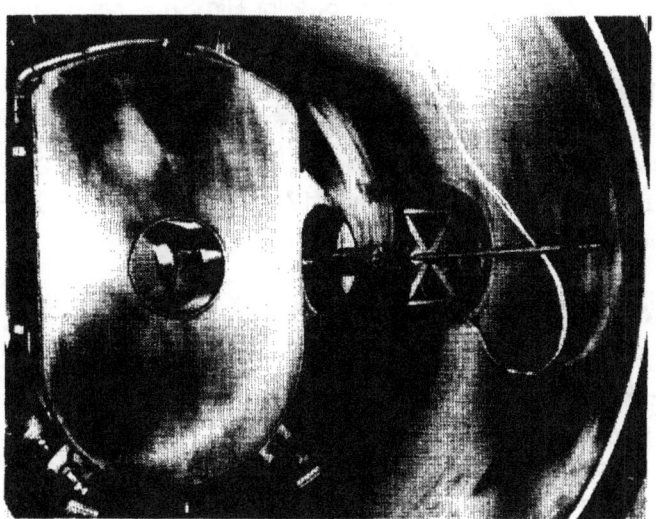

Fig. A-1. Pilot Rig superconducting magnetic mirror.

The Pilot Rig went into service for fusion research in December 1964, only 3 yr after the discovery of high-field, type II superconducting materials.[82] The Pilot Rig (Fig. A-2) is believed to be the first superconducting magnet facility to be used for high-temperature plasma physics or fusion research.

Appendix A

Fig. A-2. Pilot Rig facility.

The facility's characteristics and performance are described in Ref. 82. The Pilot Rig functioned reliably over a 13-yr period. Its performance over the first 8 yr is described in Ref. 62. This experiment was the predecessor to the EFBT, the toroidal magnetoelectric containment experiment at NASA-Lewis. The operational history of the Pilot Rig from December 2, 1964 to June 17, 1971, is presented in Table A-I.

TABLE A-I

Operational History of Pilot Rig Superconducting Magnetic Mirror Facility

Coils first operated superconducting	December 2, 1964
Coils first operated at B_{max} = 2.5 T	December 10, 1964
First experimental use	January 12, 1965
Liquid helium loadings to June 17, 1971	556
Experimental runs with magnets charged to June 17, 1971	525
Coil normalcies to June 17, 1971	107
Final operational use with plasma	December 1977

A cutaway view of the superconducting coils and dewars is shown in Fig. A-3 and a plan view in Fig. A-4.

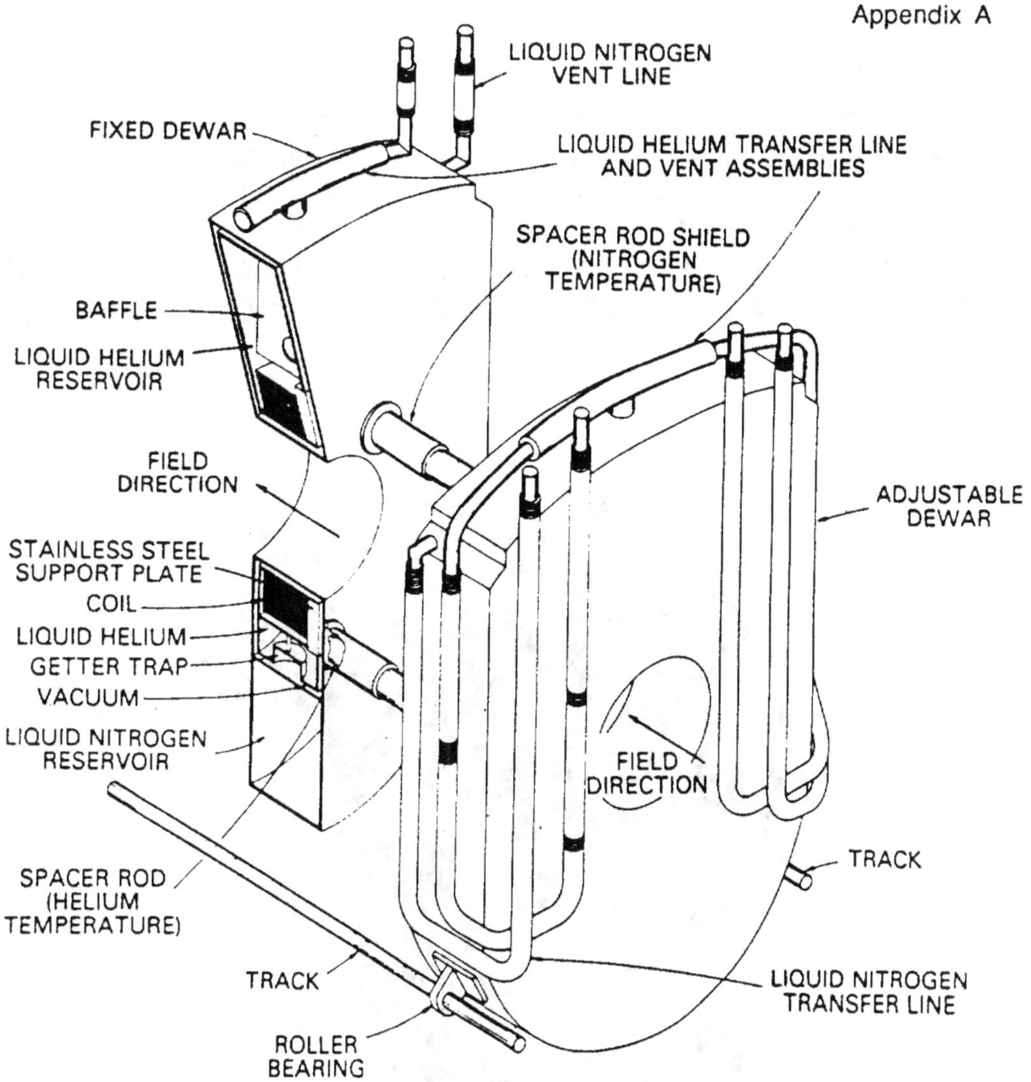

Appendix A

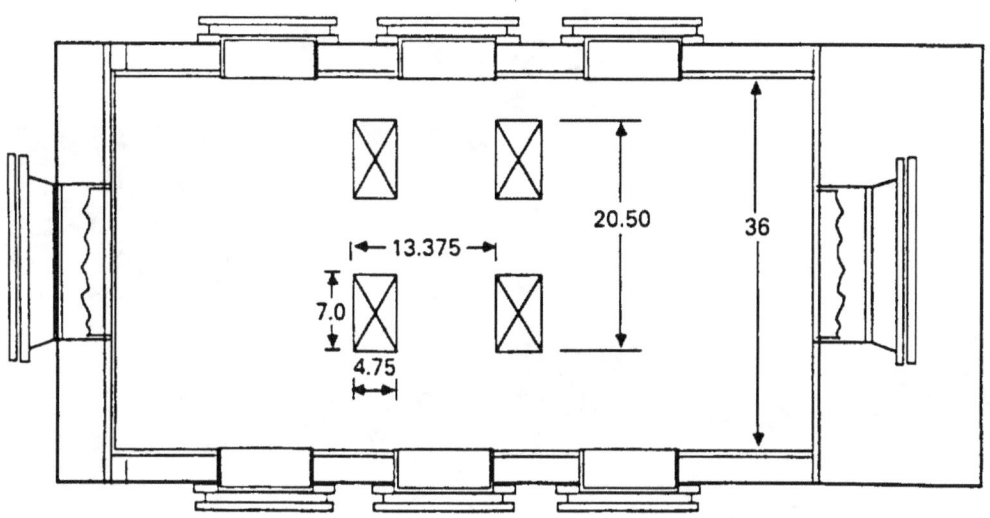

Fig. A.3. Isometric cutaway drawing in Pilot Rig superconducting coils and dewars.

Fig. A.4. Cutaway of the superconducting Pilot Rig Facility. Dimensions are given in inches.

A–7

Appendix A

The success of the superconducting Pilot Rig and of the modified Penning discharge as a magnetoelectric plasma heating and containment method led to the approval of the EFBT, the design of which is described in Refs. 29 and 62 through 66. A cutaway of the EFBT confinement concept is shown in Fig. A-5a, which depicts the inside of the facility. A photograph of the EFBT superconducting bumpy torus magnet array is shown in Fig. A-5b.

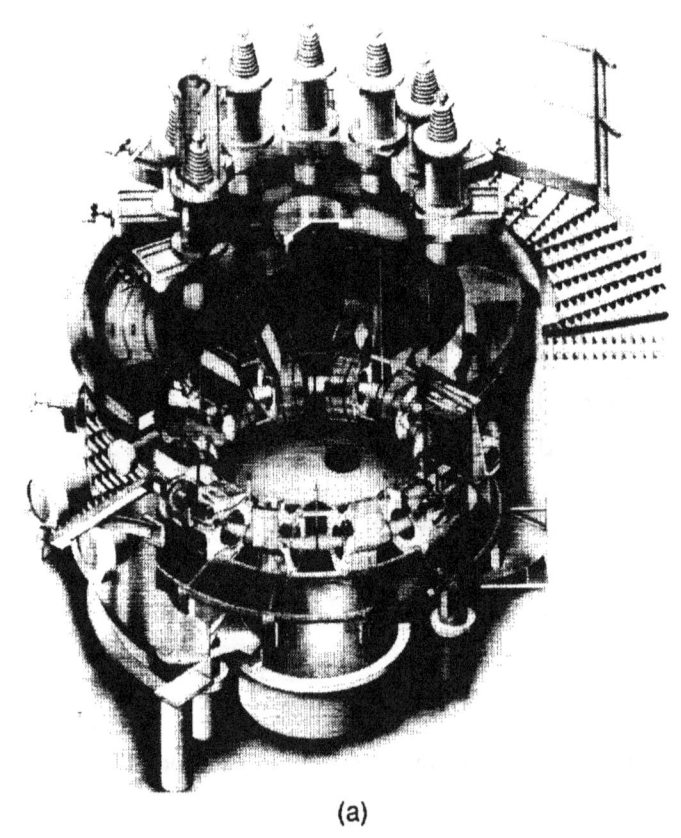

(a)

Appendix A

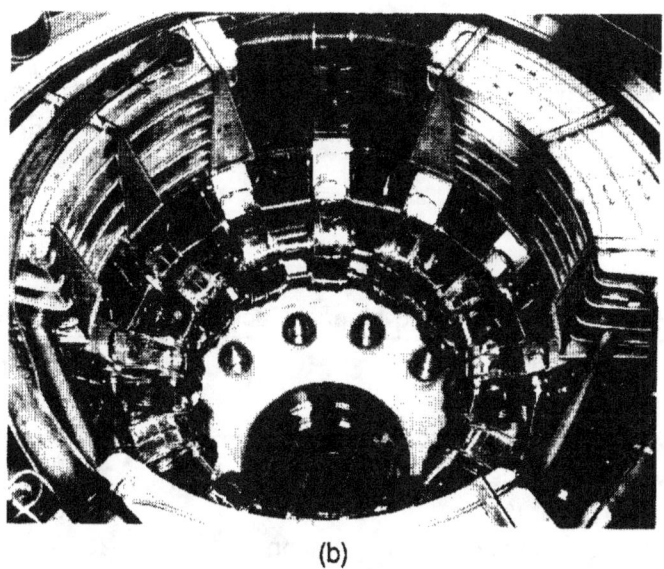

(b)

Fig. A-5. EFBT superconducting bumpy torus magnet array: (a) cutaway and (b) internal views of
the facility installation.

An external view of the facility in operation is presented in Fig. A-6.

Fig. A-6. EFBT facility in operation.

The EFBT consisted of 12 Pilot Rig plasmas bent around into a torus in such a way that no significant losses of plasma occurred along the magnetic field lines, since the individual ions and electrons circulated around the major circumference of the torus. The plasmas in both devices were magnetoelectrically contained and heated in a modified Penning discharge configuration. The EFBT commenced operation in 1972 and continued to perform reliably until it was shut down on March 31, 1978, when the program was terminated and high-temperature plasma research at NASA-Lewis concluded.

Appendix A

An intensive program of experimental research was conducted on the superconducting EFBT magnet facility throughout its 6-yr operation, as shown in Table A-II (Ref. 29).

TABLE A-II

EFBT Superconducting Magnet Facility Utilization Summary

First coil operation	April 24, 1972
First plasma	December 5, 1972
Days of operation with coils charged	436
Working days since first plasma	1337
Utilization factor (%)	33
Total hours of experimental operation	2620
Number of coil normalcies	189
Final shutdown	March 31, 1978

The data in this table represent a respectable utilization of the facility over a long period of time, which reflects highly both on the NASA professional and technical support staff responsible for this facility and on the reliability and availability of this apparatus as a superconducting magnet facility. The professional staff responsible for this facility included Walter M. Krawczonek, who was responsible for the electrical and electronic instrumentation of the facility; A. David Holmes, who was responsible for the cryogenic design, general engineering, and fabrication of the EFBT facility; and Willard D. Coles, who was responsible for the design, fabrication, and testing of the superconducting coils. The NASA-Lewis technical support staff did an excellent job in fabricating, maintaining, and repairing the superconducting EFBT magnet facility and in providing the necessary liquid helium for its operation. The overall design and characteristic features of the EFBT facility were the responsibility of J. R. Roth as was the responsibility for day-to-day operation during the 6-yr operational history summarized in Table A-II.

The basic research on high-temperature plasma physics and magnetoelectric confinement that was conducted on the EFBT facility was assisted by two contractors from academic institutions, Andrew L. Gardner of BYU, and his student R. T. Perkins,[122] who developed and put into operation a polarization diplexing microwave interferometer to measure the plasma number density in the EFBT and Edward J. Powers and his students Young C. Kim and Jae Y. Hong, from the University of Texas-Austin, who developed and applied to the EFBT plasma a diagnostic method that directly measured the radial transport of ions due to fluctuation-induced transport.[4,5,11,24,25,28-31] This diagnostic method has, since 1987, found wide application in the attempt by the fusion community to understand radial transport in tokamaks. In addition to these contractors, a number of NASA-Nuclear Regulatory Commission postdoctoral associates assisted with research on the EFBT plasma over the years. These postdoctoral associates included Glenn A. Gerdin, George X. Kambic, Raghuveer Mallavarpu, Richard W. Richardson, Chandra Singh, Chitra Sen, and Hans Persson.

The parameters of the EFBT plasma are listed in Table A-III (Ref. 29).

Table A-III
Plasma Parameters of the NASA-Lewis EFBT Experiment

Highest plasma densities
 $n_e = 3.1 \times 10^{12}/cm^3$, average
 $n_{e,\,max} = 6.2 \times 10^{12}/cm^3$, $\tau_p = 2.52$ ms, on axis

Highest particle confinement time
 $\tau_p = 6.0$ ms, $n_{e,\,max} = 1 \times 10^{12}/cm^3$

Highest simultaneous $n_{e,\,max}\,\tau_p$
 $n_{e,\,max}\,\tau_p = 1.6 \times 10^{10}$ s/cm^3

Ion kinetic temperatures in deuterium gas
 For the above conditions, $360 \leq T_i \leq 520$ eV
 Highest ever, $T_i = 2500$ eV

Electron kinetic temperatures
 For above conditions, $2 \leq T_e \leq 10$ eV
 Highest ever observed, $T_e \approx 150$ eV

The highest values of the individual fusion parameters (ion density, kinetic temperature, and particle containment time) are indicated, as well as the best simultaneous combination of the Lawson parameter (the product of electron number density and containment time) and ion kinetic temperature. The highest electron number density shown in Table A-III is that on the axis of the plasma, and the average value is about half the value on the axis.

The simultaneously observed plasma parameters in Table A-III for the NASA-Lewis EFBT experiment are compared in Table A-IV with the other U.S. bumpy torus experiment, the Oak Ridge National Laboratory's (ORNL) Elmo Bumpy Torus (EBT), which went into service in 1973 and was shut down in the early 1980s.

Appendix A

TABLE A-IV

Performance Comparison of the EBT and EFBT Experiments

Plasma Parameter	EBT	EFBT
Average density (cm^{-3})	1 to 1.5 × 10^{12}	3.2 × 10^{12}
Hot electron temperature (keV)	200 to 400	---
Toroidal electron temperature (eV)	200	5 to 30
Particle containment time (ms)	0.1 to 0.3	2.5
Energy containment time (ms)	0.1	~1
Ion energy (eV)	20 to 50	300 to 500
B_{max} (T)	1.0	2.4
Radial electric field (V/cm)	30	1500
Type of gas	Hydrogen, D$_2$	D$_2$
Neutral pressure (Torr)	5 × 10^{-5}	3 × 10^{-5}

The magnetic containment configuration of these two experiments was similar, a bumpy toroidal configuration of individual magnetic mirrors arranged end to end in a toroidal array; the methods of plasma production and heating, however, were entirely different. The EFBT relied on magnetoelectric confinement, which heated the ions preferentially with dc electrical power supplied to biasing electrode rings in contact with the plasma. The EBT experiment relied on relativistic electron rings at the midplane of each magnetic mirror to maintain the plasma and provide magnetohydrodynamic stability. In Table A-IV, the average electron number densities, ion kinetic temperatures, and particle containment times, essentially all of the important fusion relevant parameters, were significantly higher for the NASA-Lewis EFBT than for the EBT experiment at ORNL.

The plasma parameters shown in Table A-IV represent, in each experiment, the best simultaneously measured parameters. Both experiments produced higher values of each plasma parameter at other points on its performance envelope. These maximum values for the EFBT experiment are shown on Table A-III. The EFBT experiment did not employ electron cyclotron resonance heating and therefore had no relativistic electrons in the plasma. As a result, the EFBT plasma operated at low beta, unlike the EBT experiment, where the values of the plasma stability index, under some conditions, approached $\beta = 0.50$ as a result of the presence of hot electrons in the steady state. This demonstration of stable, steady-state, high-beta toroidal confinement was one of the most significant results of the EBT experiment.

The simultaneous values of the Lawson parameter $n\tau$, and the ion kinetic temperature of the EFBT experiment in the mid-1970s were equal to those of the mainline toroidal confinement program 15 yr earlier. This milestone was achieved, however, with a far smaller input of resources than was required for the mainline approaches. The radial electric field shown for the two experiments in Table A-IV was the ambipolar electric field resulting from the plasma potential of the EBT plasma; in the case of the NASA-Lewis EFBT, the very large radial electric fields were externally imposed by the negative multikilovolt bias on an electrode in contact with the plasma.[21,29] These very strong radially inward electric fields on the EFBT plasma not only created a toroidal electrostatic potential well for ions, thus assisting their magnetoelectric confinement, but also provided ion heating by E/B drift. This E/B drift resulted in high azimuthal drift velocities that thermalized through violent electrostatic plasma turbulence to the high ion kinetic temperatures shown. It was a characteristic feature of the EFBT plasma that the magnetoelectric E/B mechanism heated the ions preferentially, since the E/B drift velocity in the azimuthal direction was the same for both electrons and ions. This resulted in the imposed dc electric field depositing its energy preferentially in the heavier species.

In these magnetic containment experiments at NASA-Lewis, both the Pilot Rig and EFBT produced steady-state neutrons when deuterium was the operating gas.[29] Experiments demonstrated that the ion energies were Maxwellian[29] and that the plasma densities and temperatures were high enough to produce the observed neutron fluxes. A photograph of a steady-state plasma in the Pilot Rig is shown in Fig. A-7.

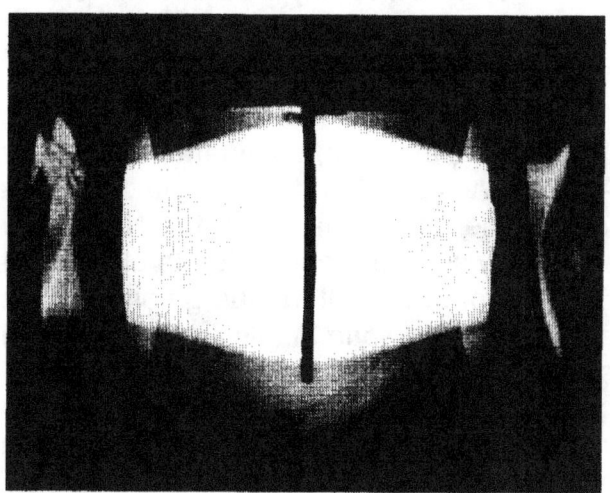

Fig. A-7. Steady-state D-D plasma in the Pilot Rig.

The observation of steady-state neutron production from D-D reactions in the Pilot Rig in 1967 is believed to be among the first in which steady-state neutron production was observed in a magnetically confined mirror plasma. In experiments on the EFBT plasma in the mid-1970s, steady-state neutron production from D-D reactions was observed at a level of 100 μW, at a time when the dc power input to the plasma was 100 kW, yielding a gain for those conditions of 10^{-9}. This level of neutron production was the maximum allowable by considerations of radiological safety.

The NASA-Lewis EFBT approach to plasma confinement and heating is referred to as "magnetoelectric confinement." This confinement and heating technique was unique and differed from any pursued by the DOE or any other country doing fusion research. Only since 1988 has the mainline tokamak program shown an interest in external biasing of toroidal tokamak plasmas in order to achieve magnetoelectric confinement and improve their densities and confinement times.

Other magnetic containment concepts rely solely on the magnetic field for plasma confinement, while the EFBT approach uses both electric and magnetic fields for containment and heating. Gross confinement is provided by the bumpy torus magnetic field (but gross confinement can be provided by other toroidal magnetic field geometries, such as the tokamak). Containment and heating of the plasma are assisted by radially inward electric fields that are imposed by a negative biasing electrode. This negative bias provides electrostatic containment of ions in the plasma while at the same time heating ions and electrons by E/B drift. During the NASA-Lewis experiments, the combination of negative plasma bias and the background magnetic field was shown to have a major beneficial effect on plasma density and confinement time.[22,27] These findings were later corroborated in negative bias experiments on the Macrotorr tokamak at University of California-Los Angeles and in more recent experiments in which tokamak plasmas have been negatively biased in the mainline "tokamak transport initiative" program.

NASA-Lewis developed magnetoelectric confinement sufficiently well to demonstrate the kinetic temperatures, confinement times, and number densities noted in Tables A-III and A-IV, and to do so in the steady state. Because of the magnetoelectric confinement principle used, the NASA-Lewis EFBT team was able to achieve higher Lawson parameters for its plasma than any other bumpy torus experiment in the world; this effort was among the top alternate magnetic confinement experiments in the world in this respect. As indicated in Table A-III, the highest plasma parameters achieved in the EFBT experiment, at the boundary of its performance envelope and not simultaneously achieved, are electron number densities on axis of $6.2 \times 10^{12}/cm^3$; the highest particle containment time was 6.0 ms, and the highest ion kinetic temperatures were 2500 eV for deuterium and 3500 eV for helium. In deuterium, the highest simultaneous Lawson parameter $n\tau$ was 1.6×10^{10} s/cm^3.

A major issue in any fusion experiment is radial transport and the resulting confinement time scaling. One of the major accomplishments of the EFBT team was identifying fluctuation-induced transport as the dominant mechanism by which particles escaped from the inside to the outside of the plasma[24,28,29] That transport research was collaboratively accomplished by J. R. Roth and Edward J. Powers. They also observed that an inward radial electric field produced fluctuating azimuthal electric fields, and this electrostatic turbulence could transport ions radially inward against the density gradient.[24,25] The scaling law for the particle containment time was derived from first principles and later confirmed experimentally. On the equally important subject of plasma heating, the magnetoelectric heating mechanism resulted in up to 45% of the dc input power to the plasma appearing in the ion population, with values between 20 and 30% being typical. The superconducting bumpy torus magnet facility continued to operate satisfactorily until it was shut down on March 31, 1978.

The most significant results of the LeRC Pilot Rig and EFBT work are summarized below.

The Pilot Rig was first operated in January 1965 with a steady-state plasma.[39,82] The initial research[36-38,40,43] with the Pilot Rig aimed to determine the magnetic field strength and geometry required to conserve the magnetic moment in mirror devices. The light emitted and charged-particle efflux from the plasma were found to occur in periodic pulses, with a frequency proportional to the square root of the neutral particle and electron number density product. These oscillations were proven to result from periodic solutions to the continuity equations for charged and neutral particles in the plasma, a previously unreported mechanism.[35,42,44-50,52]

The modified Penning discharge in the Pilot Rig was characterized by violent electrostatic turbulence, the properties of which were measured in an attempt to understand the observed Maxwellianization of the ion energy distribution.[41,51,53,54] Research to understand the magnetoelectric ion heating mechanism in the Pilot Rig plasma revealed two distinct spokes rotating in the sheath between the anode ring and the plasma. These spokes consisted, respectively, of ions and electrons with drift velocities that differed by a factor of 10. The ion heating mechanism was discovered, described by a theory, and subsequently confirmed experimentally. The results included analytical expressions for the ion and electron spoke rotation frequency, the ion kinetic temperature, and the efficiency of the ion heating process.[55-60]

The results of the EFBT facility's superconducting coil shakedown tests are reported in Refs. 62 through 66.

Electrode rings, located at one or more of the sector midplanes of the EFBT's bumpy torus magnetic field, biased the toroidal plasma to high potentials. The radial electric fields raised the ions to kilovolt energies by E/B drift. These ions were thermalized to Maxwellian distributions by processes similar to those of the modified Penning discharge in the Pilot Rig, and with electron temperatures <10% of the ion temperature.[15-22] The heating efficiency was estimated to range between 5 to 45%.

The crossed electric and magnetic fields in the EFBT plasma were used to form ion and electron spokes as in a smooth-bore magnetron. The ion spokes were in phase around the toroidal plasma's major circumference. The radial potential profile was measured with an ion beam probe in the Pilot Rig modified Penning discharge, and radial profiles of relative number density and electron

Appendix A

temperature were obtained spectroscopically from the toroidal EFBT plasma.[3,12-14,32-34] The strong radial electric fields – up to 20 kV/cm – had a profound effect on the plasma density and the particle confinement time. When the electric field pointed inward to push ions into the plasma, the number densities and confinement times were a factor of >10 higher than when the polarity was reversed and the field pointed outward. A team of contractors from the University of Texas, under the direction of Edward J. Powers, worked with NASA-Lewis staff under J. R. Roth to identify fluctuation-induced transport as the dominant radial transport mechanism and to show that the plasma transport is radially into the plasma when the electric field pointed inward.[11,24-25,27-31]

The spectroscopic and rf emission data on plasma density permitted the development of scaling laws for the total plasma current drawn by the power supply, the ion kinetic temperature, and the plasma number density as functions of the magnetic field, the background neutral gas pressure, and the electrode voltage.[19,22, 3,26,29] These scaling laws were valid for at least one order of magnitude over the independent and dependent variables.

In collaboration with Andrew L. Gardner of BYU, NASA-Lewis developed an interferometer based on the plasma-induced phase change between the ordinary and extraordinary modes of propagation to obtain reliable electron number density measurements.[22,122] The thermal expansion effects from steady-state operation made this difficult to accomplish using conventional microwave interferometry.

RF HEATING AND MAGNETIC MIRROR RESEARCH

Research on plasma heating and diagnostics was accomplished using a strong in-house analytical and experimental team that included Gerald W. Englert, Henry J. Hettel, Richard A. Krajcik, Roman Krawec, Milton R. Lauver, Carl F. Monnin, Richard W. Patch, George M. Prok, Warren D. Rayle, John J. Reinmann, Eli Reshotko, George R. Seikel, Donald R. Sigman, Aaron Snyder, Clyde C. Swett, and Richard R. Woollett. Their plasma science contributions include the following, which illustrate the broad scope and acceptance of the research results.

In experiments on steady-state ICRH, it was proven that a symmetric Faraday shield eliminated the inefficiency of electrostatic coupling of rf power when inductively producing ICRH of plasmas.[83-105] This scheme was adopted by Princeton Plasma Physics Laboratory (PPPL) personnel for application in their stellarator.

It was demonstrated that the effects of finite electron mass had to be included in the theory to obtain analytical solutions for the coupling of ion cyclotron waves into plasmas with radial density gradients.[100,101] This work was utilized by the PPPL, where it was compared with experimental measurements. The PPPL staff further extended the theory.

A partial "divertor" was invented to remove escaping unburned fuel and reaction products from a toroidal reactor in such a way as to produce a unidirectional beam of charged particles for use in space propulsion or for direct energy conversion.[78,79,132] An independent experiment by the British at the Culham Laboratory successfully demonstrated this concept. The NASA work is reviewed in Ref. 141.

A simple formula was obtained from first principles for calculating electron cyclotron radiation losses from a plasma.[76] The results accounted for both plasma and reflective wall parameters. This work, which is used in fusion reactor energy balance calculations, has been widely used by researchers in the field.

The first comprehensive plasma model was formulated and analyses were developed to obtain plasma ion temperature from cx neutral particle spectrometer and optical spectroscopy measurements.[109,110,118-121] At the time, the highest known product of steady-state plasma density ($2 \times 10^{13}/cm^{-3}$) and kinetic temperature (several kilo-electron-volts) for fusion research had been produced in the Superconducting Magnetic Mirror Apparatus (SUMMA) facility. Ion kinetic temperatures of several kilo-electron-volts were documented with line broadening via optical spectroscopy and with cx neutral particle mass/energy analysis. These were the first published (April 1974) empirical scaling relations for the temperature and density of a hot ion plasma experiment.[107,108,111-117]

This group had been concerned initially with steady-state ICRH. Both theoretical and experimental work were involved. The early ICRH experimental work was accomplished using water cooled magnet facilities in the 1960s and early 1970s. The SUMMA was the last facility (Fig. A-8) used for the steady-state plasma heating research at NASA-Lewis. It initially went into service in 1974 (Refs. 80 and 81).

Appendix A

Fig. A-8. SUMMA facility.

The magnetic field requirements and facility magnet configuration were designed to ensure that SUMMA would provide maximum versatility for experimental plasma physics research. The design and fabrication of SUMMA was coordinated among NASA, industry, and DOE. The design, fabrication, and installation of the SUMMA facility were done under the direction of Milo C. Swanson of the Engineering Design Division at NASA-Lewis. Key concerns addressed in the design were the low-temperature (4.2 K) structural properties of the metals involved and the minimization of heat leaks into the helium dewars. Swanson wrote the final specifications and contracted with the Nuclear Division of the Union Carbide Corporation (UCC-ND) to fabricate all the magnet dewar hardware and the cryogenic piping. At UCC-ND, a detailed design was produced by a group led by James E. Brewer. Out of this work came a unique design for low-heat-leak straps to support the heavy magnets that were contained in the helium dewars. The straps were made of fiberglass epoxy, which provided both the required strength and the necessary low thermal conductance. Charles R. Nichols, a section head in the Electromagnetic Propulsion Division at NASA-Lewis, brought the facility into operation. Stan Obloy, another section head at NASA-Lewis, was responsible for the design and procurement of the magnet power supplies and other electrical systems in SUMMA. Lawrence Nagy and Steven Posta were responsible for electrical and electronic instrumentation for the SUMMA facility and the plasma physics experiments.[80,81]

The SUMMA was the largest superconducting magnet facility used for plasma research at NASA-Lewis. It was characterized by high magnetic fields (designed for 8.6 T at the mirrors) and a large-diameter working bore (51-cm diam) with room-temperature access. The goal of the plasma research program in SUMMA was to produce steady-state plasmas of fusion reactor densities and temperatures, but not confinement times. The program included electrode development to produce a hot, dense, large-volume, steady-state plasma, and it included diagnostic development to document the plasma properties. SUMMA and its hot ion plasma were ideally suited to develop advanced plasma diagnostic methods. Two such methods whose requirements were well matched to SUMMA were (a) heavy ion beam probing to measure the plasma space potential and (b) submillimetre wavelength laser Thompson scattering to measure the local ion temperature and electron number density. Two NASA university grants were established to identify major requirements for developing these two diagnostic techniques.[98,123]

SUPERCONDUCTING MAGNET RESEARCH

At the formation of NASA in 1958, Wolfgang E. Moeckel, recognizing the potential of superconducting magnets for advanced propulsion devices in space, organized the Magnetics and Cryophysics Branch within his division. At his suggestion, Gerald W. Englert, then in the Plasma Physics Branch under Eli Reshotko, set out to study the feasibility of adapting such magnets to a thermonuclear rocket.

The Magnetics and Cryogenic Branch under James Laurence, and later Gerald Brown, was a leader in the design, fabrication, and testing of high-field-strength cryogenic and superconducting magnets. It supported and conducted basic research on the fabrication of copper stabilized superconducting wire. Willard D. Coles, a section head, was an outstanding magnet design engineer. He specified the materials and designed the windings for both the SUMMA and bumpy torus facilities and was responsible for acceptance testing. In the case of the SUMMA magnets, he established wire winding procedures and designs to prevent wire movement that could cause the magnets to go normal under very high-field/high-current-density operation. SUMMA's requirements led to the development of copper-stabilized, multifilament, niobium-titanium (Nb-Ti) superconducting wire. The first industrial production run of copper-stabilized Nb-Ti was for SUMMA. Windings in the high-field regions of SUMMA were fabricated from Nb_3Sn superconductive ribbon.

SUMMA was the world's largest warm-bore superconducting magnet facility for fusion research when it went into operation in 1974. It permitted investigations into several new technological areas including stabilized superconductors, superconducting magnet control systems, cryogenic heat transfer designs, and process control systems. SUMMA's requirements led to the development of a copper-stabilized, multifilament, Nb-Ti superconductor. The materials and magnet winding techniques used in the SUMMA magnet were the forerunners of the materials and techniques used in dozens of magnets produced since that time. SUMMA's 51-cm diam warm bore, designed for operation at 9 T, was the most advanced high-field superconducting magnet system up to that time.[80] The SUMMA facility was recognized by DOE as

Appendix A

making a unique contribution to the technology of superconducting magnets — an essential component of the ultimate success of controlled fusion energy.

SPACE POWER AND PROPULSION SYSTEM RESEARCH

The concepts by which fusion propulsion could be implemented using the EFBT were investigated by NASA-Lewis.[78,79] No actual experiments were conducted using these concepts before the program was terminated, since the EFBT hardware development work was directed toward plasma confinement.

The EFBT propulsion concept selected is shown in Fig. A-9.

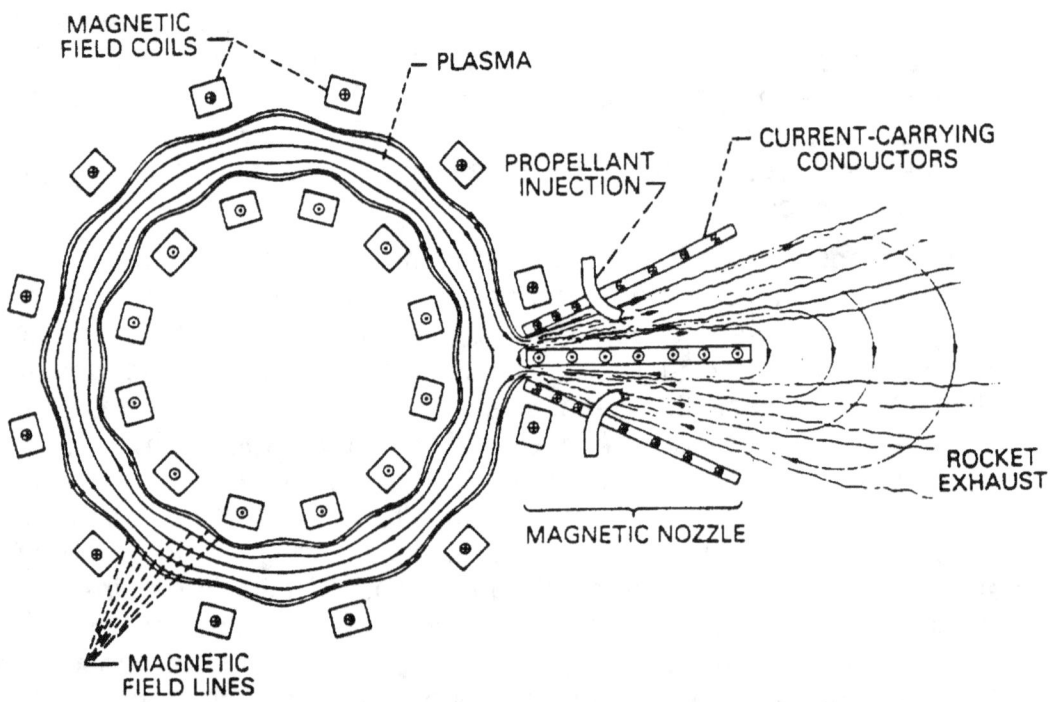

Fig. A-9. Magnetic divertor nozzle for a torroidal fusion rocket engine concept.

The fusion plasma is confined in a bumpy toroidal magnetic field generated by superconducting magnets. As the plasma reacts, it is gradually lost by outward diffusion along a minor radius. If this diffusion were to continue, the particles would hit the walls. The concept, however, incorporates a divertor and a magnetic nozzle assembly to collect the particles before they strike the walls and manipulate them into a unidirectional exhaust beam. Propellant is then added to this beam and the resulting mixture expands in the magnetic nozzle.[78,79,136] Development of the means to produce thrust was an important part of this program. John J. Reinmann invented the partial divertor nozzle for toroidal machines.[132] Fig. A-10 shows the magnetic divertor nozzles and a toroidal fusion rocket propulsion system concept.

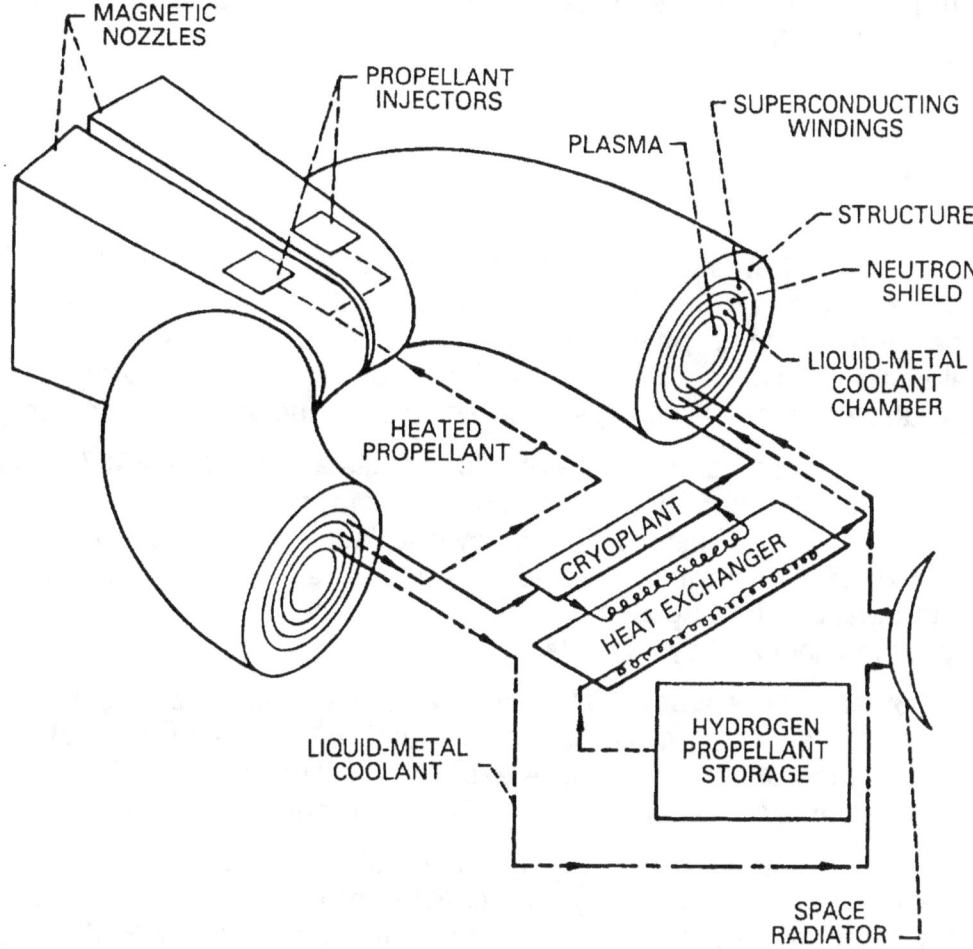

Fig. A-10. Direct toroidal fusion rocket propulsion system concept.

As mentioned earlier, Gerald W. Englert studied the feasibility of adapting superconducting magnets to a thermonuclear rocket and adding propellant to the escaping plasma to achieve optimum rocket exhaust velocity.[67-75]

MISSION ANALYSIS

The development of quickly implemented mission analysis techniques was accomplished at NASA-Lewis by Wolfgang E. Moeckel.[130-131] The Mission Analysis Branch performed the detailed trajectory and mission studies, using methods that are still in use today. Interplanetary travel using fusion energy, as well as interstellar missions,[138,139] was analyzed. It was concluded that D-^{3}He was the fusion fuel of choice for space application, although at the time the

Appendix A

presence of ^{3}He on the lunar surface was unknown. Analyses showed that considerable weight savings could be attained with superconducting magnets serving a dual role as magnetic nozzle and plasma confinement mechanism.[77,78,136]

SUMMARY

The NASA-Lewis program on the space applications of fusion energy, which ran from 1958-1978, produced new data and discoveries in high-temperature plasma physics research, and first-of-a-kind technological developments. These achievements have contributed to the progress of high-temperature plasma physics research and the technology of fusion energy.

Among the important plasma physics results achieved at NASA-Lewis is the discovery of the continuity equation plasma oscillation in the Pilot Rig plasma.[35,42,45-50] This low-frequency oscillation in the electron and neutral number density of partially ionized plasmas is related to the phenomenon of moving striations in dc gaseous discharges[50] and the fluctuations of predator and prey in ecological systems.[49]

Another discovery at NASA-Lewis was the experimental observation of the geometric mean plasma oscillation in the EFBT plasma,[6,7] which was later explained theoretically by Igor Alexeff of the University of Tennessee and published after termination of the NASA-Lewis program in 1978.

Another contribution to the mainline effort in fusion energy was the development of a diagnostic system capable of measuring fluctuation-induced transport and the application of this diagnostic system to the EFBT plasma. In this plasma, it was demonstrated that fluctuation-induced transport was the dominant radial transport mechanism and that, with a high negative bias of the toroidal plasma, the fluctuation-induced transport of ions could proceed radially inward against the density gradient. These early results on fluctuation-induced transport, made between 1975 and 1978, were recently replicated on tokamak plasmas at the University of Texas at Austin, and on the Advanced Toroidal Facility (ATF) at ORNL. The hardware and software used in these recent experiments in the mainline fusion program were originally developed for application to the EFBT plasma at NASA-Lewis. It is probably significant that three entirely different types of toroidally confined plasmas, the EFBT, tokamaks, and the ATF, all appear to be dominated by fluctuation-induced transport in their outer regions.

Other significant plasma-related measurements made during the NASA-Lewis program include the observation of steady state neutron production from thermonuclear D-D reactions in the Pilot Rig as early as 1967 and, after 1974, from the EFBT plasma. Another significant plasma-related accomplishment at NASA-Lewis was the demonstration of steady-state plasma heating by ICRH

Appendix A

and the later production of steady-state, high-density ($>10^{13}/cm^3$) hot ion (several kilo-electron-volts), neutron-producing D-D plasmas in a magnetoelectric, *E/B*, burnout-type plasma.

On the level of fusion-related technology, the NASA-Lewis program produced a number of firsts in the world fusion effort. The Pilot Rig superconducting magnetic mirror facility at NASA-Lewis went into service in December 1964. It was the first superconducting magnet facility ever to be used for high-temperature plasma or fusion-related research. This facility was reliable and was extensively used for a series of investigations from 1965 through 1972, when it was put on standby status to allow concentration on EFBT research. The Pilot Rig facility continued to operate reliably and without degradation of performance until it was finally shut down in December 1977 as a result of the termination of the NASA-Lewis program.

The superconducting bumpy torus magnet facility, built for EFBT research, was the first superconducting magnet facility to generate a toroidal magnetic field. This magnet facility went into operation in April 1972. Its reliability was such that it was used for about one-third of the working days between its initial operation in 1972 and its final shutdown on March 31, 1978. Another major advance in superconducting magnetic mirror facilities was the SUMMA, which provided a 51-cm diam room-temperature bore with a magnetic field up to 8 T on the magnetic axis. This facility was operated from 1974 through 1977, and it was shut down on termination of the NASA fusion program. Other contributions to superconducting magnet technology made during the NASA-Lewis program were the development of repeatedly demountable fittings suitable for use at liquid helium temperatures and in a vacuum system; development of fiberglass-epoxy straps to provide a low heat loss method for supporting the large SUMMA magnet coils in their helium dewars; development of superconducting wire and cables for high-field superconducting coils; and development of the hardware required for the steady-state application of ICRH power to high-temperature plasmas.

ACKNOWLEDGMENTS

The authors would like to thank John J. Reinmann and Gerald W. Englert of NASA-Lewis for providing information on aspects of the NASA-Lewis program for which they had responsibility.

Appendix A

REFERENCES

1. *Plasmas and Magnetic Fields in Propulsion and Power Research*, Conf., Cleveland, Ohio, October 16, 1969, NASA SP-226, National Aeronautics and Space Administration (1969).

2. G. A. GERDIN, "Radio Frequency Studies in the NASA Lewis Bumpy Torus," NASA TM X-71567, National Aeronautics and Space Administration (1974).

3. G. A. GERDIN, "Spoke Wavenumbers and Mode Transitions in the NASA Lewis Bumpy Torus," NASA TM X-71624, National Aeronautics and Space Administration (1974).

4. Y. C. KIM, W. F. WONG, E. J. POWERS, and J. R. ROTH, "Extension of the Coherence Function to Quadratic Models," *Proc. IEEE*, **67**, 428 (1979).

5. W. M. KRAWCZONEK, J. R. ROTH, C. E. BOYD, J. Y. HONG, and E. J. POWERS, "A Data Acquisition and Handling System for the Measurement of Radial Plasma Transport Rates," NASA TM 78849, National Aeronautics and Space Administration (1977).

6. R. MALLAVARPU, and J. R. ROTH, "Investigation of Possible Lower Hybrid Emission from the NASA Lewis Bumpy Torus Plasma," NASA TM X-73689, National Aeronautics and Space Administration (1977).

7. R. MALLAVARPU, "Lower Hybrid Emission Diagnostics on the NASA Lewis Bumpy Torus Plasma," NASA TM 73858, National Aeronautics and Space Administration (1977).

8. R. MALLAVARPU and J. R. ROTH, "Microwave Radiation Measurements near the Electron Plasma Frequency of the NASA Lewis Bumpy Torus Plasma," NASA TM 78940, National Aeronautics and Space Administration (1978).

9. H. PERSSON, "Time-Dependent Outgassing and Impurities in the NASA Lewis Bumpy Torus," NASA TM X-71639, National Aeronautics and Space Administration (1974).

10. H. PERSSON, "Ion Beam Probing of Electrostatic Fields," NASA TM X-79120, National Aeronautics and Space Administration (1979).

11. E. J. POWERS, Y. C. KIM, J. Y. HONG, J. R. ROTH, and W. M. KRAWCZONEK, "A Fluctuation-Induced Plasma Transport Diagnostic Based Upon Fast-Fourier Transform Spectral Analysis," NASA TM-38932, National Aeronautics and Space Administration (1978).

12. R. W. RICHARDSON, "Spectroscopic Results in Helium from the NASA Lewis Bumpy Torus Plasma," NASA TM X-71569, National Aeronautics and Space Administration (1974).

13. R. W. RICHARDSON, "Effect of Anode Ring Arrangement on the Spectroscopic Characteristics of the NASA Lewis Bumpy Torus Plasma," NASA TM X-71636, National Aeronautics and Space Administration (1974).

14. R. W. RICHARDSON, "Determination of Electron Temperature in a Penning Discharge by the Helium Line Ratio Method," NASA TM X-71677, National Aeronautics and Space Administration (1975).

15. J. R. ROTH, R. W. RICHARDSON, and G. A. GERDIN, "Initial Results from the NASA Lewis Bumpy Torus Experiment," NASA TM X-71468, National Aeronautics and Space Administration (1973).

16. J. R. ROTH, "Ion Heating and Containment in the NASA Lewis Bumpy Torus Plasma," NASA TM X-71630, National Aeronautics and Space Administration (1974).

17. J. R. ROTH, G. A. GERDIN, and R. W. RICHARDSON, "Characteristics of NASA Lewis Bumpy Torus Plasma Generated with Positive Applied Potentials," NASA TN D-8114, National Aeronautics and Space Administration (1976).

18. J. R. ROTH and G. A. GERDIN, "Characteristics of the NASA Lewis Bumpy Torus Plasma Generated with High Positive or Negative Applied Potentials," NASA TN D-8211, National Aeronautics and Space Administration (1976).

19. J. R. ROTH, "Preliminary Scaling Laws for Plasma Current, Ion Kinetic Temperature, and Plasma Number Density in the NASA Lewis Bumpy Torus Plasma," NASA TM X-73434, National Aeronautics and Space Administration (1976).

20. J. R. ROTH, G. A. GERDIN, and R. W. Richardson, "Characteristics of the NASA Lewis Bumpy Torus Plasma Generated with Positive Applied Potentials," *IEEE Trans. Plasma Sci.*, **PS-4**, 166 (1976).

21. J. R. ROTH and G. A. GERDIN, "Characteristics of the NASA Lewis Bumpy Torus Plasma Generated with High Positive or Negative Applied Potentials," *Plasma Phys.*, **19**, 423 (1977).

22. J. R. ROTH, "Factors Affecting the Ion Kinetic Temperature, Number Density, and Residence Times in the NASA Lewis Bumpy Torus Plasma," NASA TN D-8466, National Aeronautics and Space Administration (1977).

23. J. R. ROTH, "Optimization of Confinement in a Toroidal Plasma Subject to Strong Radial Electric Fields," NASA TM X-73690, National Aeronautics and Space Administration (1977).

24. J. R. ROTH, W. M. KRAWCZONEK, E. J. POWERS, J. Y. HONG, and Y. C. KIM, "Inward Transport of a Toroidally Confined Plasma Subject to Strong Radial Electric Fields," *Phys. Rev. Lett.*, **40**, 1450 (1977).

Appendix A

25. J. R. ROTH, W. M. KRAWCZONEK, E. J. POWERS, J. Y. HONG, and Y. C. Kim, "Inward Transport of a Toroidally Confined Plasma Subject to Strong Radial Electric Fields," NASA TM X-73800, National Aeronautics and Space Administration (1977).

26. J. R. ROTH, "A Model for Particle Confinement in a Toroidal Plasma Subject to Strong Radial Electric Fields," NASA TM X-73814, National Aeronautics and Space Administration (1977).

27. J. R. ROTH, "Effects of Applied DC Radial Electric Fields on Particle Transport in a Bumpy Torus Plasma," *IEEE Trans. on Plasma Sci.*, **PS-6**, 158 (1978).

28. J. R. ROTH, W. M. KRAWCZONEK, E. J. POWERS, Y. C. KIM, and J. Y. HONG, "Ion Confinement and Transport in a Toroidal Plasma with Externally Imposed Radial Electric Fields," NASA TP 1411, National Aeronautics and Space Administration (1979).

29. J. R. ROTH, "Ion Heating and Containment in an Electric Field Bumpy Torus (EFBT) Plasma," *Nucl. Instrum. Methods*, **207**, 271 (1983).

30. C. M. SINGH, W. M. KRAWCZONEK, J. R. ROTH, J. Y. HONG, Y. C. KIM, and E. J. POWERS, "Low Frequency Fluctuation Spectra and Associated Particle Transport in the NASA Lewis Bumpy Torus Plasma," NASA TP 1258, National Aeronautics and Space Administration (1978).

31. C. M. SINGH, W. M. KRAWCZONEK, J. R. ROTH, J. Y. HONG, and E. J. POWER, "Fluctuation Spectra in the NASA Lewis Bumpy Torus Plasma," NASA TP 1257, National Aeronautics and Space Administration (1978).

32. G. X. KAMBIC, "Heavy Ion Beam Probe Measurements of Radial Potential Profiles in the Modified Penning Discharge," NASA TM X-71643, National Aeronautics and Space Administration (1974).

33. G. X. KAMBIC, "Determination of the Radial Potential Profile in the Modified Penning Discharge with a Heavy Ion Beam Probe," *IEEE Trans. Plasma Sci.*, **PS-4**, 1 (1976).

34. G. X. KAMBIC and W. M. KRAWCZONEK, "A Heavy Ion Beam Probe System for Investigation of a Modified Penning Discharge," NASA TM X-3485, National Aeronautics and Space Administration (1977).

35. L. D. NICHOLS and J. R. ROTH, "Closed-Form Approximate Solutions to the Oscillatory Plasma Continuity Equations," NASA TM X-1944, National Aeronautics and Space Administration (1970).

36. J. R. ROTH, "Nonadiabatic Motion of a Charged Particle in an Axisymmetric Magnetic Barrier," *Phys. of Fluids*, **7**, 536 (1964).

37. J. R. ROTH, "Nonadiabatic Particle Losses in Axisymmetric and Multipolar Magnetic Fields," NASA TN D-3164, National Aeronautics and Space Administration (1965).

38. J. R. ROTH, "Optimization of Adiabatic Magnetic Mirror Fields for Controlled Fusion Research," NASA TM X-1251, National Aeronautics and Space Administration (1966).

39. J. R. ROTH, "Modification of Penning Discharge Useful in Plasma Physics Experiments," *Rev. Sci. Instrum.*, **37**, 1100 (1966).

40. J. R. ROTH, "Experimental Investigation of Single Interaction Nonadiabatic Losses from Axisymmetric Magnetic Mirrors," *Phys. Fluids*, **9**, 2538 (1966).

41. J. R. ROTH, "Plasma Stability and the Bohr-Van Leeuwen Theorem," NASA TN D-3880, National Aeronautics and Space Administration (1967).

42. J. R. ROTH, "New Mechanism for Low-Frequency Oscillation in Partially Ionized Gases," *Phys. Fluids*, **10**, 2712 (1967).

43. J. R. ROTH, "Correlation of Magnetic Moment Variation in Axisymmetric and Multipolar Magnetic Mirrors," *Plas. Phys.*, **10**, 809 (1968).

44. J. R. ROTH, "Periodic, Small-Amplitude Solutions to the Spatially Uniform Plasma Continuity Equations," NASA TN D-4472, National Aeronautics and Space Administration (1968).

45. J. R. ROTH, "Experimental Observation of Oscillations Described by the Continuity Equations of Slightly Ionized Deuterium, Neon, and Helium Gas," NASA TN D-4950, National Aeronautics and Space Administration (1968).

46. J. R. ROTH, "Experimental Observation of Continuity Equation Oscillations in Slightly Ionized Deuterium, Neon, and Helium Gas," *Plas. Phys.*, **11**, 763 (1969).

47. J. R. ROTH, "Experimental Observation of Low Frequency Oscillations Described by Plasma Continuity Equations," *Phys. Fluids*, **12**, 260 (1969).

48. J. R. ROTH, "Possible Applications of the Continuity-Equation Plasma Oscillation to Pulsars and Other Periodic Astrophysical Phenomena," NASA TN D-5078, National Aeronautics and Space Administration (1969).

49. J. R. ROTH, "Periodic Small-Amplitude Solutions to Volterra's Problem of Conflicting Populations and Their Application to Plasma Continuity Equations," *J. Math. Phys.*, **10**, 1412 (1969).

50. J. R. ROTH, "Theory of Moving Striations Based on the Continuity-Equation Plasma Oscillation," NASA TM X-52633, National Aeronautics and Space Administration (1969).

Appendix A

51. J. R. ROTH, "Experimental Observation of Quasi-Linear Mode Coupling in a Confined, Hot-Ion Plasma," NASA TM X-52718, National Aeronautics and Space Administration (1969).

52. J. R. ROTH, "Astrophysical Implications of the Continuity Equation Plasma Oscillation," *Nature*, **226**, 626 (1970).

53. J. R. ROTH, "Experimental Study of Spectral Index, Mode Coupling, and Energy Cascading in a Turbulent, Hot-Ion Plasma," NASA TM X-52919, National Aeronautics and Space Administration (1970).

54. J. R. ROTH, "Experimental Study of Spectral Index, Mode Coupling, and Energy Cascading in a Turbulent, Hot-Ion Plasma," *Phys. Fluids*, **14**, 2193 (1971).

55. J. R. ROTH, "Origin of Hot Ions Observed in a Modified Penning Discharge," NASA TM X-67956, National Aeronautics and Space Administration (1971).

56. J. R. ROTH, "Ion Heating Mechanism in a Modified Penning Discharge," NASA TN D-6985, National Aeronautics and Space Administration (1972).

57. J. R. ROTH, "Origin of Hot Ions Observed in a Modified Penning Discharge," *Phys. Fluids*, **16**, 231 (1973).

58. J. R. ROTH, "Hot Ion Production in a Modified Penning Discharge," *IEEE Trans. Plasma Sci.*, **1**, 34 (1973).

59. J. R. ROTH, "Energy Distribution Functions of Kilovolt Ions in a Modified Penning Discharge," *Plasma Phys.*, **15**, 995 (1973).

60. J. R. ROTH, "Energy Distribution Functions of Kilovolt Ions Parallel and Perpendicular to the Magnetic Field of a Modified Penning Discharge," NASA TN D-7167, National Aeronautics and Space Administration (1973).

61. C. SEN, "Probe Studies in a Modified Penning Discharge," NASA TM X-73631, National Aeronautics and Space Administration (1976).

62. J. R. ROTH, A. D. HOLMES, T. A. KELLER, and W. M. KRAWCZONEK, "A 12-Coil Superconducting 'Bumpy Torus' Magnet Facility for Plasma Research," *Proc. Applied Superconductivity Conf.*, Annapolis, Maryland, May 1-3, 1972, p. 361 (1972).

63. J. R. ROTH, A. D. HOLMES, T. A. KELLER, and W. M. KRAWCZONEK, "A 12-Coil Superconducting 'Bumpy Torus' Magnet Facility for Plasma Research," NASA TM X-68063, National Aeronautics and Space Administration (1972).

64. J. R. ROTH, A. D. HOLMES, T. A. KELLER, and W. M. KRAWCZONEK, "Performance of a 12-Coil Superconducting 'Bumpy Torus' Magnet

Facility," NASA TM X-68165, National Aeronautics and Space Administration (1972).

65. J. R. ROTH, A. D. HOLMES, T. A. KELLER, and W. M. KRAWZCONEK, "Characteristics and Performance of a 12-Coil Superconducting 'Bumpy Torus' Magnet Facility for Plasma Research," NASA TN D-7353, National Aeronautics and Space Administration (1973).

66. J. R. ROTH, A. D. HOLMES, and T. A. KELLER, "Performance of a 12-Coil Superconducting 'Bumpy Torus' Magnet Facility," *Proc. Technology of Controlled Thermonuclear Fusion Experiments and the Engineering Aspects of Fusion Reactors*, Austin Texas, November 20-22, 1972, p. 409, American Nuclear Society (1974).

67. G. W. ENGLERT, "High-Energy Ion Beams Used to Accelerate Hydrogen Propellant Along Magnetic Tubes of Flux," NASA TN D-3656, National Aeronautics and Space Administration (1966).

68. G. W. ENGLERT, "Application of Superconducting Magnets to a Thermonuclear Rocket," presented at National Superconductivity Information Mtg., Upton, New York, November 11, 1966.

69. G. W. ENGLERT, "Simulation of the Fokker-Planck Equation by Random Walks of Test Particles in Velocity Space with Application to Magnetic Mirror Systems," NASA TN D-5671, National Aeronautics and Space Administration (1970).

70. G. W. ENGLERT, "Effects of Coordinate Space Phenomena on End Losses From and Distributions Inside Magnetic Mirror Systems," Nucl. Fusion, **10**, 361 (1970).

71. G. W. ENGLERT, "Random Walk Theory of Elastic and Inelastic Time Dependent Collisional Processes in an Electric Field," *Z. Naturforschung*, **26**, 836 (1971).

72. G. W. ENGLERT, "Physical Interrelation Between Fokker - Planck and Random Walk Models with Application to Coulomb Interactions," *Appl. Sci. Res.*, **25**, 201 (1971).

73. G. W. ENGLERT, "Random Walk Study of Electron Motion in Helium in Crossed Electromagnetic Fields," NASA TN D-6648, National Aeronautics and Space Administration (1972).

74. G. W. ENGLERT, "Study of Thermonuclear Propulsion Using Super-Conducting Magnets," *Proc. 3rd Conf. Engineering Aspects of Magnetohydrodynamics*, Rochester, New York, March 28, 1962.

75. G. W. ENGLERT, "Towards Thermonuclear Rocket Propulsion," *New Scientist*, **16**, 16 (Oct. 4, 1962).

76. R. KRAJCIK, "The Effect of a Metallic Reflector Upon Cyclotron Radiation," *Nucl. Fusion*, **13**, 7 (1973).

Appendix A

77. J. J. REINMANN and W. D. RAYLE, "Deuterium-Helium-3 Fusion Power Balance Calculations," NASA TM X-2280, National Aeronautics and Space Administration (1971).

78. J. R. ROTH, W. D. RAYLE, and J. J. REINMANN, "Technological Problems Anticipated in the Application of Fusion Reactors to Space Propulsion and Power Generation," *Proc. Intersociety Energy Conversion Engineering Conf.*, Las Vegas, Nevada, September 21-25, 1970 Vol. 1, p. 2 (1970).

79. J. R. ROTH, W. D. RAYLE, and J. J. REINMANN, "Technological Problems Anticipated in the Application of Fusion Reactors to Space Propulsion and Power Generation," NASA TM X-2106, National Aeronautics and Space Administration (1970).

80. J. J. REINMANN, M. C. SWANSON, C. R. NICHOLS, S. J. OBLOY, L. A. NAGY, and F. J. BRADY, "NASA Superconducting Magnetic Mirror Facility," NASA TM X-71480, National Aeronautics and Space Administration (1973); see also in *Proc. 5th Symp. Engineering Problems of Fusion Research*, Princeton, New Jersey, November 6-9, 1973, Publication 73 CHO843-3NPS, Institute of Electrical and Electronics Engineers (1973).

81. J. J. REINMANN, M. R. PATCH, M. R. LAUVER, G. W. ENGLERT, and A. SNYDER, "Summa Hot-Ion Plasma Heating Research at NASA Lewis Research Center," NASA TM X-71840, National Aeronautics and Space Administration (1975).

82. J. R. ROTH, D. C. FREEMAN, and D. A. HAID, "Superconducting Magnet Facility for Plasma Physics Research," *Rev. Sci. Instrum.*, **36**, 1481 (1965).

83. H. J. HETTEL, R. KRAWEC; G. M. PROK; and C. C. SWETT, "Enhancement of Ion Cyclotron Waves in Hydrogen-Helium Mixtures," NASA TN D-4271, National Aeronautics and Space Administration (1968).

84. R. KRAWEC, "Radial Density and Temperature Profiles at the Ion Cyclotron Wave Resonance Point," NASA TM X-52159, National Aeronautics and Space Administration (1966).

85. R. KRAWEC, "Radial Density and Temperature Profiles at the Ion Cyclotron Wave Resonance Point," Paper 66-158, American Institute of Aeronautics and Astronautics (1966).

86. R. KRAWEC, "Steady-State Composition of a Low-Density Nonequilibrium Hydrogen Plasma," NASA TN D-3457, National Aeronautics and Space Administration (1966).

87. R. KRAWEC, "General Operating Characteristics of a Back-Streaming Direct-Current Plasma Generator," NASA TN D-4604, National Aeronautics and Space Administration (1968).

88. R. KRAWEC, "Effect of an Aperture on Measurement of the Axial Distribution Function in a Magnetically Confined Plasma," NASA TN D-5746, National Aeronautics and Space Administration (1970).

89. C. F. MONNIN and G. M. PROK, "Comparison of Gryzinski's and Born's Approximations for Inelastic Scattering in Atomic Hydrogen," NASA TN D-2903, National Aeronautics and Space Administration (1965).

90. C. F. MONNIN and G. M. PROK, "Comparison of Gryzinski and Born Cross Sections for the Metastable 2s State of Atomic Hydrogen," NASA TN D-3838, National Aeronautics and Space Administration (1967).

91. C. F. MONNIN and G. M. PROK, "Energy Transfer and Ion Cost in a Hydrogen Plasma," NASA TN D-5319, National Aeronautics and Space Administration (1969).

92. C. F. MONNIN and J. J. REINMANN, "Stability of Two-Fluid Wheel Flows with an Imposed Uniform Axial Magnetic Field," NASA TN D-5372, National Aeronautics and Space Administration (1969).

93. G. M. PROK and C. A. MCLEAN, "Intensity and Intensity Ratio of Principal Singlet and Triplet Lines of Molecular Hydrogen," NASA TN D-2522, National Aeronautics and Space Administration (1964).

94. G. M. PROK, "Two-Step Process to Increase Atomic Hydrogen Ion Production in Low-Pressure Radiofrequency Discharge," NASA TN D-2919, National Aeronautics and Space Administration (1965).

95. G. M. PROK and C. F. MONNIN, "Energy Required for Proton Production by Electron Impact in Mixtures of Atomic and Molecular Hydrogen," NASA TM X-52344, National Aeronautics and Space Administration (1967).

96. G. M. PROK, C. F. MONNIN, and H. J. HETTEL, "Estimation of Electron Impact Excitation Cross Sections of Molecular Hydrogen," NASA TN D-4004, National Aeronautics and Space Administration (1967).

97. G. M. PROK, C. F. MONNIN, and H. J. HETTEL, "Molecular Hydrogen Inelastic Electron Impact Cross Sections - A Semiclassical Method," *J. Quant. Spectros. Radia. Transfer*, **9**, 361 (1969).

98. J. J. REINMANN, M. R. PATCH, M. R. LAUVER, G. W. ENGLERT, and A. SNYDER, "Summa Hot-Ion Plasma Heating Research at NASA Lewis Research Center," NASA TM X-71840, National Aeronautics and Space Administration (1975).

99. D. R. SIGMAN, "Radiofrequency Power Transfer to Ion-Cyclotron Waves in a Collision-Free Magnetoplasma," NASA TN D-3361, National Aeronautics and Space Administration (1966).

Appendix A

100. D. R. SIGMAN and J. J. REINMANN, "Ion Cyclotron Wave Generation in Uniform and Nonuniform Plasma Including Electron Inertia Effects," NASA TN D-4058, National Aeronautics and Space Administration (1967).

101. D. R. SIGMAN and J. J. REINMANN, "Power Transfer to Ion Cyclotron Waves in a Two Ion Species Plasma," NASA TM X-1481, National Aeronautics and Space Administration (1967).

102. D. R. SIGMAN and J. J. REINMANN, "Power Coupling and Wave Fields of Ion-Cyclotron Waves in a Finite Length System," NASA TM X-52494, National Aeronautics and Space Administration (1968).

103. D. R. SIGMAN, "Calculations of Ion-Cyclotron Wave Properties in Hot Plasmas," NASA TM X-52719, National Aeronautics and Space Administration (1969).

104. D. R. SIGMAN, "Some Limitations on Ion-Cyclotron Wave Generation and Subsequent Ion Heating in Magnetic Beaches," NASA TM X-2263, National Aeronautics and Space Administration (1971).

105. D. R. SIGMAN, "A Comparison of Coupling Efficiencies for a Stix Coil and an $m = 1$ Coil," NASA TM X-2547, National Aeronautics and Space Administration (1972).

106. G. W. ENGLERT, "Trajectories of Charged Particles in Radial Electric and Uniform Axial Magnetic Fields," *IEEE Trans. on Plasma Science*, **PS-7**, 2 (1979).

107. R. KRAWEC, G. M. PROK, and C. C. SWETT, "Evaluation of Two Direct-Current Methods of Plasma Production for Use in Magnetic Mirror Experiments," NASA TN D-2862, National Aeronautics and Space Administration (1965).

108. M. R. LAUVER, "Effect of Anode-Cathode Geometry on Performance of the HIP-1 Hot Ion Plasma," NASA TP 1201, National Aeronautics and Space Administration (1978).

109. R. W. PATCH, D. E. Voss, and J. J. Reinmann, "Ion and Electron Temperatures in the SUMMA Mirror Device by Emission Spectroscopy," NASA TM X-71635, National Aeronautics and Space Administration (1974).

110. R. W. PATCH and M. R. LAUVER, "Ion Temperatures in HIP-1 and Summa From Charge-Exchange Neutral Optical Emission Spectra," NASA TM X-73471, National Aeronautics and Space Administration (1976).

111. J. J. REINMANN, M. R. LAUVER, R. W. PATCH, S. J. POSTA, A. SNYDER, and G. W. ENGLERT, "Hot Ion Plasma Heating Experiments in Summa," NASA TM X-71559, National Aeronautics and Space Administration (1974).

112. J. J. REINMANN, M. R. LAUVER, R. W. PATCH, S. J. POSTA, A. SNYDER, and G. W. ENGLERT, "Hot Ion Plasma Heating Experiments in SUMMA," *IEEE Trans. Plasma Sci.*, **PS-3**, 16 (1975).

113. J. J. REINMANN and R. W. LAYMAN, "Neutron Monitoring and Electrode Calorimetry Experiments in the HIP-1 Hot Ion Plasma," NASA TM X-3525, National Aeronautics and Space Administration (1975).

114. J. J. REINMANN, M. R. LAUVER, R. W. PATCH, R. W. LAYMAN, and A. SNYDER, "Hot Ion Plasma Production in HIP-1 Using Water-Cooled Hollow Cathodes," NASA TM X-71852, National Aeronautics and Space Administration (1975).

115. J. J. REINMANN and R. W. LAYMAN, "Neutron Monitoring and Electrode Calorimetry Experiments in the HIP-1 Hot Ion Plasma," NASA TM X-3525, National Aeronautics and Space Administration (1977).

116. D. R. SIGMAN and J. J. REINMANN, "Steady-State Hot-Ion Plasma Produced by Crossed Electric and Magnetic Fields," NASA TM X-2783, National Aeronautics and Space Administration (1973).

117. D. R. SIGMAN and J. J. REINMANN, "Parametric Study of Ion Heating in a Burnout Device (HIP-1)," NASA TM X-3033, National Aeronautics and Space Administration (1974).

118. G. W. ENGLERT, J. J. REINMANN, and M. R. LAUVER, "Interpretation of Neutral Particle Analyzer Measurements on Plasma Having Azimuthal Drift," *Plasma Phys.*, **17**, 609 (1975).

119. G. W. ENGLERT, R. W. PATCH, and J. J. REINMANN, "Model for Interpreting Doppler Broadened Optical Line Emission Measurements on Axially Symmetric Plasmas," *Plasma Phys.*, **20**, 451 (1978).

120. R. KRAWEC, "Electronic Analog for Performing Abel Inversion," *Rev. Sci. Instrum.*, **39**, 402 (1968).

121. R. W. PATCH, "Method for Decomposing Observed Line Shapes Resulting From Multiple Causes; Application to Plasma Charge-Exchange-Neutral Spectra," *J. Quan. Spectrosc. Radiat. Transfer*, **22**, 273 (1979).

122. R. T. PERKINS, "Interpreting the Plasma Phase Shift of a Polarization-Diplexing Microwave Interferometer," M.S. Thesis, Department of Physics and Astronomy, Brigham Young University (1977).

123. H. C. PRADDAUDE and P. WOSKOBOINIKOW, "Design of a Submillimeter Laser Thomson Scattering System for Measurement of Ion Temperature in Summa," NASA CR-2974, National Aeronautics and Space Administration (1978).

124. J. R. ROTH and M. CLARK, "Analysis of Integrated Charged Particle Energy Spectra from Gridded Electrostatic Analyzers," NASA TN D-4718, National Aeronautics and Space Administration (1968).

125. J. R. ROTH and M. CLARK, "Analysis of Integrated Charged Particle Energy Spectra from Gridded Electrostatic Analyzers," *Plasma Phys.*, **11**, 131 (1969).

126. J. R. ROTH and W. M. KRAWCZONEK, "Paired Comparison Tests of the Relative Signal Detected by Capacitive and Floating Langmuir Probes in Turbulent Plasma from 0.2 to 10 MHz," NASA TM X-52914, National Aeronautics and Space Administration (1970).

127. J. R. ROTH and W. M. KRAWCZONEK, "Paired Comparison Tests of the Relative Signal Detected by Capacitive and Floating Langmuir Probes in Turbulent Plasma from 0.2 to 10 MHz," *Rev. Sci. Instrum.*, **42**, 589 (1971).

128. J. C. EVVARD, "Wheel-Flow Gaseous-Core Reactor Concept," NASA TN D-2951, National Aeronautics and Space Administration (1965).

129. S. H. MASLEN, "Fusion for Space Propulsion," *IRE Trans. Military Electronics*, **MIL-3**, 52 (1959).

130. W. E. MOECKEL, "Propulsion Systems for Manned Exploration of the Solar System," *Astronaut. Aeronaut.*, **8**, 66 (1969).

131. W. E. MOECKEL, "Comparison of Advanced Propulsion Concepts for Deep Space Exploration," *J. Spacecr. Rockets*, **9**, 863 (1972).

132. J. J. REINMANN, "Fusion Rocket Concepts," NASA TM X-67826, National Aeronautics and Space Administration (1971).

133. J. R. ROTH, "A Preliminary Study of Thermonuclear Rocket Propulsion," Paper 59-944, American Rocket Society (1959).

134. J. R. ROTH, "A Preliminary Study of Thermonuclear Rocket Propulsion," *J. Bri. Interplanet. Soc.*, **18**, 99 (1961).

135. J. R. ROTH, "An Open Cycle Life Support System for Manned Interplanetary Spacecraft," *J. Spacecr. Rockets*, 3, 1971 (1966).

136. J. R. ROTH, W. D. Rayle; and J. J. Reinmann, "Fusion Power for Space Propulsion," *New Scientist*, **54**, 125 (1972).

137. J. R. ROTH, "Alternative Approaches to Plasma Confinement," *IEEE Trans. Plasma Sci.*, **PS-6,** 294 (1978).

138. D. F. SPENCER, "Fusion Propulsion System Requirements for an Interstellar Probe," JPL TR 32-397, Jet Propulsion Laboratory (1963).

139. D. F. SPENCER, "Fusion Propulsion for Interstellar Missions," *N.Y. Acad. of Sci. Ann.*, **140**, 407 (1966).

140. G. H. MILEY, "Evaluation of Performance of Select Fusion Experiments and Projected Reactors," NASA CR-3043, National Aeronautics and Space Administration (1978).

114. G. H. MILEY, *Fusion Energy Conversion*, American Nuclear Society, La Grange Park, Illinois (1976).

APPENDIX B

AN ALTERNATE STRATEGY FOR LOW SPECIFIC POWER REACTORS POWERING INTERPLANETARY SPACECRAFT, BASED ON EXPLOITING LASERS AND LUNAR RESOURCES

G. Logan
Lawrence Livermore National Laboratory
Livermore, CA

INTRODUCTION

A key requirement which establishes the minimum electric propulsion system's specific power performance for manned Mars missions is the maximum allowable radiation dose to the crew during the long transits between Earth and Mars. Penetrating galactic cosmic rays and secondary neutron showers give about 0.1 rem/day dose rate, which only massive shielding (e.g., a meter of concrete) can reduce significantly. With a humane allowance for cabin space, the shielding mass could be large enough to prohibitively escalate the propellant consumption required for reasonable trip times.

One solution that has been proposed is the use of permanently cycling spaceships with transfer vehicles which avoid acceleration and deceleration of large shielding mass, but which constrain round trip periods to long 4-year cycles. A more desirable alternative is to develop sufficient propulsion system performance for sufficiently short trip times that maximum dose limits not be exceeded. Such dose limits are not yet promulgated for space travel, but for reference, the US limits routine doses to nuclear plant workers to 5 rem/year, and 25 rem for one-time accident exposures. If the latter dose limit is applied to astronauts, the round trip time must be less than 250 days (0.7 year) at a dose rate of 0.1 rem/day. Then, for the Mars mission requirements discussed in Section 2.0, the minimum specific power for a 1,000 MT initial vehicle mass and 0.7-year round trip travel time is found from Fig. B-1 to be 0.33 kWe/kg.

Appendix B

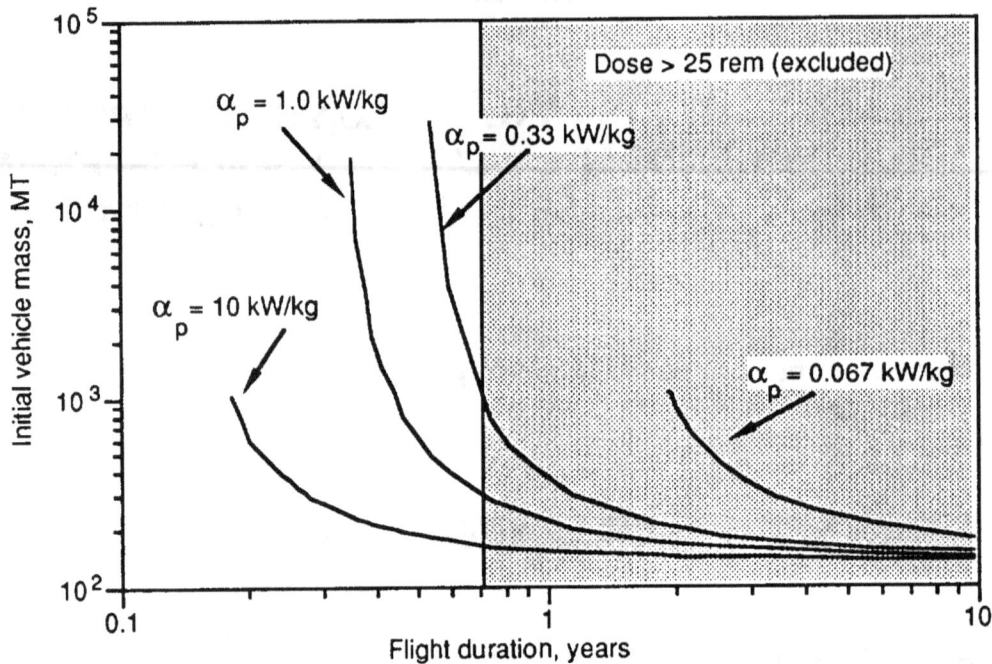

Fig. B-1. Initial vehicle mass variations with flight duration for a Manned Mars Mission.

The corresponding total mission Δv, specific impulse, and propellant consumption curves are indicated in Figs. B-2, B-3, and B-4, respectively. Dose limits less than 25 rem would require higher specific power capability than 0.33 kWe/kg.

Appendix B

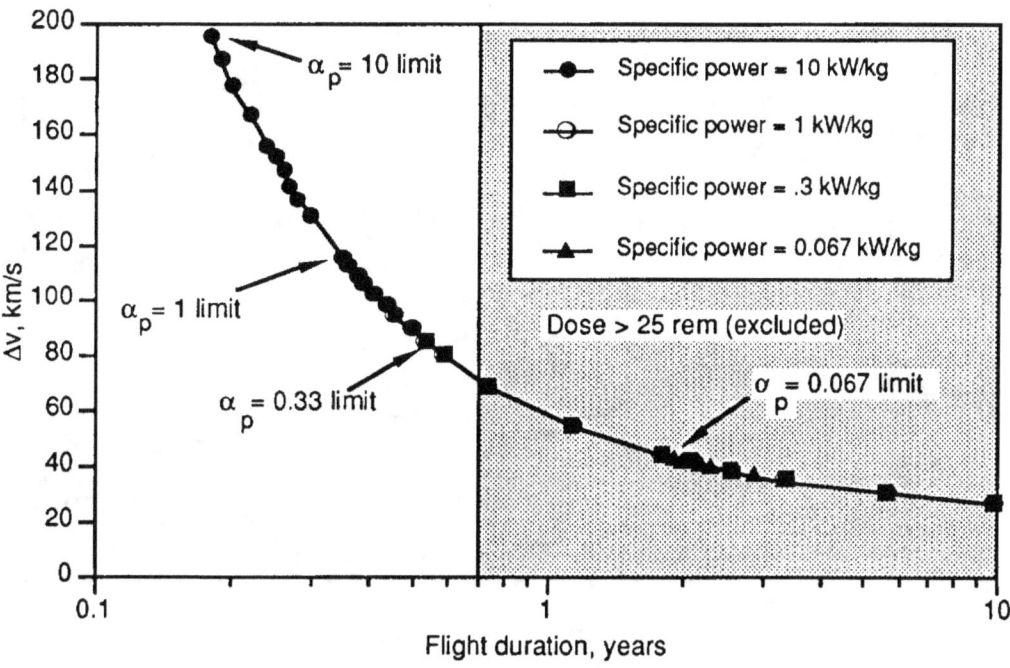

Fig. B-2. Vehicle velocity variations with flight duration for a Manned Mars Mission.

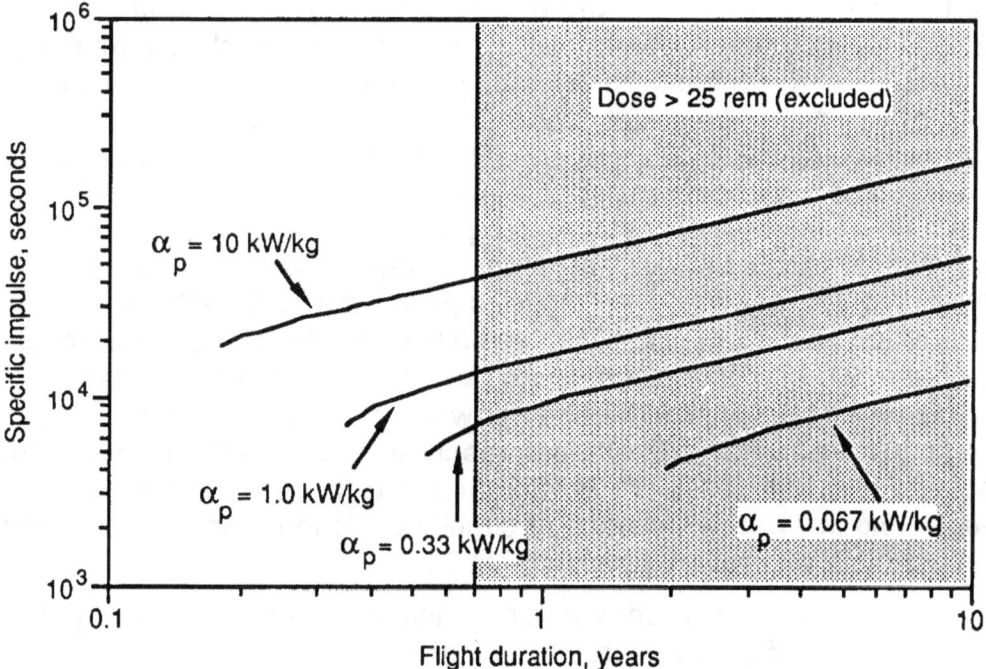

Fig. B-3. Specific impulse variations with flight duration for a Manned Mars Mission.

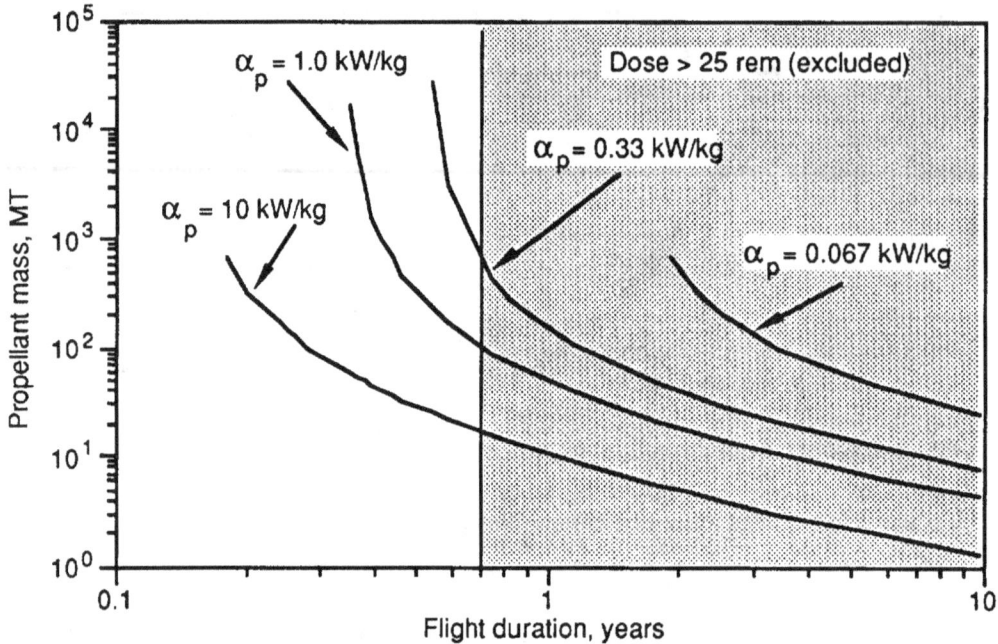

Fig. B-4. Propellant mass variations with flight duration for a Manned Mars Mission.

Given the present state of knowledge about solar, fission, and fusion candidates for spacecraft power, we cannot say that such minimum specific power values can be assured with any candidate, although, with various degrees of optimism, we might say that such a performance level might be reached with advances in technology. Rather than have the fate of important Mars and other manned interplanetary missions depend solely on the achievement of such threshold specific powers, it would be prudent to seek other paths to achieve such missions, even if fission or fusion reactor developments turn out with lower specific power than 0.33 kWe/kg. One such concept, which I dub "LASERPATH," would site lower specific power reactors at a lunar base and use their electricity to power large free-electron-lasers (FEL). These lasers in turn would remotely power lower mass spacecraft at much higher specific powers. Given that reactors at 0.33 kWe/kg were indeed available, it could still be more advantageous to base them on the moon for laser-powering the vehicles instead of directly powering them on-board with the same reactor specific power, provided that:

(1) the laser conversion efficiency were sufficiently high at sufficiently short wavelengths,

(2) the specific power of laser-driven photovoltaics for the vehicle's electrical power system were sufficiently greater than 0.33 kWe/kg, and

(3) a large fraction of the lunar-based reactors and lasers could be constructed from indigenous lunar materials.

Appendix B

The following description of the LASERPATH concept and comparisons of LASERPATH powered cases with on-board reactor-powered cases are not an attempt to fully substantiate these assumptions, but rather they serve to illustrate possibilities that warrant further study.

LUNAR-BASED FREE ELECTRON LASERS AND TRANSMITTERS

There have been several previous assessments of laser space power transmission,[1,2] but since these studies were completed, the recent advent of FEL's in the US SDI program and in the Japanese Center for Science and Technology Development at Osaka appear much more promising to meet the desired characteristics for lunar-based laser power transmission:

(1) 100 megawatt-level high average power,

(2) high conversion efficiency (20 to 40%),

(3) high specific power (≥1 kWe/kg), and

(4) tunability to any desired wavelength.

The last characteristic is important to match $h\nu$ to the optimum quantum energy above the bandgap of the vehicle photovoltaic receiver to achieve high photovoltaic power density and conversion efficiency described in the next section.

Fig. B-5 illustrates the basic components of one type of FEL, called Induction-Linac FEL, or IFEL, which is under development at the Lawrence Livermore National Laboratory.

Appendix B

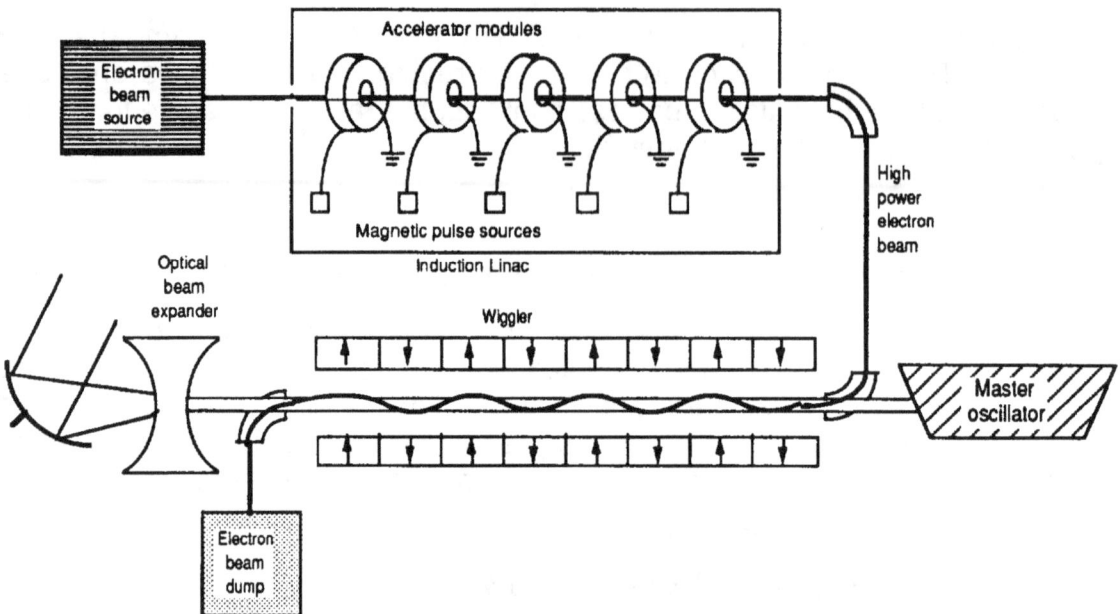

Fig. B-5. Schematic representation of the IFEL system, including the Linac, a wiggler, its driving source, and the output transmission system.

A 40% conversion efficiency at peak powers of a gigawatt have been recently demonstrated in a microwave IFEL at LLNL, and experiments at much shorter wavelengths are under way. Another basic type of FEL driven by an RF Linac is under development at Los Alamos National Laboratory and also in Japan. Both approaches accelerate an electron beam to high energies and pass the beam through a series of alternating transverse magnetic fields, called a wiggler, as shown in Fig. B-5. Provided a certain relationship between the electron energy, input light wavelength, wiggler field, and wiggler wavelength are satisfied, the periodic transverse motion of the electron beam in the wiggler field amplifies the input light intensity. Because the gain medium in the FEL is simply a bunch of free electrons traveling in a vacuum magnetic field, and because there need be no window in the space vacuum between the wiggler and reflective (Cassegrain-type) transmitter optics, there is no fundamental limit on the laser intensity set by the breakdown of materials, nor any constraints on the wavelength set by any atomic optical transitions. Thus, in principle, the power density in an FEL can be quite high, and the wavelength can be adjusted to any desired value.

Instead of being limited by the gain medium, the maximum IFEL power density would be set by cooling of the dielectric and ferromagnetic materials used in the magnetic pulse sources and accelerator modules of Fig. B-5. With typical stored energy densities in a single pulse of a hundred joules per kilogram, the "intrinsic" IFEL specific power, α_{IFEL}, in these accelerators active media depends on the pulse repetition rate, F_{rep}:

$$\alpha_{IFEL} \text{ (intrinsic)} \approx 0.1 \text{ kJ/kg} \times F_{rep}(\text{Hz}).$$

Appendix B

With solid state switching, the upper limit on F_{rep} set by cooling is currently expected to be 10 to 20 kilohertz. Thus, intrinsic $\alpha_{IFEL} > 10^3$ kWe/kg are possible. Of course, the overall IFEL system specific power will be much lower due to structure, wigglers, power supplies and space radiators for cooling. Thus, the maximum system α_{IFEL} might not be much larger than 1 kWe/kg. At this specific power, the laser system mass will be dominated by the structure, power supplies, and radiators.

LUNAR-BASED TRANSMITTER

The transmitter to direct the laser beam out of the FEL to the spacecraft photovoltaic receiver millions of kilometers away needs to be very large, both to limit diffraction losses discussed later on and to allow adequate cooling at the high beam power levels envisioned. To achieve diffraction-limited beam quality, the favored approach is to subdivide a large aperture transmitter into many smaller mirror segments. Each segment would be a thin hexagonal wafer, supported and adjusted by a set of three small, computer controlled electromagnetic or piezoelectric actuators. In this way, arbitrarily large phased optical transmitter arrays could be constructed at a moderate areal mass of about 40 kg/m^2. Balancing beam losses using optical coatings with radiative cooling would limit average beam intensity to about 100 kW/m^2. This corresponds to a transmitter specific power of 2.5 kW(beam)/kg, 10% as much mass as the IFEL at 1 kWe/kg and 25% efficiency.

The adaptive optics would control the beam phase front to within a small fraction of a laser wavelength, correct for thermal and gravitational distortions, and provide a small angular range of electronic beam steering. The beam would most likely be directed to an adaptive relay mirror at a high synchronous lunar orbit and then be redirected to track the spacecraft. The spacecraft receiver would best be a large diameter, parabolic foil collector ($\leq 10^{-2}$ kg/m^2 areal mass, $D_r \sim 1000$ m diameter), which concentrates the laser beam onto a smaller ($D_f \sim 100$ m) photovoltaic array at much higher areal mass (≤ 1 kg/m^2). The characteristics of this photovoltaic array is discussed next, and then the allowed laser transmission range versus laser power will be estimated.

PHOTOVOLTAIC RECEIVER CHARACTERISTICS

It is well known that photovoltaic conversion efficiencies with spectrally narrow laser light can be much higher than with solar radiation, much of the latter spectrum falling uselessly outside the semiconductor band gap.[3] A promising photovoltaic candidate for a Mars LASERPATH mission is a thin diamond film semiconductor, now under development at several laboratories. With a 5 eV band gap energy, E_b, a high conversion efficiency (e.g. 70%) might be achieved with UV laser wavelengths of 100 to 200 nm.[4] Furthermore, the conversion

Appendix B

efficiency should remain high up to higher temperatures, allowing more waste heat radiation off the wafer backsides.

For 100 μ thin film photovoltaic array at 1 kg/m² areal mass (including structure) and an equal total foil collector mass, a specific power of 10 kWe/kg would require 30 kW/m² average laser intensity on the photovoltaics (300 W/m² on the foil collector) to produce 20 kWe/m² of photovoltaic area. The waste heat radiated would be 10 kW_{th}/m² off the back side, giving an equilibrium photovoltaic temperature of 670K.

An important consideration for manned missions is reliability, with backups to system failure, if possible. During a 4-month, one-way flight duration mission, there may be sufficient time to repair a failed LBR or laser if the failure is experienced prior to the midpoint of the flight. This can be accomplished using the lunar base infrastructure or even shipping up spare parts from Earth. Building-in redundancy such as an extra reactor and laser, also helps. But ultimately, if all else fails, a LASERPATH system has an emergency backup energy source, albeit with less <u>usable</u> power – the sun. In principle, the large foil collectors envisioned could also deliver solar radiation to the photovoltaics, up to the limit imposed by photovoltaic temperature limits and waste heat radiation. For diamond photocells, the bandgap accepts only a slice of the less intense solar UV spectrum, so the solar conversion efficiency would probably be low, perhaps only a couple of percent. Nonetheless, since the concentrated solar flux and operating temperature can be higher, the diamond photovoltaic array might still have a solar output of electricity comparable to conventional silicon solar cells ($\approx$ 0.2 kWe/m²). With emergency solar power, a LASERPATH vehicle could limp home, provided sufficient propellant reserves exist at the time of the laser failure. The astronauts would receive a higher radiation dose with a longer solar-powered trip home, but they would survive.

RANGE OF LASER-POWER TRANSMISSION

Now that we have determined laser intensities at the lunar-base transmitter (100 kW/m²) and on the photovoltaic array (30 kW/m² within D_f, the collector focus), we can determine a relationship between average laser power, P_L, and range, R, between the transmitter and the foil receiver (collector), provided we specify the ratio of foil receiver diameter to photovoltaic (focus) diameter, D_r/D_f:

$$D_r = \frac{D_r}{D_f}D_f = \frac{D_r}{D_f}\left[\frac{4}{\pi}\left(\frac{0.9 P_L(W)}{3\times 10^4 (W/m^2)}\right)\right]^{1/2} \quad (1)$$

$$D_f = \left[\frac{4}{\pi}\left(\frac{0.9 P_L(W)}{3\times 10^4 (W/m^2)}\right)\right]^{1/2} \quad (2)$$

Now, diffraction relates the product, D_r/D_f, to the range, R, and the laser wavelength, λ, according to:

$$\frac{D_r}{D_f} = 2.44R \times \lambda = 220P_L \text{ (MW)} \qquad (3)$$

where we have used Eq. 1 and 2. The results are plotted in Fig. B-6 for various wavelengths, λ.

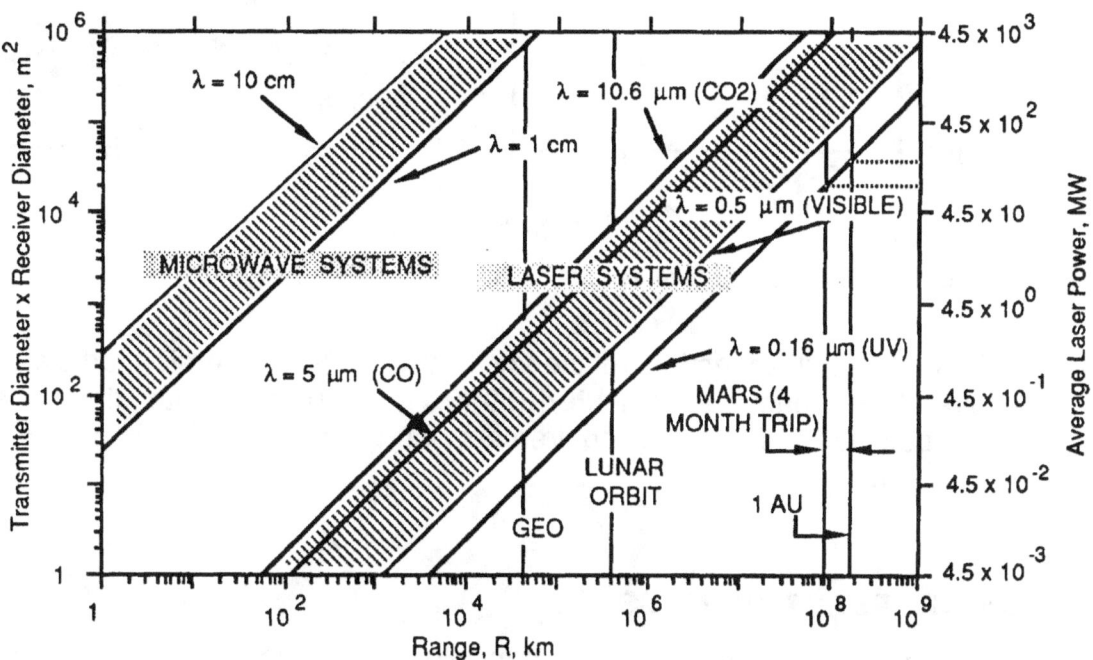

Fig. B-6. Percent of transmitter and receiver aperture (left scale) and average laser power (right scale) as a function of range between transmitter and receiver. Transmitter laser intensity = 100 kW/m^2; receiver laser transmitter = 300 W/m^2.

We see from Fig. B-6 that, for the short UV wavelengths we assume, a Mars LASERPATH mission can be achieved with 200 MW laser power. Longer wavelength lasers require either more power, or several laser stations enroute, to decrease the range requirement. Eventually, for regular manned shuttles supporting a permanent base, it would be advantageous to install at least one additional reactor and UV laser on the Martian moon Phobos.

From the mission requirements plotted in Figs. B-1 to B-4 and from Fig. B-7, we see that an on-board power source of 129 MWe is required for a 10 kW$_e$/kg specific power propulsion system to complete the mission within 0.7 year round trip travel time. That mission flight duration will limit the dose to 25 rem (background exposure).

Appendix B

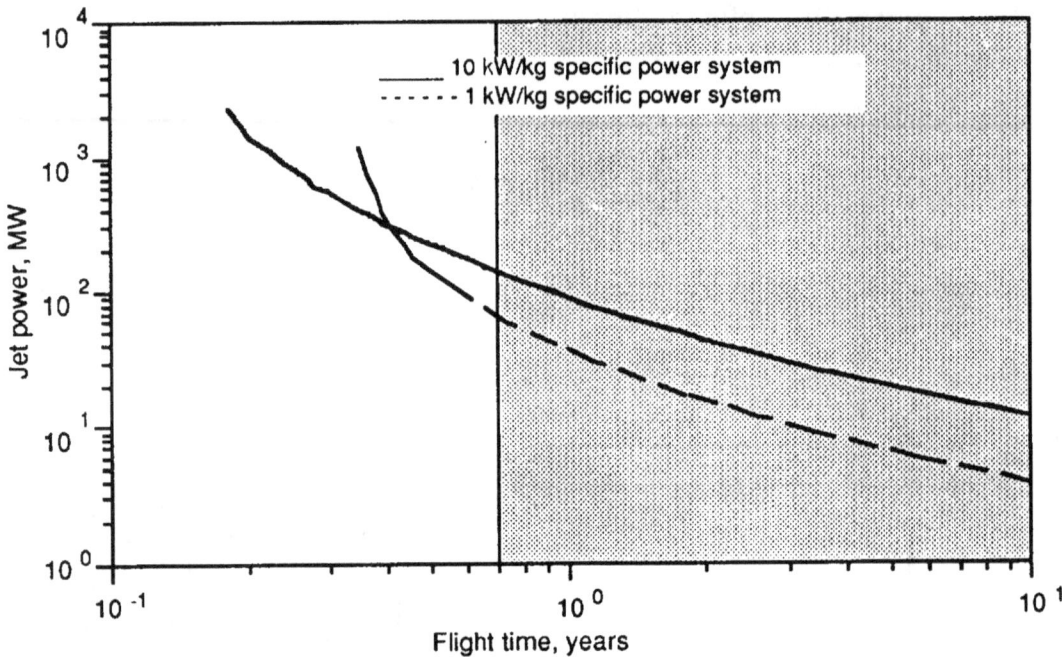

Fig. B-7. Flight power level requirements.

With a 90% foil collection efficiency and a 70% conversion efficiency, the required laser beam power is $129/[(0.9)(0.7)] \approx 200$ MW. Thus, there is a good match between the mission requirements and the LASERPATH power system performance.

LUNAR REACTOR MASS

Finally, we can address the performance requirements for lunar-based reactors (or other power sources) to power the Mars LASERPATH mission. Such lunar-based power sources could in principle be fission or fusion reactors, or even large solar-power stations. In any case, we inquire whether or not the propellant and vehicle mass savings made possible by laser driven photovoltaics could offset the greater reactor or power source mass incurred by the inefficiency of laser conversion in the LASERPATH scheme. Taking our IFEL laser example with a conversion efficiency of 25%, the 200 MW laser power output demands an 800 MWe lunar-based power source, ten times the 80 MWe required for an on-board power source with the 0.33 kWe/kg specific power necessary to meet the same 0.7 year round trip mission (see Fig. B-3). If one assumes the vehicles are reusable (but with a spare vehicle), one could compare the sum of the vehicles' power/propulsion system mass and the total propellant consumed for say, 10 round trips (20 years, given the 2-year Earth-Mars synodic period), with the corresponding sum in the LASERPATH case plus the added mass of the laser where M_{laser} = 800 MWe/(0.9 kWe/kg), MT including the transmitter optics, laser power conditioning, cooling, and supporting structural mass) and the added mass of the reactor, $M_{reactor} = 800$ MW_e/α_r, MT. Such comparisons are presented in Table B-1, for two on-board power sources characterized by $\alpha_r = 0.33$ kWe/kg (the minimum required for the mission – case 1), and $\alpha_r = 1$ W_e/kg (case 2), to represent the aspiration of more advanced fusion-powered vehicles, to be compared with two LASERPATH examples (cases 3 and 4) characterized by lunar-reactor specific powers of 0.33 kWe/kg and 0.067 kWe/kg, respectively.

TABLE B-1. Case Comparisons of Propulsion System Mass: On-Board Reactors versus Lunar-Based Reactors + Laser Transmission.

Parameter	Case 1 On-Board Reactor, $\alpha_r = $ 0.33 kWe/kg	Case 2 On-Board Reactor, $\alpha_r = $ 1 kWe/kg	Case 3 Lunar-Based Reactor, $\alpha_r = $ 0.33 kWe/kg	Case 4 Lunar-Based Reactor, $\alpha_r = $ 0.067 kWe/kg
Manned Mars Vehicle (1) Power, (2) Specific Power	80 MWe 0.33 kWe/kg	115 MWe 1 kWe/kg	129 MWe[a] 10 kWe/kg	129 MWe 10 kWe/kg
Two-Vehicle Power System Mass (MT)	480 (264)[b]	115 (94)	26 (25)	26 (25)
Propellant (MT) for 10 round trips	6300 [1890][c]	908 [272]	160 [48]	160 [48]
Lunar-based Reactor Mass (MT)	NA	NA	2400 (1320)	12,000 (600)
Lunar-Base Laser + Transmitter Mass (MT)	NA	NA	800 (722)	880 (722)
Total Pwr/Prop/Sys[d] Mass (MT)	6780	1023	3466	13,066
Total Pwr/Prop/System Mass (MT), Non-Lunar Origin, if $f_{mp} = 0$	(6564)	(1002)	(2227)	(1507)
Total Pwr/Prop/Sys Mass (MT), Non-Lunar Origin, if $f_{mp} = 0.7$	[2154]	[366]	[2115]	[1395]

Assumptions: 133 ton payload, 250 day round-trip travel time, (25 rem round trip dose), 2 vehicles (one for standby), specific power of lunar-based laser + optics system = 0.9 kWe/kg lunar-mass-utilization factors, f_m: $f_m = 0.02$ for $\alpha = 10$, $f_m = 0.18$ for $\alpha = 1$, $f_m = 0.45$ for $\alpha = 0.33$, $f_m = 0.95$ for $\alpha = 0.067$; propellant $f_{mp} = 0$ or 0.7 (as indicated).

Appendix B

    a Foil concentrators + photovoltaic array for vehicle power (case 3 and 4).
    b Figures in parenthesis subtract mass of lunar origin $(1 - f_m) \times$ component mass.
    c Figures in bracket adjusted by $(1 - f_m) = 0.3$ factor for propellant.
    d Includes vehicle power systems for 2 vehicles, propellant for 10 trips, and lunar-based reactors and lasers, where appropriate.

As the specific detail of optimized lunar reactor designs is beyond the scope of this work, I seek to characterize such reactors by specifying only their specific power. The lunar reactor case 3 with 0.33 kWe/kg is chosen to compare with case 1, having the same specific power for an on-board reactor which can barely meet the mission requirement. The lunar reactor case 4 with 0.067 kWe/kg is chosen to illustrate what happens with a specific power no better than the target SP-100 nuclear fission units, which <u>cannot</u> meet the Mars mission as on-board reactors (at least with <25 rem round-trip dose constraints).

LUNAR MASS UTILIZATION

Normally, one compares total mass between competing space power systems meeting the same mission since transportation costs to LEO could likely dominate over terrestrial material and fabrication costs for >10^3 ton space systems. When that is the case, the unit costs of very different materials and fabrications tend to be closer to the same transportation costs per unit mass. This is even more likely to be the case for lunar space systems, if transport from the Earth to the Moon were required.

The NASA Office of Exploration sponsored studies of possible use of lunar materials for space development and ways to manufacture various commodities and structures on the moon, and the Office of Aeronautics and Space Exploration has established the Center for the Utilization of Local Planetary Resources at the University of Arizona. For example, heavy radiation shielding might be made of lunar concrete. Iron-nickel micrometeorite particles collected from the lunar soil might provide steel structures, and traces of low-atomic-number solar-wind gases trapped in the finer lunar dust can be outgassed by heating (H_2, H_2O, He, CO_2, etc). Without a detailed design, one cannot determine what fraction, f_m, of a given lunar system, such as a reactor, could be made of indigenous lunar materials. However, <u>if</u> a substantial fraction of reactor systems, which might be dominated by structures, shielding, transformer iron in power supplies, etc. <u>could</u> be made of lunar materials; and, furthermore, <u>if</u> such a fraction were different for different types of reactors (as is likely to be the case), then the important comparison between competing propulsion systems would be the total mass <u>minus</u> any lunar-origin mass, i.e., the mass portion that must be transported from Earth. This assumes that the unit cost of Earth-origin mass significantly exceeds the unit cost of lunar-origin mass, which would be the case if the total lunar mass of each type produced were a large multiple of the initial investment of lunar mining and manufacturing equipment mass. If the lunar production equipment mass were not negligible, it could be included as an effectively smaller lunar mass utilization factor, f_m.

I will not attempt to fully justify the f_m values assumed in Table B-I, which are picked primarily to illustrate how the impact of large f_m fractions might change the comparative system economics of the various cases. I inserted just a tiny bit of logic to the f_m assumptions. For the f_m values pertaining to power generation and conversion (reactors and lasers), I suppose that f_m can in general increase with decreasing specific power on the argument that, the higher the specific power, the narrower the choice of materials which can reach the higher performance levels, and the more likely such specialty materials would have to be transported from Earth. Thus, I chose $f_m = 0.02$ for $\alpha = 10$ kW/kg, $f_m = 0.18$ for $\alpha = 1$ kW/kg, $f_m = 0.45$ for $\alpha = 0.33$ kW/kg, and $f_m = 0.95$ for $\alpha = 0.067$ kW/kg, for either reactors or lasers, which reflects this tendency, although the actual values are arbitrary. I would like to mention, at least in the case of magnetic fusion, of which I am most familiar, that $f_m = 0.95$ is not obviously impossible to achieve. At 800 MWe and $\alpha_r = 0.067$ kW/kg, a 12,000 ton D-^{3}He tokamak (5) might consist of:

    (1) 4000 tons superconducting magnets consisting of 3400 tons of iron-nickel steel structure,

    (2) 300 tons of aluminum stabilizer,

    (3) 300 tons of superconducting wire,

    (4) 3000 tons of steel neutron shielding,

    (5) 2000 tons of blankets (which could be a simple, helium cooled, ferritic steel structure),

    (6) 2700 tons of heat injection space radiators (mainly low-pressure steel tubing), and

    (7) 300 tons of solid-state microwave rectenna convertors.

If meteorite-derived steel can be used, there would be essentially only 600 tons of superconductor and rectenna converters to import from Earth.

As for the vehicle propellant, I assumed two different values for $f_{mp} = 0$ and 0.7, to illustrate the impact of using imported propellant ($f_{mp} = 0$), such as argon or sodium, or using lunar-derived propellant, such as hydrogen. Most electric-powered plasma thrusters would run on either heavy noble gases, alkali metals, mercury, or cesium, none of which are likely to be lunar indigenous, due to their intrinsic volatility. Although hydrogen is difficult to use in electric thrusters, and difficult to store for long periods, these problems might be overcome in the future. The hydrogen exists only in trace amounts in lunar soil, so f_{mp} should not be too close to unity when accounting for the hydrogen extraction, liquefaction, and storage equipment mass.

CONCLUSIONS AND RECOMMENDATIONS

From the results in Table B-1 we can draw some conclusions, some of which are more qualitative than quantitative until more analysis is done:

(1) The rationale for LASERPATH hinges mainly on how high a specific power fission, fusion, or solar power systems can be developed for powering manned vehicles. If, for example, sufficiently advanced fusion reactors could achieve $\alpha_r = 1$ kWe/kg, then it would be best to pursue the conventional approach with the reactor carried on-board. If, however, $\alpha_r \ll 0.33$ kWe/kg, then a mission with less than 0.7 year travel time and 25 rem doses cannot be achieved at all with on-board reactors, and in this case the LASERPATH approach might meet the mission requirement with lower specific mass reactors and with comparable total mass investment as if $\alpha_r = 0.33$ kW$_e$/kg reactors were available.

(2) The viability of LASERPATH depends on the development of advanced photovoltaics, adaptive transmitter optics, and efficient free-electron-lasers, all of which appear to be promising but remain to be demonstrated at the performance levels needed. NASA should encourage and participate in such developments, as a hedge against the uncertainty of reactors reaching the high specific powers required for on-board power systems.

(3) The actual commitment of mass transport from Earth to establish lunar power reactors and lasers might be heavily influenced by the availability and suitability of lunar materials in their construction. NASA should sponsor a study, in conjunction with the ongoing lunar resource studies, to explore the different degrees to which different lunar power sources – fission, fusion, and solar – can utilize lunar materials, and in so doing, encourage innovative thinking from reactor designers to more fully exploit lunar materials, i.e., reoptimize the reactor designs for the lunar base development.

(4) As the duty factor required for Mars missions every two years is low ($\approx 35\%$), an investment in a lunar LASERPATH system could be utilized for a variety of other space enterprises in between flights, further leveraging the investment.

REFERENCES

(1) R. D. Arno, J. S. MacKay, and K. Nishioka, "Applications Analysis of High Energy Lasers," NASA TM X-62, NASA AMES, 142 (March 1972).

(2) W. J. Schafer Associates, Inc., "A Study to Survey NASA Laser Applications," Feb. 1978, in "Space Laser Transmission Studies," by M. D. Williams and E. J. Conway, editors, NASA Conference Publication 2214, (1982).

(3) G. H. Walker and J. H. Heinbockel, "Photovoltaic Conversions of Laser Power to Electrical Power," NASA TM 89041, NASA Langley (Sept. 1988).

(4) B. G. Logan, "Initiative for the 21st Century: Advanced Space Power and Propulsion Based on Lasers," Seminar given at NASA Lewis Research Center, April 26, 1988. Lawrence Livermore National Laboratory, UCRL-98520 (preprint).

(5) G. L. Kulcinski, G. A. Emmert, J. P. Blanchard, L. El-Guebaly, H. Y. Khater, J. F. Santarius, M.E. Sawan, I. N. Sviatoslavsky, L. J. Wittenberg, and R. J. Witt, "Apollo-An Advanced Fuel Fusion Power Reactor for the 21st Century," University of Wisconsin Fusion Technology Institute, Report UWFDM-780 (Oct. 1988).

APPENDIX C

CHARACTERISTIC PLASMA DEVELOPMENT REQUIREMENTS FOR GENERIC SPACE FUSION REACTORS

G. Logan
Lawrence Livermore National Laboratory
Livermore, CA

INTRODUCTION

Here we examine basic plasma characteristics and requirements for fusion space reactors of three generic types: inertial-confinement-fusion (ICF), magnetically-insulated and inertially-confined-fusion (MICF), and magnetic-confined-fusion (MCF). We consider both D-T and D-^{3}He fuel cycles, using D-T as a spark-plug for initiating the D-^{3}He burn in the latter cases. Plasma characteristics are determined on the common basis that sufficient charged plasma energy yield from a fusion burn pulse be at least 10 times the electrical energy consumed to start the reactor, taking into account the electrical efficiency, η_a, of the auxiliary plasma heating system in each case and the efficiency, η_c, of coupling that energy input to get the plasma to ignition temperature. For ICF, we assume a laser-driven implosion with $\eta_a = 0.1$ and $\eta_c = 0.15$; and for MICF, we assume a high-velocity compact torus accelerator, e.g., RACE, as the driver with $\eta_a = 0.5$ and $\eta_c = 0.9$. For MCF, we take a generic torus with an aspect ratio $R/a = 3.3$ with $\eta_a = 0.5$ and $\eta_c = 0.9$ for some unspecified form at plasma heating but with a similar efficiency as for MICF.

CONFINEMENT

The basic plasma confinement times in the three different generic approaches to fusion are:

(a) ICF:
$$\tau_i = \frac{a}{4\bar{v}_i}, \text{ (s)} \tag{1}$$

where τ_i is the characteristic fuel ion confinement time during the burn at peak compression, "a" is the radius of the laser-compressed fuel mass, and $\bar{v}_i$ is the average ion speed as given by:

Appendix C

$$\bar{v}_i = 4.4 \times 10^5 \left[\frac{T_i(\text{keV})}{\bar{A}_i} \right]^{1/2}, \quad (\text{m/s}) \tag{2}$$

the mean ion velocity at the ion (burn) temperature, T_i, and average ion mass number, $\bar{A}_i$, in a.m.u.

B. MICF:
$$\tau_i = \frac{a}{0.7(p/\rho_w)^{1/2}}, \quad (\text{s}) \tag{3}$$

where "a" is the radius of the spherical, heavy-metal tamper shell (hollow cannonball), "p" is the plasma pressure:

$$p = Cn_i T_i(\text{keV}) \times 1.6 \times 10^{-16}, \quad (\text{pascals}) \tag{4}$$

$C = 2$ for D-T, 2.5 for D-^{3}He, and n_i is the fuel ion density:

$$n_i \equiv n_D + n_T, \text{ or } n_D + n_{^3\text{He}} \quad (\text{ions/m}^3), \tag{5}$$

and ρ_w is the MICF shell mass density (17,000 kg/m³ for gold, for example).

C. MCF:
$$t_i = 5\tau_E, \quad (\text{s}) \tag{6}$$

where τ_E is the cross-field thermal conductivity loss time for plasma energy, given by

$$t_E = \frac{a^2}{4\chi_\perp}, \quad (\text{s}) \tag{7}$$

with $\chi_\perp$ (min) ≈ 0.1 m²/sec taken as the lowest achieved thermal diffusivity measured in MCF experiments to date. The ICF and MICF plasma volumes, V_p, are computed as spheres $4/3\pi a^3$ (m³), and the MCF plasma volume is a torus of $2\pi^2 R a^2 K$ (m³) with an aspect ratio of $R/a = 3.3$ and an elongation of $K = 2$.

IGNITION AND BURN

We use the following simple statement for ignition:

$$\tfrac{1}{4} n_i^2 <\sigma v> E^* > \frac{n_i \tfrac{3}{2} T_i + n_e \tfrac{3}{2} T_e}{t_E} + P_{rad}, \tag{8}$$

where the left hand side of inequality (8) is the charged fusion product heating power per unit volume, with $E^* = E_\alpha = 3520$ keV in the case of D-T, and $E_\alpha + p = 18{,}300$ keV in the case of D-^{3}He. The first term in the right hand side of inequality (8) is the plasma power loss by transport (e.g., expansion in the ICF and MICF cases, with $\tau_E \sim \tau_i$, and by cross-field thermal conductivity with $\tau_E = \tau_i/5$ in the MCF case), and P_{rad} is the radiation loss, mostly soft-x-ray bremsstrahlung at high densities and betas needed for space reactors. The ideal ignition temperatures are that for which Eq. (8) is satisfied, neglecting the transport term, T_{ign} (ideal) ≈ 5 keV for D-T and 35 keV for D-^{3}He. The reaction parameter, $<\sigma v>$ (tritium), climbs more rapidly with temperature than does P_{rad} for tritium above T_{ign} (ideal), so that, to satisfy Eq. (8) including transport, one generally needs to heat the fusion plasma to an average initial temperature, T_{ign}, somewhat higher than T_{ign} (ideal), say 8-10 keV for D-T depending on the impurity level. ICF is a special case in that properly designed targets can ignite with average T_{ign} well below T_{ign} (ideal) by surrounding a hotter core ("spark plug") with cold compressed fuel, using alpha particles from the ignited core to heat the cold surrounding fuel. For D-^{3}He ignition, which requires heating to a higher temperature, it is generally advantageous to first ignite a D-T plasma and then use the D-T alpha particle heating to heat additional D-^{3}He fuel to ignition. In the ICF case, D-^{3}He ignition can occur within one pellet by compressing the additional cold D-^{3}He fuel around a small D-T core. In the MICF and MCF D-^{3}He cases, a D-T plasma is first ignited and then is refueled to a higher total density by additional cold D-^{3}He fuel.

Once ignition is achieved, the burn temperature is assumed to rise to an optimum T_i for maximum fusion energy gain, generally where the quantity, $<\sigma v>/T^2$, is a maximum.

PRESSURE RELATED CONSIDERATIONS

For the MICF and ICF cases, there is a magnetic field energy that must be established by the auxiliary plasma heating system in addition to the plasma energy, $E_{ign} = V_p (3/2) C n_i T_{ign}$. The additional magnetic energy is

$$E_M \approx E_{ign} \left(\frac{V_m}{V_p}\right)\left(\frac{1}{\beta}\right), \text{ (joules)} \tag{9}$$

where beta is defined as

Appendix C

$$\beta = \frac{p}{B^2/2\mu_o},\tag{10}$$

$\mu_o = 4p \times 10^{-7}$ h/m, **B** in tesla. For MICF, the magnetic field must only reduce cross-field thermal conductivity losses below the wall expansion losses. It need not fully support the plasma pressure, i.e., $\beta \gg 1$ is possible, and so $E_M \ll E_{ign}$ and can be neglected. For good measure, we assume $\beta = 1$ for the MICF case here, driven by a compact torus plasma accelerator which contains a stronger, embedded magnetic flux and which results in $\beta = 1$ at stagnation with $V_m \ll V_p$ (relatively thin flux layer). For MCF, the field must fully support the plasma pressure, so β is limited to less than unity. MCF candidates (Tokamaks, Reversed Field Pinches, Field-Reversed Configurations, stellarators, etc.) have widely different β limits, so we merely report the required magnetic field for $\beta = 1$ and $\beta = 0.06$. We do not go further to describe the required magnet performance for these MCF space reactor candidates except to note that since waste heat rejection carries a severe penalty in space, cooling cryogenic superconducting magnets would be a lesser problem for these cases than a copper one, assuming there is space for a thin shield. Perhaps advances in new high temperature superconductors will meet this challenge, but we leave this subject for future work.

For a given plasma pressure, "p," the fusion plasma power density scales as

$$\tfrac{1}{4}n_i^2 <\sigma v> E_{fus} \approx \frac{p^2 <\sigma v>}{T_i^2},\tag{11}$$

so that there is an optimum ion (burn) temperature, T_i, depending on the temperature dependence of $<\sigma v>$. Fig. C-1 shows the variation of this reaction parameter for D-T, D-D, and D-^{3}He fuels.

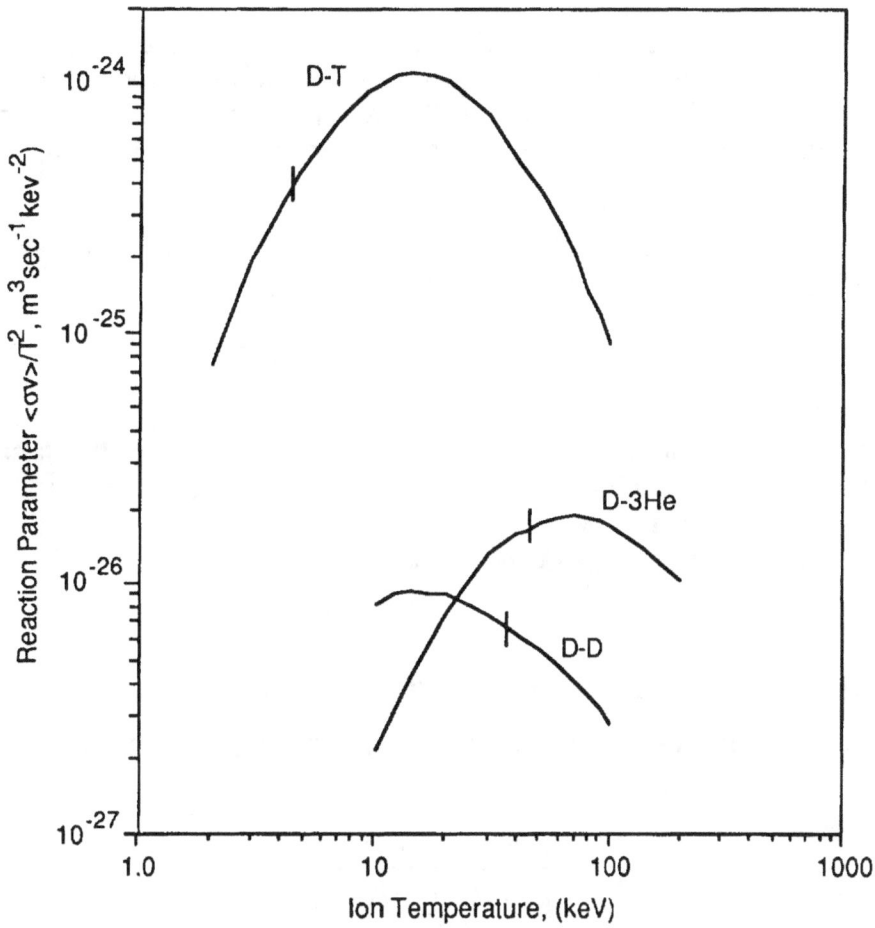

Fig. C-1. Reaction parameter for deuterium-tritium, deuterium-helium-3, and deuterium-deuterium.

For D-T the maximum occurs at $T_i = 15$ keV and $<\sigma v>_{D-T} = 2.7 \times 10^{-22}$ m³/sec. For D-³He the maximum occurs at $T_i = 60$ keV and $<\sigma v>_{D^3He} = 6.8 \times 10^{-23}$ m³/sec. The maximum $<\sigma v>/T_i^2$ for D-³He is lower than for D-T by a factor of about 50. However, the charged particle yield for D-³He, minus the x-ray bremsstrahlung losses, constitutes 65% of the fusion yield while the same for D-T is only 19%. Therefore, if we consider charged plasma energy more useful for space reactors than neutron or x-ray energy, the ratio of maximum useful power density at a given pressure for D-T to that at D-³He would be 15 for the case of MCF. This factor of 15 lower reactivity for useful charged-particle power with D-³He, as compared to D-T, can be compensated (in view of Eq. 11) by raising the pressure (density) by a factor of $(15)^{1/2} = 3.8$.

In ICF, some of the neutron energy and x-ray bremsstrahlung can be absorbed within the compressed fuel and also in non-fuel material shells which augment plasma thrust for propulsion. About 20% of the neutron energy and nearly all of the x-ray bremsstrahlung energy can be converted in-situ to useful plasma thrust in suitable ICF designs. Likewise, at least 20% of the neutron energy and nearly all x-ray energy can be converted in the metal tamper shell used to aid confinement in MICF targets. When these effects are taken into account, the effective charged (useful) fraction of fusion energy yield is

$$F_c = \frac{E_{charged,\ eff}}{E_{fusion}} = 0.4 \text{ for D-T,}$$

$$= 0.9 \text{ for D-}^3\text{H} \qquad (12)$$

for ICF and MICF cases. The $f_c = 0.9$ for D-^{3}He arises from the fact that there are side neutron-producing D-D reactions and neutrons from using D-T for spark-plug ignition of D-^{3}He. For MCF, $f_c = 0.19$ for D-T and 0.65 for D-^{3}He because neutrons and x-rays do not absorb in the plasma.

REQUIREMENTS FOR HIGH FUSION GAIN

Here we determine the required fractional fuel burnup, f_b, where

$$f_b = \frac{1/2(n_i t_i)<\sigma v>}{1+1/2(n_i t_i)<\sigma v>}, \qquad (13)$$

to achieve a minimum useful fusion gain, defined as an effective charged particle energy release at least 10 times the electrical energy consumed to heat the fuel to ignition in each case. First, we define the ideal fusion gain

$$G_{ideal} = \frac{E_{fusion}}{2\left[\frac{3}{2}T_i + \frac{n_e}{n_i}\frac{3}{2}T_e\right]} f_b, \qquad (14)$$

where E_{fus} is the total (neutron + ion) energy released per fusion reaction, which consumes 2 fuel ions of mean energy, $(3/2 T_i)$, and the associated electron energy $(n_e/n_i)(3/2\ T_e)$. G_{ideal} is the total energy gain for a fraction, f_b, of fuel burned divided by the minimum fuel ion and electron energy invested to heat the fuel to ignition. Thus, for D-T:

$$G_{ideal} = \frac{17{,}600 \text{ keV}}{6T_{ign}} f_b, \qquad (15)$$

where we have neglected impurity species so that $n_e = n_i$, and where we took $T_i = T_e = T_{ign}$ in keV. Since the temperature can climb during the burn, the $<\sigma v>$ in the expression (13) for f_b can be calculated at a higher temperature, i.e., the Ti for optimum $<\sigma v>/T^2$. For D-^{3}He, the corresponding expression for G_{ideal} is:

$$G_{ideal} = \frac{18,300 \text{ keV}}{7.5 \overline{T}_{ign}} f_b, \qquad (16)$$

where the 7.5 factor arises because $n_e = 1.5\, n_i$ with 50% deuterium, 50% doubly-charged helium. In fusion ICF experiments, the fusion gain is often quoted in terms of fusion yield divided by the driver (laser) energy incident on the target, $G_{ideal}\, \eta_c$, which takes into account the coupling efficiency, η_c. Now suppose that we adopt a performance target for generic fusion space reactors that the useful plasma energy generated per pulse be at least 10 times the electrical energy consumed to ignite the plasma which takes into account the electrical auxiliary system efficiency η_a. Then we require a figure-of-merit, G_{fom}, where

$$G_{fom} \equiv (G_{ideal}\, \eta_c)\, f_c\, \eta_a \geq 10, \qquad (17)$$

or an ideal fusion gain, G_{ideal}, where

$$G_{ideal} \geq \frac{10}{\eta_c \eta_a f_c}. \qquad (18)$$

By substituting Eq. (15) for G_{ideal} we see that

$$f_b(\min) \geq \frac{3.4 \times 10^{-3}\, \overline{T}_{ign}}{\eta_c \eta_a f_c}. \qquad (19)$$

for D-T, and

$$f_b(\min) \geq \frac{4.1 \times 10^{-3}\, \overline{T}_{ign}}{\eta_c \eta_a f_c}. \qquad (20)$$

Appendix C

for D-^{3}He. Substitution of the result for the minimum required f_b(min) into Eq. (13) gives a minimum required ($n_i\tau_i$) product:

$$n_i t_i(\min) \geq \frac{2}{<\sigma v>}\left(\frac{f_b \min}{1-f_b \min}\right). \qquad (21)$$

and this, in turn, leads to a minimum required fuel ion confinement time, τ_i, by dividing $(n_i\tau_i)_{\min}$ by the maximum fuel ion density, n_i. The limits on the maximum fuel ion density are not precisely known in the ICF and MICF cases, but the values for n_i used in Table C-1 reflect current projections of maximum compression (fuel convergence ratio) in ICF.

TABLE C-1. Characteristic plasma development requirements for generic fusion space reactors.

Parameter	ICF Laser driver		MICF Compact torus driver		MCF Generic torus: R/A = 3.3	
	D-T	D-^{3}He	D-T	D-^{3}He	D-T	D-^{3}He
Fuel ignition temp., T_{ign}, (keV) (with D-T spark plug)	1	2	8	10	8	10
Burn temp., T_i, (keV)	15	60	15	60	15	60
Fuel ion density, n_i, (cm^{-3})	6×10^{26}	2.4×10^{27}	10^{21}	6×10^{21}	6×10^{14}	2.4×10^{15}
Plasma pressure, p, (bar)	3×10^{13}	6×10^{14}	4.8×10^7	1.4×10^9	29	580
Magnetic Field B(T),						
• at $\beta = 1$	NA	NA	3.5×10^3	1.9×10^4	2.7	12
• at $\beta = 0.06$	NA	NA	NA	NA	11	49
f_c = $E_{charged}/E_{fusion}$	0.4	0.9	0.4	0.9	0.19	0.65
Auxiliary efficiency, η_a	0.1	0.1	0.5	0.5	0.5	0.5
Coupling efficiency, η_c	0.15	0.15	0.9	0.9	0.9	0.9
Fuel burnup fraction, f_b	0.57	0.31	0.15	0.10	0.30	0.44
Fuel confinement time, τ_i, (s)	1.6×10^{-11}	2.2×10^{-11}	1.3×10^{-6}	5.3×10^{-7}	5	5
$n_i \tau_i$ product, (cm^{-3} s)	9.8×10^{15}	1.3×10^{16}	1.3×10^{15}	3.2×10^{15}	3×10^{15}	1.3×10^{16}
Energy/fuel confinement ratio	1	1	1	1	0.2	0.2
Plasma radius, a, (cm)	7.2×10^{-3}	4.8×10^{-3}	1.4	1.9	63	63
Plasma ignition energy, E_{ign} (MJ)	0.45	1.1	45	90	76	76
Driver energy, E_{ign}/η_c, (MJ)	3	7.3	50	100	84	84
Electrical input $E_{ign}/(\eta_c \eta_a)$, (MJ)	30	73	100	200	168	168
Fusion gain, ($G_{ideal} \eta_c$)	250	111	56	220*	100	400*
Useful plasma output energy, ($G_{ideal} f_c E_{ign}$), (MJ)	300	730	1000	20,000*	1680	30,400*
G_{fom} = ($G_{ideal} \eta_c$)$f_c \eta_a$ charged MJ/electrical MJ	10	10	10	100*	10	180*

* Assumes an initial D-T plasma re-fueled with D-^{3}He.

The current wisdom in MICF puts n_i at a few percent of solid D-T density, so the tamper shell can initially be loaded with a solid D-T layer and still be largely hollow to allow the driver energy input. The maximum n_i in MCF could be limited either by βB^2 due to some limit on beta or B, but here we take the long view that future magnet advances will allow sufficiently high B fields that n_i will be limited instead by fundamental surface heat flux limits due to neutron and x-ray bremsstrahling heating of the first wall. The densities chosen in Table C-1

Appendix C

for the MCF cases are the maximum that might be allowed by surface heat flux considerations.

Once the minimum τ_i is determined in this way, the formulas for τ_i -- i.e., Eq. (1), (3) and (7) for the cases of ICF, MICF, and MCF respectively -- can then be used to determine the minimum plasma radius "a" at maximum density to achieve the desired gain G_{ideal} and G_{fom}. Then the plasma volume, V_p, for each case can be computed, the initial plasma energy, E_{ign}, where:

$$E_{ign} = V_p C (\tfrac{3}{2}) \overline{T}_{ign} \qquad (22)$$

that must be supplied, the driver energy E_{ign}/η_c required to "light the match," and finally, the driver electrical energy consumed, $E_{ign}/(\eta_c \eta_a)$, to start-up the plasma. In the case of MCF, the plasma could in principle be continuously refueled, provided a continuous removal of alpha ash were possible without extinguishing the ignited plasma, and then the fusion gain for such a plasma maintained in a steady-state operation might be arbitrarily high. In any case, the same driver energy and electrical stored energy are still required for start-up. The fusion gains given in Table C-1 for ICF assume a pulsed burn to the burn-up fractions given, and any steady-state case would still require the same confinement ($n_i \tau_i$), size "a," and start-up energy, E_{ign}.

CONCLUSIONS

Start-up energy is an important parameter for space reactors because it implies a requirement of storing and converting such energy and a minimum start-up system mass with components having desired ratios of stored energy per unit mass. Such electrical and auxiliary start-up systems are different in each case, and their mass to provide the ignition energies given in Table C-1 must be determined for future designs. The energies given in Table C-1 are characteristic; more detailed designs may raise or lower the values some. The predictions of this simple model for D-T are roughly consistent with more detailed designs -- the LLNL Laboratory Microfusion Facility parameters for ICF, the MICF design by Hasegawa, and the Compact Ignition Tokamak design (CIT) for MCF. Since the model reflects a consistent level of detail and methodology, and a common constraint of $G_{fom} = 10$ for D-T, the results of Table C-1 are informative by providing a comparison of the performance between ICF, MICF, and MCF. The required driver electrical energies for start-up -- 30 MJ, 100 MJ, and 168 MJ -- respectively, go in the order of increasing start-up energy with decreasing plasma density, demonstrating the advantage of high density and pressure for fusion. However, the energy/mass ratio for the different drivers available to each fusion case may decrease with the higher density cases because the energy has to be delivered in shorter times with higher density, raising the peak power required from the auxiliary systems. Thus, future designs may turn out to show that higher start-up energies delivered over longer pulses than in ICF, might still result in lower overall system mass. Regardless of which fusion approach is used, Table C-1 indicates the auxiliary systems required for ignition will not likely be so small that average power output ($G_{ideal} f_c E_{ign}$) × (pulse repetition rate) can likely be much less than 100 MW. It is anticipated that high pulse repetition rates (>1 Hz) may be needed to achieve an average driver system specific power $\alpha > 1$ kW/kg, a value considered to be useful for solar system space travel. Thus, it appears likely that all fusion candidates will be suitable only for large spacecraft with missions requiring power levels of 100 MW or more.

As seen in Table C-1, D-T reactors can have a much smaller energy output than burning D-^{3}He because D-^{3}He requires higher burn temperatures and $n_i \tau_i$ products and, thus, higher plasma energy, E_{ign}, to get high gain ($G_{fom} > 10$). However, D-^{3}He fuel is available from extraterrestrial sources while tritium is not, and long term storage of tritium with its decay heat is a formidable problem. Fortunately, one can mitigate the problem of igniting D-^{3}He by using a small amount of D-T as a "spark-plug." In ICF, the driver energy is expended to compress the additional D-^{3}He fuel as well as the D-T, whereas in MICF and MCF, the magnetic confinement should be sufficient to allow refueling with D-^{3}He without much refueling energy required after igniting and burning D-T to an initial plasma energy gain of 10. In the D-^{3}He MICF case of Table C-1, for example, the initial D-T burn provides 1000 MJ of plasma energy to ignite the injected D-^{3}He fuel mass which then burns to raise the plasma gain another factor of 10 for an overall $G_{fom} = 10 \times 10 = 100$.

This higher gain results in higher output energies (20 GJ), of course, which may limit the application to very large spacecraft; but this scheme would keep the driver energy reasonable small. Otherwise, without the initial D-T gain, the driver would need to supply all of the 1000 MJ of plasma energy needed to ignite the D-^{3}He. Thus, we see a progression of sizable fusion output from D-T reactors to an even larger output from D-^{3}He reactors which can exploit extraterrestrial ^{3}He fuel.

www.ingramcontent.com/pod-product-compliance
Lightning Source LLC
Chambersburg PA
CBHW081714170526

45167CB00009B/3579